THE GROWTH PLATE

Biomedical and Health Research

Volume 54

Earlier published in this series

Vol. 21. N. Yoganandan, F.A. Pintar, S.J. Larson and A. Sances Jr. (Eds.), Frontiers in Head and Neck Trauma
Vol. 22. J. Matsoukas and T. Mavromoustakos (Eds.), Bioactive Peptides in Drug Discovery and Design: Medical Aspects
Vol. 23. M. Hallen (Ed.), Human Genome Analysis
Vol. 24. S.S. Baig (Ed.), Cancer Research Supported under BIOMED 1
Vol. 25. N.J. Gooderham (Ed.), Drug Metabolism: Towards the Next Millennium
Vol. 26. P. Jenner (Ed.), A Molecular Biology Approach to Parkinson's Disease
Vol. 27. P.A. Frey and D.B. Northrop (Eds.), Enzymatic Mechanisms
Vol. 28. A.M.N. Gardner and R.H. Fox, The Venous System in Health and Disease
Vol. 29. G. Pawelec (Ed.), EUCAMBIS: Immunology and Ageing in Europe
Vol. 30. J.-F. Stoltz, M. Singh and P. Riha, Hemorheology in Practice
Vol. 31. B.J. Njio, A. Stenvik, R.S. Ireland and B. Prahl-Andersen (Eds.), EURO-QUAL
Vol. 32. B.J. Njio, B. Prahl-Andersen, G. ter Heege, A. Stenvik and R.S. Ireland (Eds.), Quality of Orthodontic Care – A Concept for Collaboration and Responsibilities
Vol. 33. H.H. Goebel, S.E. Mole and B.D. Lake (Eds.), The Neuronal Ceroid Lipofuscinoses (Batten Disease)
Vol. 34. G.J. Bellingan and G.J. Laurent (Eds.), Acute Lung Injury: From Inflammation to Repair
Vol. 35. M. Schlaud (Ed.), Comparison and Harmonisation of Denominator Data for Primary Health Care Research in Countries of the European Community
Vol. 36. F.F. Parl, Estrogens, Estrogen Receptor and Breast Cancer
Vol. 37. J.M. Ntambi (Ed.), Adipocyte Biology and Hormone Signaling
Vol. 38. N. Yoganandan and F.A. Pintar (Eds.), Frontiers in Whiplash Trauma
Vol. 39. J.-M. Graf von der Schulenburg (Ed.), The Influence of Economic Evaluation Studies on Health Care Decision-Making
Vol. 40. H. Leino-Kilpi, M. Välimäki, M. Arndt, T. Dassen, M. Gasull, C. Lemonidou, P.A. Scott, G. Bansemir, E. Cabrera, H. Papaevangelou and J. Mc Parland, Patient's Autonomy, Privacy and Informed Consent
Vol. 41. T.M. Gress (Ed.), Molecular Pathogenesis of Pancreatic Cancer
Vol. 42. J.-F. Stoltz (Ed.), Mechanobiology: Cartilage and Chondrocyte
Vol. 43. B. Shaw, G. Semb, P. Nelson, V. Brattström, K. Mølsted and B. Prahl-Andersen, The Eurocleft Project 1996-2000
Vol. 44. R. Coppo and Dr. L. Peruzzi (Eds.), Moderately Proteinuric IgA Nephropathy in the Young
Vol. 45. L. Turski, D.D. Schoepp and E.A. Cavalheiro (Eds.), Excitatory Amino Acids: Ten Years Later
Vol. 46. I. Philp (Ed.), Family Care of Older People in Europe
Vol. 47. H. Aldskogius and J. Fraher (Eds.), Glial Interfaces in the Nervous System – Role in Repair and Plasticity
Vol. 48. H. ten Have and R. Janssens (Eds.), Palliative Care in Europe – Concepts and Policies
Vol. 49. T. Reilly (Ed.), Musculoskeletal Disorders in Health-Related Occupations
Vol. 50. R. Busse, M. Wismar and P.C. Berman (Eds.), The European Union and Health Services
Vol. 51. G. Lebeer (Ed.), Ethical Function in Hospital Ethics Committees
Vol. 52. J.-F. Stoltz (Ed.), Mechanobiology: Cartilage and Chondrocyte, Vol. 2
Vol. 53. In production

ISSN: 0929-6743

The Growth Plate

Edited by

Irving M. Shapiro

Department of Orthopaedic Surgery,
Thomas Jefferson University
Philadelphia, PA USA

Barbara Boyan

Department of Orthopaedics and Biochemistry,
University of Texas Health Sciences Center
San Antonio, TX USA

H. Clarke Anderson

Department of Pathology and Laboratory Medicine
University of Kansas Medical Center
Kansas City, KS USA

IOS
Press

Ohmsha

Amsterdam • Berlin • Oxford • Tokyo • Washington, DC

ISBN 1 58603 240 2 (IOS Press)
ISBN 4 274 90506 3 C3047 (Ohmsha)
Library of Congress Control Number: 2002104882

Publisher
IOS Press
Nieuwe Hemweg 6B
1013 BG Amsterdam
The Netherlands
fax: +31 20 620 3419
e-mail: order@iospress.nl

Distributor in the UK and Ireland
IOS Press/Lavis Marketing
73 Lime Walk
Headington
Oxford OX3 7AD
England
fax: +44 1865 75 0079

Distributor in the USA and Canada
IOS Press, Inc.
5795-G Burke Centre Parkway
Burke, VA 22015
USA
fax: +1 703 323 3668
e-mail: iosbooks@iospress.com

Distributor in Germany, Austria and Switzerland
IOS Press/LSL.de
Gerichtsweg 28
D-04103 Leipzig
Germany
fax: +49 341 995 4255

Distributor in Japan
Ohmsha, Ltd.
3-1 Kanda Nishiki-cho
Chiyoda-ku, Tokyo 101-8460
Japan
fax: +81 3 3233 2426

PRINTED IN THE NETHERLANDS

Preface

Skeletal development and growth is now emerging as one of the most exciting areas of current research activity. Evidence generated by an ever growing number of genetic studies indicate that growth is regulated by a large number of genes and that interference with their expression can have catastrophic effects on the health of the whole organism. With the realization that multiple regulatory pathways exist, work is now focusing on identification of those signals that control the activity of the cells in the epiphyseal growth plate.

One advance that has served to catalyze analysis of the processes that regulate growth plate activity has been the use of culture systems that model events that occur in vivo. In these culture systems, chondrocytes recapitulate many of the changes that are seen in the epiphyseal growth plate itself as chondrocytes mature and become terminally differentiated. Ongoing studies have shown that these models can be used very effectively to probe and identify local environmental signals and to assess how these signals influence gene expression. Outcomes from these studies have also impacted on the large number of conditions that cause growth plate anomalies, malformations, hypomineralizations and dysplasias.

To understand the clinical problems linked to abnormalities of child growth and to address the wide spread interest in epiphyseal chondrocyte biology, a number of leading researchers gathered in San Antonio, Texas in June 2001 for the first International Conference on the Growth Plate. This group of individuals examined the regulation of craniofacial growth and mineralization, skeletal morphogenesis, developmental regulation of the cell cycle, angiogenesis, apoptosis, the function of collagenous and non-collagenous proteins, the role of growth factors in the cartilage matrix, and the interaction of matrix vesicles, matrix proteins and mineral components during the calcification process. The book, which is a compilation of many of these presentations, has been designed to provide an update on the current state of the field and at the same time to review major topics in growth plate and chondrocyte research.

The organizers wish to thank the workshop participants, and the chapter authors for their expertise and their dedication. We wish to thank Dr. H.I. Roach, University Orthopaedics, General Hospital, Southampton, England and Dr. T. Aizawa, Department of Orthopaedic Surgery, Tohoku University School of Medicine, Sendai, Japan for the cover photograph. A special thanks to Ms. Linda A. Keller for her organizational skills and sympathetic understanding of the problems of hosting an international conference. We also wish to thank the companies and organizations listed on the following pages for their help with funding this memorable meeting.

Irving M. Shapiro
Barbara D. Boyan
H. Clarke Anderson

Philadelphia 2002

Dedication

In Recognition of

S. Yousuf Ali, Ph.D.,
David Howell, M.D.,
Fujio Suzuki, Ph.D.

for their contributions to the study of the growth plate

Sponsors

Biomet/EBI
Biora
Center for the Enhancement of the Biology/Biomaterials Interface
Howard Hughes Medical Institute
Mission Pharmacal
National Institute of Aging
National Institute of Arthritis and Musculoskeletal and Skin Diseases
National Institute of Child Health and Human Development
National Institute of Dental and Craniofacial Research
National Institute of Diabetes and Digestive and Kidney Diseases
Shriners Hospital for Children
The University of Texas Health Science Center at San Antonio

Contents

Indian Hedgehog and Retinoids Orchestrate Multiple Growth Plate Functions in Developing Long Bones: The Growth Plate as a Highly Interactive Structure

Maurizio Pacifici[1], Chiara Gentili[2], Melinda Yin[1], Masahiro Iwamoto[3], Motomi Enomoto-Iwamoto[4], William R. Abrams[1], and Eiki Koyama[1]
[1]Department of Anatomy and Histology, School of Dental Medicine, University of Pennsylvania, Philadelphia, PA 19104-6003; [2]Laboratorio Differenziamento Cellulare, Istituto Nazionale Ricerca sul Cancro, Centro Biotecnologie Avanzate, Genova, Italy; and Departments of [2]Oral Anatomy and Developmental Biology and [3]Biochemistry, Osaka University Faculty of Dentistry, Osaka 565, Japan

Abstract. Previous studies from this and other laboratories have indicated that *Indian hedgehog* (IHH) and retinoids play important signaling roles in growth plate function and long bone development. In the present chapter, we present highlights from our previous studies, particularly those describing the role of IHH in intramembranous bone collar development and osteogenic cell differentiation, and the need for retinoid signaling in chondrocyte maturation and hypertrophy and endochondral ossification. In new studies using immunohistochemical procedures, we show that IHH is not limited to the prehypertrophic zone of growth plate where it is produced, but is present also in the hypertrophic zone and perichondrial osteogenic layers, indicating long-range diffusion and action by IHH. Using recombinant IHH we found that the factor is a mitogen for chondrocytes. New studies on retinoids reveal that endogenous retinoids present in perichondrial tissues may diffuse into the growth plate and promote chondrocyte maturation starting along the chondro-perichondrial border. Evidence with chondrocyte cultures indicates that retinoids regulate IHH gene expression, which is normally down-regulated during transition from prehypertrophic to hypertrophic cartilage. The new data presented here reinforce our proposal that IHH- and retinoid-dependent signaling pathways are important orchestrators of multiple steps and processes central to long bone development. They also suggest a new view of the growth plate. Rather than being a structure made of independent zones, the growth plate would be a highly interactive and interdependent structure in which events in each zone are influenced by, dependent on and coordinated with, events in flanking zones and events in perichondrial tissues.

Introduction

Long bone formation is a multi-step process (Thorogood, 1983; Hinchcliffe and Johnson, 1990). It initiates with the emergence of mesenchymal cell condensations at specific times and sites that are patterned by the concerted action of the zone of polarizing activity, apical ectodermal ridge and dorsal ectoderm. The condensed cells differentiate into chondrocytes that produce characteristic cartilage matrix components and give rise to readily identifiable cartilaginous elements. The chondrocytes within each element become organized in growth plates and progress through the resting, proliferative, prehypertrophic, hypertrophic and mineralizing phases of maturation. Once formed, the hypertrophic mineralized cartilage is invaded by bone, marrow and vascular progenitor cells from adjacent perichondrial tissues and replaced by endochondral bone and marrow. In addition, perichondrial cells give rise to an intramembranous bone collar surrounding the elements, which is critical for determining diameter and shape of the shaft (Fell, 1925). Maturation, hypertrophy, blood vessel invasion and ossification first occur in the diaphyseal region and then spread toward the opposing epiphyses with increasing developmental time. Thus, long bone formation includes and depends on (a) multiple and topographically restricted events within the cartilaginous elements, (b) progression of chondrocytes through the maturation process in the growth plate, and (c) related events in perichondrial tissues. It is not well understood at present how all these processes and events are set, regulated and coordinated.

Studies have indicated that the signaling molecules *Indian hedgehog* (IHH) and retinoids have critical roles in long bone formation. IHH belongs to the powerful *hedgehog* family of secreted signaling proteins and is exclusively expressed in prehypertrophic chondrocytes in the growth plate of long bone anlagen (Bitgood and McMahon, 1995; Vortkamp et al., 1996; Koyama et al., 1996). In contrast, the hedgehog cell surface receptor *Patched*-1 and the hedgehog-responsive nuclear factor GLI are strongly expressed in growth plate zones flanking the prehypertrophic zone as well as in perichondrial tissue surrounding the IHH-expressing prehypertrophic cells (Vortkamp et al., 1998). These and other findings suggested that IHH is a major regulator of chondrocyte behavior in the growth plate, inhibits maturation, and determines the overall number of chondrocytes entering and completing the maturation process with the aid of perichondrium-derived PTHrP (Vortkamp et al., 1996; 1998). Additional work from our group has indicated that IHH may have other important roles in long bone development. We were the first to report that the perichondrial tissue adjacent to the IHH-expressing prehypertrophic chondrocytes is the site of initiation of intramembranous bone collar development (Koyama et al., 1996). We then found that treatment of osteoprogenitor cell lines with recombinant IHH induces differentiation into osteoblast-like cells (Nakamura et al., 1997). These and other findings led us to propose that IHH is an osteo-inductive factor and directs intramembranous ossification along the outer perimeter of developing long bones (Koyama et al., 1996, 1999; Nakamura et al., 1997). In very good agreement with our proposal, St-Jacques et al. (1999) have reported recently that in IHH-null mice there is no ossification in the limbs; interestingly, the IHH-null long bone elements remain cartilaginous, and contain disorganized growth plates with much fewer proliferative chondrocytes and more numerous and dispersed hypertrophic chondrocytes. In sum, IHH appears to have multiple roles in long bone development.

With regard to retinoids, their involvement in skeletogenesis was first suggested by nutrition studies over 4 decades ago (Walbach and Hegsted, 1952). Since then, such connection has been substantiated by work on retinoic nuclear receptors. The receptors comprise two subfamilies, the retinoic acid (RA) receptors RARα, RARβ and RARγ, and the retinoid receptors RXRα, RXRβ and RXRγ (Chambon, 1994; Mangelsdorf et al., 1994). During limb skeletal development, RARα and RARγ are first expressed broadly; with time, RARγ becomes expressed preferentially in prechondrogenic condensations, RARα remains diffuse, and RARβ becomes restricted to perichondrium (Dolle et al., 1994). Gene inactivation studies have shown that loss of a single RAR gene usually causes minor to no skeletal defect, whereas double gene inactivation, such as double null mutants of RARα and RARγ, produces serious skeletal abnormalities (Mendelsohn et al., 1994). Past and recent work from our group has provided more specific and detailed insights into the roles of retinoid signaling in skeletogenesis. One of our initial findings was that prehypertrophic chondrocytes isolated from developing skeletal elements and maintained in standard culture conditions appeared to be unable to fully mature into hypertrophic mineralizing chondrocytes. The cells, however, promptly did so following treatment with physiologic doses of natural retinoids, such as all-*trans*-retinoic acid or 9-*cis*-retinoic acid (Iwamoto et al., 1994). We went on to show that retinoids play a similar role in long bone development in vivo (Koyama et al., 1999). We found that the emergence of hypertrophic chondrocytes is invariably accompanied by a marked upregulation of RARγ gene expression, that endogenous retinoids are present in the developing cartilaginous elements, and that pharmacological interference with retinoid signaling and action blocks the terminal phases of chondrocyte maturation and endochondral ossification.

In the present chapter, we present data from our previous studies to illustrate the roles of IHH and retinoid signaling in long bone development. In addition, we present new data on IHH distribution in the growth plate in vivo and IHH effects on chondrocyte proliferation, and data on the role of retinoids in IHH gene expression and chondrocyte hypertrophy. The observations reinforce our proposal that IHH- and retinoid-dependent signaling pathways are important regulators of chondrocyte maturation and ossification during long bone development. They also suggest a new view of the growth plate as a highly interactive structure, in which events in each zone would be influenced by, dependent on and coordinated with, events in flanking zones and events in perichondrial tissues.

IHH and Osteogenesis in Developing Long Bones

The first clue that IHH has a role in intramembranous bone collar development came from a study we reported a few years ago (Koyama et al., 1996). The study focused on the question of how morphogenesis of long bones is regulated and in particular on how the diaphysis comes to acquire its characteristic cylindrical and elongated configuration compared to the three-dimensionally complex epiphyses. It had been reported at the time that the powerful morphogenetic factor *Sonic hedgehog* (SHH) is expressed in the zone of polarizing activity (ZPA) located in the posterior part of the limb (Riddle et al., 1993). The ZPA has a critical role in patterning the prechondrogenic mesenchymal condensations during limb development. Thus, we reasoned that SHH itself or another member of its family (i.e. IHH) may be expressed in chondrocytes and may have a role in regulating morphogenesis of long bones. To approach this question, we monitored the expression of SHH, IHH and other relevant genes in developing long bones in chick embryo limbs. We found that IHH gene expression was first turned on in the incipient diaphysis of early long bone cartilaginous anlagen present in Day 6-6.5 chick embryo; at this stage, the anlagen were quite primitive and it was hard to precisely establish the maturation stage of the diaphyseal IHH-expressing chondrocytes. Once the anlagen had developed further and the growth plates were more discernable (Fig. 1A), it became apparent that IHH expression characterized prehypertrophic chondrocytes (Fig. 1B, star). The IHH-expressing prehypertrophic chondrocytes were surrounded by a thin intramembranous bone collar that was recognizable histologically (Fig. 1A, arrow) and was characterized by very strong gene expression of type I collagen (Fig. 1C, arrow) and staining by alizarin red (Fig. 1D, arrow). There was no bone collar, no strong type I collagen gene expression and no alizarin red staining in perichondrial tissues adjacent to proliferating or more immature epiphyseal chondrocytes which did not express IHH (Figs. 1A-1D, arrowheads). As shown later in this chapter (see Fig. 7), there was selective expression of the bone-characteristic matrix protein osteopontin in bone collar flanking IHH-expressing prehypertrophic chondrocytes, but no osteopontin expression in perichondrium flanking IHH-negative chondrocytes (see Koyama et al., 1999 for further findings). Thus, our data clearly showed that expression of IHH coincides with formation of a bone collar, indicating that the two events may be causally related.

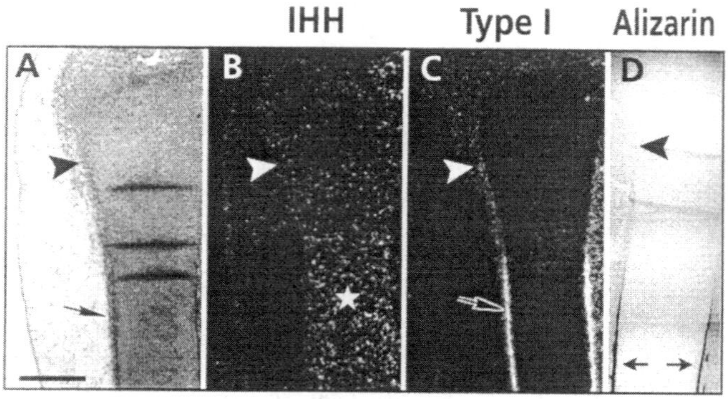

Figure 1. Analyses of Day 8.5 chick embryo ulna. Longitudinal sections were examined by: phase microscopy (A); in situ hybridization with IHH (B) and type I collagen (C) cDNA probes; and histochemical staining with alizarin red (D). Arrows in A, C and D point to the intramembranous bone collar forming around the IHH-expressing prehypertrophic chondrocytes (star in B). Arrowheads point to lack of collar formation around the preceding proliferative zone. Bar, 200 μm.

If indeed IHH induces bone collar formation, IHH produced by prehypertrophic chondrocytes would have to diffuse away from the cells and reach the appropriate perichondrial layers. To test this prediction, we prepared monospecific rabbit antibodies to avian IHH and used them to determine the distribution of IHH in the growth plates of long bone anlagen at different stages of chick embryo development (Yin et al., 2001). We first examined early long bone anlagen, in which bone collar development has just started and the metaphyseal-diaphyseal cartilaginous portion has not yet become fully hypertrophic. We found that IHH was clearly associated with prehypertrophic and early hypertrophic chondrocytes (Fig. 2A). Indeed, the protein was also present in the inner layers of perichondrial tissues flanking the IHH-rich chondrocytes (Fig. 2A, arrow) but was undetectable in the outer layers (Fig. 2A, arrowhead). This distribution is particularly interesting because the inner perichondrial layers are osteogenic, while the outer layers serve mechanical roles (Pechak et al., 1986; Gigante et al., 1996). When we examined later developmental stages, we found that IHH had similar distribution patterns (Fig. 2B-2C). However, it was also clearly detectable in fully hypertrophic chondrocytes as well as along the chondro-endochondral bone border (Figs. 2B-2C, arrows).

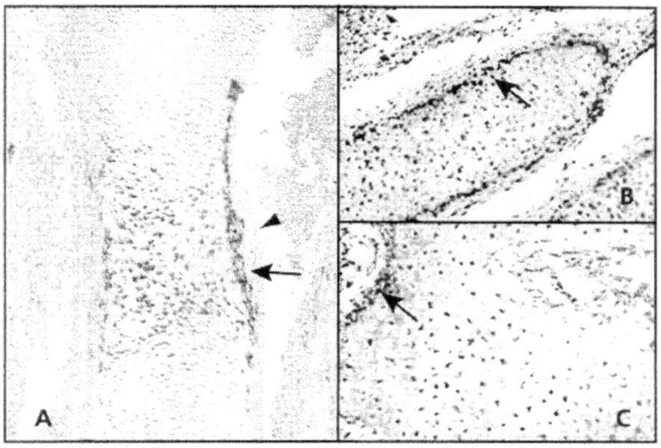

Figure 2. IHH distribution. Longitudinal frozen sections of (A) Day 8.5 and (B-C) Day 9.5 chick embryo ulnas were processed for immunohistochemical detection of IHH, using monospecific antibodies to a synthetic IHH peptide. Arrow and arrowhead in (A) point to the IHH-positive and IHH-negative perichondrial tissue layers, respectively. Arrows in B and C point to strong signal at sites along the chondro-endochondral bone border.

To determine whether the perichondrium-associated IHH is functional and has a direct and necessary role in intramembranous collar formation, we performed in vivo experiments using cyclopamine, a powerful hedgehog protein antagonist (Incardona et al., 1998). Cyclopamine-filled beads were microsurgically implanted next to the metaphyseal-diaphyseal region Day 6 chick embryo humerus; embryos were reincubated for 36-48 hrs and were then processed for in situ hybridization analysis of bone collar formation. We should point out that since the beads were placed on one side of the humerus, they created a drug concentration gradient (Eichele et al., 1984) with the near side of the humerus (closest to the beads) receiving more drug levels than the far side. We found that in control mock-treated embryos, expression of the bone marker osteopontin was detectable on each side of Day 7.5 humerus, reflecting ongoing intramembranous collar formation all around the metaphysis-diaphysis (Figs. 3A-3B, arrows). In contrast, in the cyclopamine-implanted embryos, osteopontin expression was undetectable on the near side of humerus close to the beads (Figs. 3C-3D, arrowhead) but was detectable on the far side (Fig. 3D, arrow). Differential collar formation in near versus far side of the humerus was confirmed by histochemical staining with alizarin red (not shown).

Taken together, the above data demonstrate that IHH is a product of prehypertrophic chondrocytes, is able to move away from its site of synthesis and reach the inner layers of perichondrial tissues as well as flanking growth plate zones, and induces differentiation of osteoprogenitor cells directly (Nakamura et al., 1997). The data also provide an explanation for the strong and selective expression of *Patched-1* in perichondrial cells surrounding IHH-expressing prehypertrophic chondrocytes seen in long bone anlagen in vivo (Vortkamp et al., 1996), pointing to the occurrence of positive feedback loops between IHH-producing chondrocytes and *Patched*-expressing perichondrial cells. They correlate well with data and conclusions in the recent report on IHH-null mice in which there is no ossification in the limb (St-Jacques et al., 1999).

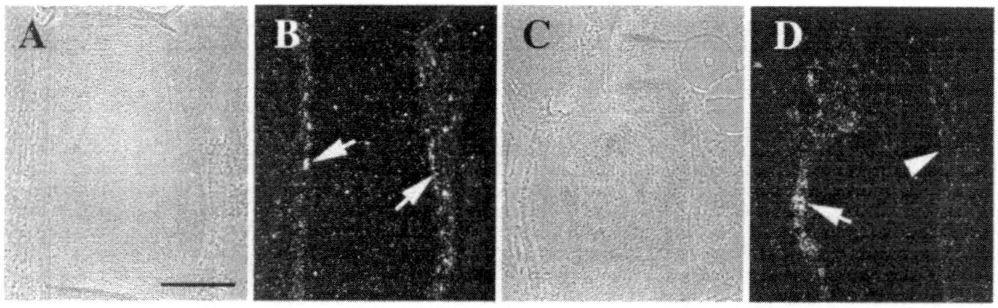

Figure 3. Cyclopamine effects on bone collar formation determined by in situ hybridization analysis of osteopontin gene expression. A-B: control Day 7.5 humerus displaying osteopontin transcripts all around its diaphysis (arrows). C-D: cyclopamine-treated humerus lacking osteopontin transcripts on the near side (arrowhead) but containing them on the far side (arrow). Beads are visible on the upper right corner in C. Bar, 250 μm.

IHH Stimulates Chondrocyte Proliferation

One of the defects seen in the IHH-null mice is decreased chondrocyte proliferation . To determine whether this is a direct effect, we tested whether IHH stimulates proliferation in cultured chondrocytes. Thus, we first produced the N-terminal half of avian IHH (amino acids 24-198; N-IHH) in bacteria. Biological activity was verified by its ability to induce supernumerary digits when implanted in the anterior margin of stage 20 chick embryo wing buds (not shown). Immature proliferating chondrocytes were isolated from the caudal region of Day 17 chick embryo sterna, a convenient and popular source of this type of cells (Gibson and Flint, 1985). Cells were reared in monolayer cultures for approximately five days to recover from the isolation procedures and to adapt to the in vitro condition. The cells were then switched to medium containing 0.5% serum to reduce endogenous mitotic activity and treated with 0.5-1.0 μg/ml of N-IHH for 1, 2 and 3 days. Mitotic activity was determined by incorporation of [^{3}H]thymidine. We found that N-IHH treatment did stimulate chondrocyte proliferation (Fig. 4). The cells displayed a 25 to 30% increase in incorporation at each time point tested. The effect was consistent and highly reproducible, but was lower than that seen with known strong chondrocyte mitogens, including PTH, PTHrP or FGF-2 (not shown). The data indicate that IHH is able to stimulate chondrocyte proliferation directly. By extrapolating to the in vivo condition, IHH diffusing from the prehypertrophic to proliferative zone of the growth plate could exert a similar role and influence chondrocyte proliferation in a positive manner. This is consistent with the well established roles of hedgehog proteins in cell proliferation in a variety of developmental processes and cell types (Jensen and Wallace, 1997; Duprez et al., 1998). It is also consistent with the strong gene expression of IHH receptor *Patched-1* and hedgehog protein mediator GLI seen in the proliferative zone of mouse growth plate (Vortkamp et al., 1998).

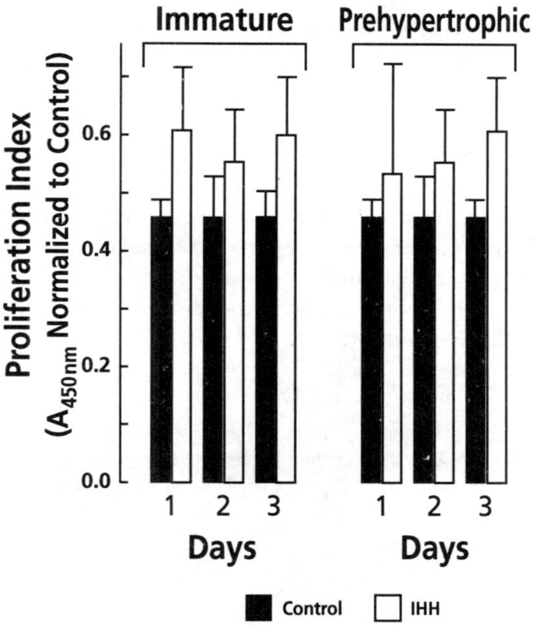

Figure 4. Histograms showing effects of recombinant IHH treatment on chondrocyte proliferation in culture.

Need for Retinoid Signaling in Endochondral Ossification

In a series of previous studies from our laboratory (Iwamoto et al., 1993, 1994 and refs. therein), we had provided evidence that retinoic signaling promotes the development of immature chondrocytes into hypertrophic chondrocytes in vitro. We found that upon induction by retinoids, the cells progress to the terminal stage of maturation and closely resemble the post-hypertrophic cells present at the chondro-osseous border in the growth plate in vivo. The phenotypic traits expressed by the retinoid-induced chondrocytes include a very large cell diameter, production of mineralization-competent matrix vesicles, ability to deposit apatitic crystals and high alkaline phosphatase activity. Since all these traits are actually needed for the transition from mineralized hypertrophic cartilage to endochondral bone, our data suggested that by inducing such traits, retinoid signaling may be required for the cartilage-to-bone transition in vivo. Thus, we carried out additional studies to obtain evidence in support of this important conclusion; these studies were reported recently (Koyama et al., 1999) and key findings are summarized next.

In a first set of experiments, we asked whether expression of retinoid nuclear receptors is upregulated in pre-hypertrophic and/or hypertrophic chondrocytes. We reasoned that such upregulation may be necessary for retinoids to act on those cells and promote terminal maturation into mineralizing post-hypertrophic chondrocytes ready for replacement by bone cells. Thus, we used in situ hybridization to monitor expression of RARα, RARβ and RARγ during long bone development in the embryonic limb. As above, we first examined newly-emerged cartilaginous anlagen in young Day 5.5 chick embryo limb; these anlagen are composed entirely of immature chondrocytes, display a still primitive morphological organization, and do not contain growth plates. We found that in these anlagen expression of RARα and RARγ was quite broad and diffuse throughout the cartilaginous tissue and RARβ expression was strong in incipient perichondrial cells. We then examined older Day 9 through Day 18 skeletal elements that display typical elongated morphologies, well defined diaphysis and epiphyses, and obvious growth plates (Fig. 5A). At these stages, RARα expression remained uniform, broad and relatively low throughout the cartilaginous tissue (Fig. 5B), while RARβ remained confined to perichondrial tissue. Interestingly, expression of RARγ was sharply and selectively upregulated in hypertrophic chondrocytes (Fig. 5C). Identity of the hypertrophic cells was based on their large cell size and location as well as strong expression of a typical marker, type X collagen (Fig. 5D). Equally interesting was the finding that there was a sharp boundary and minimal overlap between RARγ expression in hypertrophic chondrocytes and IHH expression in the preceding prehypertrophic chondrocyte zone (cfr. Figs. 5E and 5C).

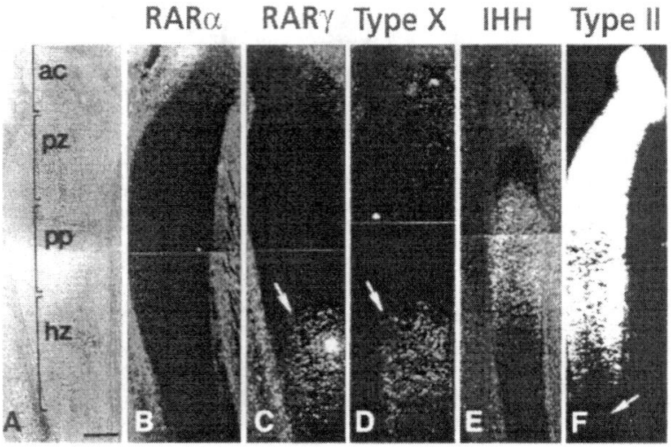

Figure 5. In situ hybridization analysis of expression of indicated genes in Day 10 chick embryo ulna. Arrows in C and D point to hypertrophic chondrocytes expressing RARγ and type X collagen. Arrow in F points to type II collagen RNA-negative post-hypertrophic chondrocytes. Ac, articular cap; pz, proliferative zone; pp, prehypertrophic zone; and hz, hypertrophic zone. Bar 185 μm.

Having shown that there is a selective upregulation of RARγ in hypertrophic chondrocytes, we carried out a second set of studies to determine whether these and/or other chondrocytes contain endogenous retinoids serving as RAR ligands. To approach this question, we used a bioassay commonly employed to determine endogenous retinoid levels in embryonic tissues; the bioassay is very sensitive, requires small amounts of tissue and is thus ideal for analyses of scarce specimens such as embryonic tissues (Wagner et al., 1992). It consists of an F9 cell line stably transfected with a retinoid sensitive RARE/β-galactosidase construct; the cell line is exposed to tissue extracts, reporter activity increases in proportion to retinoid content in the extracts, and reporter activity is finally measured biochemically or histochemically. Accordingly, we isolated whole cartilaginous elements from Day 5.5, 8.5 and 10 embryos by microsurgical procedures; for comparison, we isolated other tissues and organs from the same embryos, including the perichondrial tissues immediately adjacent to the cartilages. About 100 mg of each sample were homogenized and extracted, and extracts were added to the reporter cell line; 24 h later, cultures were stained histochemically for β-galactosidase. Standards included cultures receiving known amounts of natural retinoids, such as all-*trans*-retinoic acid or 9-*cis*-retinoic acid. We found that at each stage studied, the cartilaginous elements contained endogenous retinoids (Fig. 6, left panels). These levels were higher than those in brain but much lower than those in liver (not shown). Surprisingly and unexpectedly, extremely large amounts of retinoids were present in perichondrial tissues (Fig. 6, right panels); on a tissue wet weight basis, these amounts were comparable to those in liver. Very similar observations were made in a recent study with a transgenic mouse carrying a RARE/β-galactosidase reporter construct which is activated by endogenous retinoids (von Schroder and Heersche, 1998); the authors found that strong β-galactosidase activity (and hence high retinoid content) was present in perichondrial tissues adjacent to the prehypertrophic and hypertrophic zones of long bone growth plate as well as in hypertrophic cartilage itself.

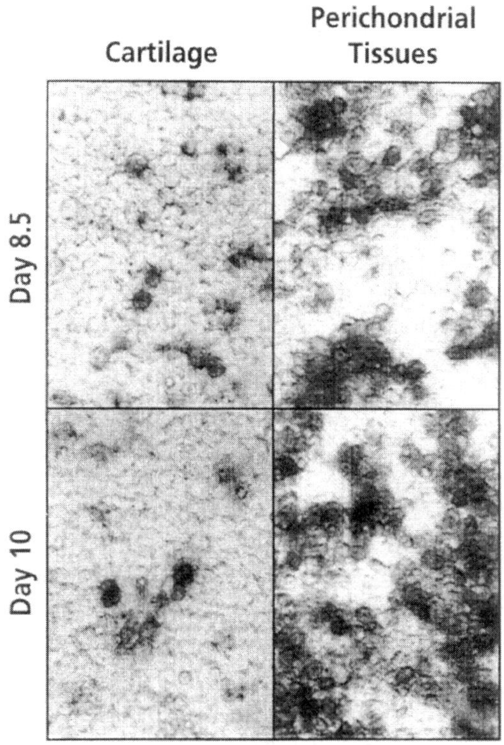

Figure 6. Bioassay of endogenous retinoid content in cartilaginous elements (left panels) and perichondrial tissues (right panels) isolated from Day 8.5 and Day 10 chick embryo limbs. Tissue extracts were used to treat F9 cells stably transfected with a β-galactosidase/RARE reporter construct, and reporter activity was determined 24 hrs later by histochemical staining.

The above data, combined with the finding of a specific RARγ upregulation in hypertrophic chondrocytes, set the stage for a third series of experiments in which we asked whether the endogenous retinoids and RARγ are actually required for chondrocyte hypertrophy and ossification in vivo. To approach this question, we implanted beads containing retinoid antagonists around newly-formed early cartilaginous elements in the chick wing and determined effects over developmental time. As pointed out above (see Fig. 3), the major advantage of this pharmacological approach is that the antagonists can be used at specific stages of development and can be placed in contact with specific skeletal elements or portions thereof. Thus, drug action and developmental consequences can be studied at the local level, minimizing the possibility that the effects are global and of a systemic nature. The RAR antagonist used was RO 41-5253 from Hoffman-LaRoche (Keidel et al., 1994) which exerts antagonist effects on all RARs. Three to 4 beads containing the antagonist were placed around the Day 4.5-5.5 humeral anlagen and embryos were examined over time. The results were dramatic. By Day 10, humerus in control embryos (implanted with beads containing vehicle) had developed normally and displayed a typical elongated morphology and size; in sharp contrast, the antagonist-treated humerus was about half the length. No effects were seen in radius and ulna, attesting to the fact that the effects were limited to the site of bead implantation and were not systemic.

Histology and in situ hybridization provided further insights into the developmental perturbations caused by the block of retinoid signaling (Fig. 7). In control humerus, the growth plate displayed normal zones of proliferating, prehypertrophic, hypertrophic and mineralizing chondrocytes; the metaphysis was surrounded by an intramembranous bone collar (Fig. 7A, arrowhead), and the diaphysis was undergoing invasion and replacement by endochondral bone and marrow (Fig. 7A, arrow). There was strong and typical gene expression of IHH in prehypertrophic chondrocytes (Fig. 7D, arrow), RARγ in hypertrophic chondrocytes (Fig. 7B, arrow), and osteopontin in endochondral bone (Fig. 7C). There was also osteopontin expression in intramembranous bone collar surrounding the IHH-expressing prehypertrophic chondrocytes (Fig. 7C, arrowhead). In sharp contrast, the antagonist-treated specimens were entirely cartilaginous and displayed no hypertrophic chondrocytes, no endochondral bone and marrow (Fig. 7E) and no expression of RARγ (Fig. 7F). Interestingly, IHH expression was not only present but seemed more extensive than control (Fig. 7H, arrows), and the metaphyseal-diaphyseal portion was surrounded by a conspicuous intramembranous bone collar (Fig. 7E, arrowhead) strongly expressing osteopontin (Fig. 7G, arrowheads). Thus, interference with retinoid signaling has very specific consequences on long bone development and prevents completion of this process. The chondrocytes can reach the prehypertrophic IHH-expressing stage but cannot pass it; likewise, the intramembranous bone collar forms but there is no formation of endochondral bone and no marrow invasion. Retinoid signaling thus appears to be required for normal progression through the terminal phases of long bone development.

Retinoid Signaling and IHH Expression

The in situ data above indicate that IHH gene expression is not only maintained in antagonist-treated skeletal anlagen, but is broader and more extensive than in control specimens. This led us to ask whether under normal circumstances retinoid signaling regulates IHH gene expression. To test this hypothesis, we carried out studies with cultured prehypertrophic-early hypertrophic chondrocytes isolated from the cephalic portion of Day 17 chick embryo sterna (Gibson and Flint, 1985). Cells were grown in monolayer for a few days in complete serum-containing medium and were then treated with 30 nM all-*trans*-retinoic acid for 2, 4 and 6 day. RNA was isolated from each culture and processed for northern blot analysis, using a cDNA probe encoding IHH. This retinoid was chosen because it is a natural retinoid and is present in the developing limb (Eichele and Thaller, 1987); the dose used is precisely within the range seen in the developing limb as well. For comparison, we determined the effects of the retinoid antagonist used above, namely RO 41-5253. Control untreated chondrocytes displayed obvious expression of IHH (Fig. 8A, lane 1). Upon treatment with all-*trans*-retinoic acid, IHH RNA levels were decreased

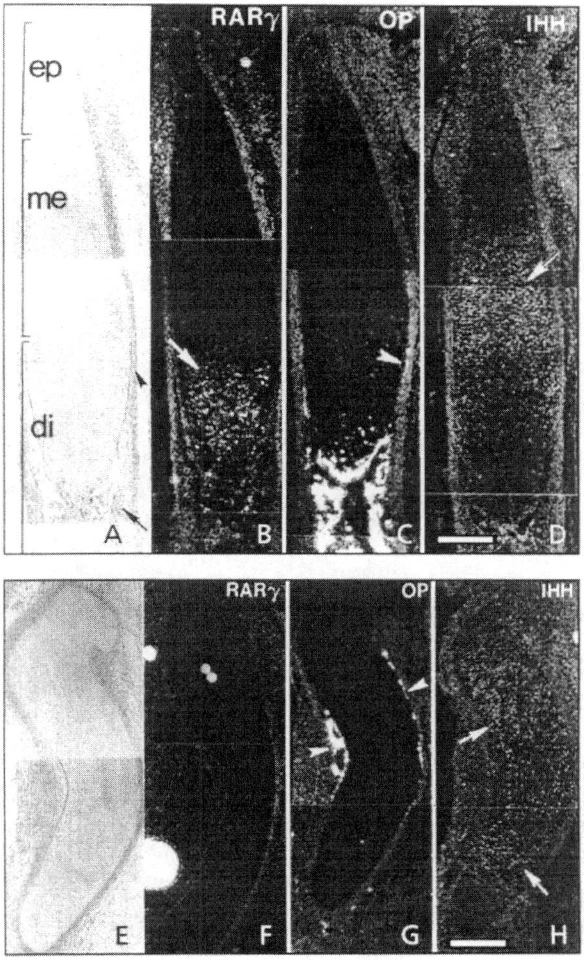

Figure 7. In situ hybridization analysis of expression of indicated genes in control Day 10 humerus (A-D) and retinoid antagonist-treated humerus (E-H). See text for details. Ep, epiphysis; me, metaphysis; and di, diaphysis. Bars, 250 μm.

markedly (Fig. 8A, lanes 2-4); on the contrary, treatment with 50 nM retinoid antagonist boosted IHH gene expression by several fold (Fig. 8A, lanes 5-7), in good correlation with the in situ data (see Fig. 7H). Clearly, retinoid signaling appears to represent a powerful and effective switch by which expression of the IHH gene is inhibited in prehypertrophic-early hypertrophic chondrocytes.

To determine the specificity of this effect, we examined in the above cultures whether treatment with all-*trans*-retinoic acid or RO 41-5253 affected expression of alkaline phosphatase (APase), a typical hypertrophic cell trait. Northern hybridization showed that treatment with all-*trans*-retinoic acid led to a powerful increase in APase gene expression (Fig. 8B, lanes 2-4) compared to control values (Fig. 8B, lane 1), while antagonist treatment decreased it (Fig. 8B, lanes 5-7). Thus, retinoid signaling down-regulates traits characteristic of prehypertrophic chondrocytes (i.e. IHH) and induces expression of hypertrophic traits (i.e. APase).

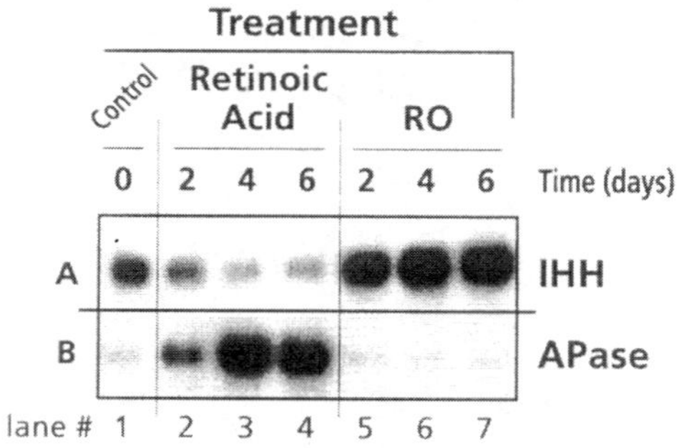

Figure 8. Northern blot analysis of IHH and APase gene expression in cultured chondrocytes. Cells were left untreated (lane 1) or were treated with all-*trans*-retinoic acid (lanes 2-4) or antagonist (lanes 5-7) for 2, 4 and 6 days.

Perichondrial Tissues as Positive Regulators of Chondrocyte Maturation

Perichondrial tissues adjacent to the prehypertrophic to hypertrophic zones of growth plate contain large amounts of endogenous retinoids, which in turn could exert a positive effect on neighboring chondrocytes and favor their maturation. To gain support for our hypothesis, we carried out the following studies.

We reasoned that if perichondrial tissues were to provide positive signals for chondrocyte maturation, hypertrophic chondrocytes should first emerge along the chondro-perichondrial border in an early developing long bone anlage, because chondrocytes in that location would be closer to the source of positive perichondrially-derived signals. Thus, we systematically examined the development of long bone anlagen between Day 7.5 and Day 9.0 of chick embryogenesis. We knew from previous observations that a Day 7.5 anlage contains chondrocytes up to the prehypertrophic stage but does not contain hypertrophic cells yet; conversely, a Day 9 anlage displays a clear hypertrophic zone in the diaphysis. Thus, we prepared longitudinal sections of limbs from Day 7.5 through Day 9 chick embryos and processed them for histology and in situ hybridization, using type X collagen as a molecular marker of chondrocyte hypertrophy. We found that the first hypertrophic type X collagen-expressing hypertrophic chondrocytes emerged on Day 8.5 of development and were indeed located along the chondro-perichondrial border; no such cells were present in the center where the distance from the border is greater (Fig. 9A- 9B, arrows). By Day 9, hypertrophic chondrocytes had formed a "zone", that is they were uniformly present from border to border (not shown).

To corroborate this finding, we implanted a single bead containing the retinoid antagonist RO 41-5253 next to the incipient diaphysis of a Day 5.5 humerus anlage and reincubated the embryos until Day 8.5. Because the antagonist-filled bead creates a concentration gradient (see above), we predicted that the antagonist would block emergence of type X collagen-expressing chondrocytes in the near side of humerus diaphysis (close to the bead) but not in the far side. In situ hybridization on longitudinal sections of Day 8.5 control and antagonist-treated humerus showed that this prediction was correct. Type X collagen-expressing chondrocytes were absent in the near side (Fig. 9D, arrowhead), but were present on the far side (Fig. 9D, arrow) Together, the data indicate that the chondro-perichondrial border serves as the initial site for emergence of hypertrophic chondrocytes. This site may thus have special pro-maturation properties, including presence of pro-maturation retinoids.

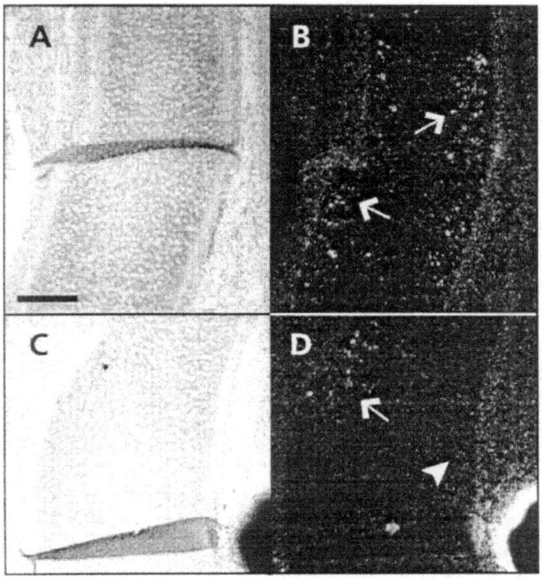

Figure 9. In situ hybridization analysis of type X collagen gene expression in Day 8.5 ulna. A-B: control ulna displaying newly-emerged type X collagen-expressing hypertrophic chondrocytes along the chondro-perichondrial border (arrows). C-D: ulna implanted with a single antagonist-filled bead (visible in the lower left corner) and displaying type X collagen transcripts only on far side (arrow) but not the near side (arrowhead) from the bead. Bar, 200 μm.

Conclusions and a Model

The data presented here provide further insights into the roles of IHH and retinoids in growth plate and long bone development. It is clear from our immunohistochemical data that IHH is not limited to the prehypertrophic zone where it is produced, but can reach surrounding zones and tissues. IHH could do so via long-range diffusion mechanisms mediated by proteoglycan molecules (Bellaiche et al., 1998) or micelle-like structures (Zeng et al., 2001). Presence of IHH in inner osteogenic layers of perichondrium, lack of collar formation following cyclopamine treatment, and IHH ability to induce differentiation of osteogenic cells in culture provide the most direct evidence to date that IHH is a temporal-spatial regulator of intramembranous bone collar formation. Its presence in the hypertrophic zone of growth plate, particularly along the chondro-osseus border, provides the first evidence that IHH could be involved in endochondral ossification as well. These findings correlate, and provide an explanation for, the previous observation that in IHH-null mice there is neither intramembranous nor endochondral ossification in the limbs (St-Jacques et al., 1999). Lastly, our data indicate that IHH simulates chondrocyte proliferation in vitro. Taken together, the data portray IHH as a global regular and coordinator of functions and events in multiple growth plate zones and adjacent tissues and structures.

With regard to retinoids, our previous data and the additional evidence presented here indicate that these signaling molecules have equally important roles in long bone development. When retinoid signaling and action are experimentally stopped by pharmacological intervention, the chondrocyte maturation process is arrested at the prehypertrophic IHH-expressing stage. In very good agreement with the data on IHH, the retinoid antagonist-mediated maturational arrest is accompanied by lack of endochondral bone but formation of an intramembranous bone collar. This indicates that the two ossification pathways are dissociable and that prehypertrophic chondrocytes can still exert their positive influence on collar formation even though downstream processes (hypertrophy and endochondral ossification) have been halted.

The new data presented indicate that retinoid signaling has two additional important roles. The first would be to regulate IHH gene expression in the growth plate. IHH transcripts are limited to the prehypertrophic zone, but the mechanisms accounting for this restricted expression pattern are still unknown. Our data suggest that retinoids may have the important role of turning off gene expression during the transition from prehypertrophic to hypertrophic chondrocytes. This would be important because misexpression of IHH or ablation of the IHH gene both have serious consequences on the growth plate and long bone formation (Vortkamp et al., 1996; St-Jacques et al., 1999). Misexpression results in suppression of chondrocyte hypertrophy and endochondral ossification, without major changes in intramembranous collar formation (Vortkampt et al., 1996). IHH gene ablation is accompanied by lack of both intramembranous and endochondral ossification and growth plate abnormalities (St-Jacques et al., 1999). Clearly, normal amounts and distribution patterns of IHH in several growth plate zones (Fig. 2) must be of paramount importance for skeletogenesis. Lastly, our in situ hybridization data show that the very first type X collagen-expressing chondrocytes emerge along the lateral border between diaphyseal cartilage and perichondrial tissues. These tissues are rich in endogenous retinoids (Fig. 6) (Koyama et al., 1999; von Schroder and Heersche, 1998) and retinoids readily induce type X collagen gene expression in cultured chondrocytes (Adams et al., 1991). It is thus conceivable that perichondrium-derived retinoids may diffuse into the adjacent cartilage tissue and turn on type X collagen gene expression. In broader terms, the retinoid-rich perichondrium surrounding the diaphyseal portion of long bone anlagen would provide a positive stimulus for completion of chondrocyte maturation.

The sequential and interrelated roles of IHH and retinoid signaling pathways in long bone development are depicted and integrated in the following working model (Fig. 10). IHH produced by prehypertrophic chondrocytes (Fig. 10, step 1) would reach adjacent perichondrial tissues as well as the proliferative and hypertrophic zones of growth plate. IHH diffusing into the proliferative zone would influence chondrocyte mitotic activity and overall maturation rates together with other powerful mitogens, such as PTHrP (step 2) (Vortkamp et al., 1996). IHH diffusing into perichondrium would induce osteogenesis and formation of the intramembranous bone collar (Fig. 10, step 3). IHH diffusing through the hypertrophic zone would reach the chondro-marrow border and stimulate endochondral bone formation (step 4). The model prescribes also that formation of a bone collar and associated vasculature would result in increased local amounts of blood-derived retinoids (step 5); these molecules would diffuse into the growth plate and provide a positive stimulus for completion of chondrocyte maturation, including up-regulation of RARγ gene expression and down-regulation of IHH gene expression (step 6). Formation of fully mature hypertrophic mineralizing chondrocytes, combined with IHH protein diffusing from the prehypertrophic zone, would permit and favor endochondral bone formation.

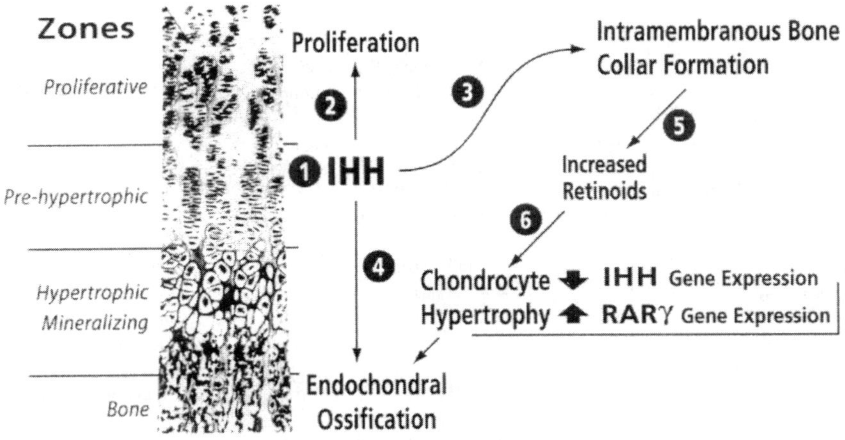

Figure 10. Model depicting the distinct but interrelated roles of IHH and retinoids in long bone development. See text for details.

The model has a major and novel implication for growth plate biology. Up until recently, the growth plate has been viewed as a structure in which each of its zones is a separate and independent entity. In other words, behavior and function of chondrocytes in each zone would not be directly influenced by those in adjacent zones. Our model and that proposed in a recent related study (Chung et al., 2001) suggest a different view, one in which the growth plate is a highly interactive structure and each zone is influenced by, and dependent upon and coordinated with, events in flanking zones and events in perichondrial tissues. The growth plate would be controlled by flows of regulatory cues going from the resting to the hypertrophic zone, from the hypertrophic zone to the resting zone, and from growth plate to perichondrial tissues. This novel view of the growth plate is not as radical as it may appear, and findings supporting it have actually appeared in the literature previously. For instance, in their elegant studies Jacenko and co-workers showed that expression of a mutant non-functional type X collagen in the hypertrophic zone of growth plates in transgenic mice has negative repercussions not only in the hypertrophic zone itself but in other zones as well (Jacenko et al., 1993; Gress and Jacenko, 2000). The proliferative zone seems to be particularly affected. Similar global growth plate changes were noted in type X collagen gene null mice (Gress and Jacenko, 2000). These examples do suggest that there may indeed be a "retrograde" flow of regulatory cues going from the hypertrophic zone back to the resting/proliferative zone and that the growth plate is indeed far more integrated than previously realized. Based on this evidence and our own data, we propose that this "retrograde" flow is of paramount importance and perhaps more crucial than a standard flow from resting to hypertrophic zone. Ongoing experiments are examining these tantalizing possibilities.

Acknowledgements

We thank our colleagues Drs. S. L. Adams and T. Kirsch who participated in studies upon which this chapter is based, and Ms. Eleanor Golden for help with experiments with cultured chondrocytes. Original work was supported by NIH grants AR 40833 and AR 45402 to M.P.

References

Adams, S.L., Pallante, K.M., Niu, Z., Leboy, P.S., Golden, E.B., and Pacifici, M. (1991). Rapid induction of type X collagen gene expression in cultured chick vertebral chondrocytes. *Exp. Cell Res.* 193, 190-197

Bellaiche, Y., The, I., and Perrimon, N. (1998). Tout-velu is a Drosophila homologue of the putative tumor suppressor EXT-1 and is needed for Hh diffusion. *Nature* 394, 85-88.

Bitgood, M.J., and McMahon, A.P. (1995). Hedgehog and Bmp genes are coexpressed at many diverse sites of cell-cell interaction in the mouse embryo. *Dev. Biol.* 172, 126-138.

Chambon, P. (1994). The retinoid signaling pathway: molecular and genetic analyses. *Semin. Cell Biol* 5, 115-125.

Chung, U.-I., Schipani, E., McMahon, A.P., and Kronenberg, H.M. (2001). Indian hedgehog couples chondrogenesis to osteogenesis in endochondral bone development. *J. Clin. Invest.* 107, 295-304.

Dolle, P., Ruberte, E., Kastener, P., Petkovich, M., Stoner, C.M., Gudas, L.J., and Chambon, P. (1989). Differential expression of genes encoding α, β and γ retinoic acid receptors and CRABP in the developing limbs of the mouse. *Nature* 342, 702-705.

Duprez, D., Fournier-Thibault, C., and Le Douarin, N. (1998). Sonic hedgehog induces proliferation of committed skeletal muscle cells in the chick limb. *Development* 125, 495-505.

Eichele, G., and Thaller, C. (1987). Characterization of concentration gradients of a morphogenetically active retinoid in the chick limb bud. *J. Cell Biol.* 105, 1917-1923.

Eichele, G., Tickle, C., and Alberts, B. (1984). Microcontrolled release of biologically active compounds in chick embryos: beads of 200-μm diameter for the local release of retinoids. *Anal. Biochem.* 142, 542-555.

Fell, H.B. (1925). The histogenesis of cartilage and bone in the long bones of the embryonic fowl. *J. Morphol. Physiol.* 40, 417-459.

Gibson, G.J., and Flint, M.H. (1985). Type X collagen synthesis by chick sternal cartilage and its relationship to endochondral development. *J. Cell Biol.* 101, 277-284.

Gigante, A., Specchia, N., Nori, S., and Greco, F. (1996). Distribution of elastic fiber types in the epiphyseal region. *J. Orthop. Res.* 14, 810-817.

Gress, C.J., and Jacenko, O. 2000. Growth plate compressions and altered hematopoiesis in collagen X null mice. J. Cell Biol. 149, 983-993

Hinchcliffe, J.R., and Johnson, D.R. (1990). *The Development of the Vertebrate Limb.* Clarendon Press, Oxford.

Incardona, J. P., Gaffield, W., Kapur, R. P., and Roelink, H. (1998). The teratogenic Veratrum alkaloid cyclopamine inhibits Sonic hedgehog signal transduction. *Development* 125, 3553-3562

Iwamoto, M., Shapiro, I.M., Yagami, K., Boskey, A.L., Leboy, P.S., Adams, S.L., and Pacifici, M. (1993). Retinoic acid induces rapid mineralization and expression of mineralization-related genes in chondrocytes. *Exp. Cell Res.* 207, 413-420.

Iwamoto, M., Yagami, K., Shapiro, I.M., Leboy, P.L., Adams, S.L., and Pacifici, M. (1994). Retinoic acid is a major regulator of chondrocyte maturation and matrix mineralization. *Microsc. Res. Tech.* 28, 483-491.

Jacenko, O., Lu Valle, P., and Olsen, B. R. 1993. Spondylometaphyseal dysplasia in mice carrying a dominant negative mutation in a matrix protein specific for cartilage-to-bone transition. *Nature* 365, 56-61.

Jensen, A.M., and Wallace, V.A. (1997). Expression of Sonic hedgehog and its putative role as a precursor cell mitogen in the developing mouse retina. *Development* 124, 363-371.

Keidel, S., LeMotte, P., and Apfel, C. (1994). Different agonist- and antagonist-induced conformational changes in retinoic acid receptors analyzed by protease mapping. *Mol. Cell Biol.* 14, 287-298.

Koyama, E., Leatherman, J.L., Noji, S., and Pacifici, M. (1996). Early chick limb cartilaginous elements possess polarizing activity and express hedgehog-related morphogenetic factors. *Dev. Dynam.* 207, 344-354.

Koyama, E., Golden, E.B., Kirsch, T., Adams, S.L., Chandraratna, R.A., Michaille, J.J., and Pacifici, M. (1999). Retinoid signaling is required for chondrocyte maturation and endochondral bone formation during limb skeletogenesis. *Dev. Biol.* 208, 375-391.

Mangelsdorf, D.J., Umesono, K., and Evans, R.M. (1994). The retinoid receptors. In "*The Retinoids: Biology, Chemistry, and Medicine*". [M.B. Sporn et al., eds.]. Raven Press, New York, pp. 319-349.

Mendelsohn, C., Lohnes, D., Decimo, D., Lufkin, T., LeLeur, M., Chambon, P., and Mark, M. (1994). Function of the retinoic acid receptors (RARs) during development. II. Multiple abnormalities at various stages of organogenesis in RAR double mutants. *Development* 120, 2749-2771.

Nakamura, T., Aikawa, T., Iwamoto-Enomoto, M., Iwamoto, M., Higuchi, Y., Pacifici, M., Kinto, N., Yamaguchi, A., Noji, S., Kurisu, K., and Matsuya, T. (1997). Induction of osteogenic differentiation by hedgehog proteins. *Biochem. Biophys. Res. Commun.* 237, 465-469.

Riddle, R.D., Johnson, R.L., Laufer, E., and Tabin, C. (1993). Sonic hedgehog mediates the polarizing activity of the ZPA. *Cell* 75,1401-1416

St-Jacques, B., Hammerschidt, M., and McMahon, A.P. (1999). Indian hedgehog signaling regulates proliferation and differentiation of chondrocytes and is essential for bone formation. *Genes Dev.* 13, 2076-2086.

Thorogood, P. (1983). Morphogenesis of cartilage. In "*Cartilage*" (B.K. Hall, ed.), vol. 2. Academic Press, New York, pp. 223-254.

von Schroder, H.P., and Heersche, J. N. M. (1998). Retinoic acid responsiveness of cells and tissues in developing fetal limbs evaluated in a RAREhsplacZ transgenic mouse model. *J. Orthop. Res.* 16, 355-364.

Vortkamp, A., Lee, K., Lanske, B., Segre, G.V., Kronenberg, H.M., and Tabin, C.J. (1996). Regulation of the rate of cartilage differentiation by Indian hedgehog and PTH-related protein. *Science* 273, 613-622.

Vortkampt, A., Pathi, S., Peretti, G.M., Caruso, E.M., Zaleske, D.J., and Tabin, C.J. (1998). Recapitulation of signals regulating embryonic bone formation during postnatal growth and in fracture repair. *Mech. Dev.* 71, 65-75.

Wagner, M., Han, B., and Jessell, T.M. (1992). Regional differences in retinoid release from embryonic neural tissue detected by an in vitro reporter assay. *Development* 116, 55-66.

Walbach, S.B., and Hegsted, D.M. (1952). Vitamin A deficiency in the duck. Skeletal growth and the central nervous system. *Arch. Pathol.* 54, 548-563.

Yin, M., Gentili, C., Koyama, E., Zasloff, M., and Pacifici, M. (2001). Antiangiogenic treatment delays chondrocyte maturation and bone formation during limb skeletogenesis. *J. Bone Min. Res.* (in press).

Zeng, X., Goetz, J.A., Suber, L.M., Scott, W.J., Schreiner, C.M., and Robbins, D.J. (2001). A freely diffusible form of Sonic hedgehog mediates long-range signalling. *Nature* 411, 716-720.

The Growth Plate
I.M. Shapiro et al. (Eds.)
IOS Press, 2002

Involvement of Cbfa1 in Chondrocyte Differentiation Maturation, and Endochondral Ossification, and the Specification of the Cartilage Phenotype

Toshihisa KOMORI[1,4], Masahiro IWAMOTO[2], Naoko KANATANI[1], Carolina YOSHIDA[1], Motomi ENOMOTO-IWAMOTO[3], and Chisato UETA[1]

[1]*Department of Molecular Medicine, Osaka University Medical School, Osaka, Japan,* [2]*Department of Oral Anatomy and Developmental Biology,* [3]*Department of Biochemistry, Osaka University Faculty of Dentistry, Suita, Osaka 565-0871, Japan,* [4] *"Form and Function", Precursory Research for Embryonic Science and Technology, Japan Science and Technology Corporation, and Department of Molecular Medicine, Osaka University, Medical School, Osaka, Japan*

Abstract. Cbfa1 (core binding factor α1)/Runx2 (runt-related gene 2) is expressed in chondrocytes as well as osteoblasts. In the growth plate, Cbfa1 expression is up-regulated in prehypertrophic chondrocytes. Cbfa1-deficient mice which display a complete lack of bone formation, owing to the maturational arrest of osteoblasts, evidence disturbed chondrocyte maturation. We analyzed the function of Cbfa1 in chondrocytes using ATDC5 cells and chick chondrocytes. In the chondrogenic cell line, ATDC5, treatment with Cbfa1 antisense oligonucleotides suppressed ATDC5 cell maturation. Retrovirally forced expression of Cbfa1 in immature chondrocytes induced chondrocyte maturation, while the dominant-negative form of Cbfa1 (DN-Cbfa1) inhibited maturation. It was concluded that Cbfa1 expression stimulates chondrocyte maturation. To further understand the roles of Cbfa1 in chondrocytes during skeletal development, we generated DN-Cbfa1 transgenic mice that overexpress Cbfa1. In the Cbfa1 transgenic mice, chondrocyte maturation and endochondral ossification was greatly accelerated, whereas endochondral ossification was completely blocked. In DN-Cbfa1 transgenic mice, the cartilages were composed of immature chondrocytes. Further, Cbfa1 transgenic mice failed to form most of their joints, while permanent cartilage was present in endochondral bone. In contrast, most of the chondrocytes in DN-Cbfa1 transgenic mice retained the permanent cartilage phenotype. These findings demonstrate that Cbfa1 plays an important role, not only in chondrocyte maturation in the process of endochondral ossification, but also in the expression of the cartilage phenotype.

Introduction

Cbfa1, also called Runx2, is a transcription factor that belongs to the runt-domain gene family [1,2]. There are three runt-domain genes (Cbfa1/Runx2, Cbfa2/Runx1, and Cbfa3/Runx3), all of which have a DNA-binding domain, runt, that is homologous with the Drosophila pair-rule gene runt. These proteins form heterodimers with the transcriptional co-activator Cbfb in vitro, and specifically recognize a consensus sequence, TGT/CGGT.

Cbfa1 has at least two isoforms with different N-terminal sequences. One product starts with a sequence (MRIPVD) encoded by exon 2 (hereafter referred to as type I Cbaf1) [3]. The other product, with a sequence (MASNS), contains sequences from exon 1 spliced to an aspartic acid residue at position 6 of the former sequence located in exon 2 (hereafter referred to as type II Cbfa1) [4-6]. Although type I Cbfa1 is upregulated earlier than type II Cbfa1 [7], both Cbfa1 isoforms are weakly expressed in proliferating chondrocytes, and their expression is upregulated during chondrocyte maturation. Terminal hypertrophic chondrocytes as well as osteoblasts show the most extensive expression of both isoforms of Cbfa1.

In Cbfa1-deficient mice, while the entire skeleton is composed of cartilage [8, 9], chondrocyte differentiation is severely disturbed [10, 11]. Parathyroid hormone/parathyroid hormone related peptide receptor (PTH/PTHrPR), Indian hedgehog (Ihh), type X collagen, and BMP-6 expression was not detected in humerus and femur, indicating that chondrocyte differentiation was blocked at the prehypertrophic chondrocyte stage [10]. However, it is possible that chondrocyte differentiation leading to endochondral ossification is largely controlled by surrounding osteoblastic and hematopoietic cells, and that a lack of Cbfa1 in chondrocytes may not be a major reason for suppression of chondrocyte maturation. In this communication we examine the mechanism of inhibition of endochondral ossification in Cbfa1-deficient mice.

Cbfa1 is a Positive Inducer of Chondrocyte Maturation

In the chondrogenic cell line, ATDC5, type I Cbfa1 expression was elevated prior to differentiation into the hypertrophic phenotype, and treatment with antisense oligonucleotides for type I Cbfa1 reduced type X collagen expression [7]. Retrovirally forced expression of either type I or type II Cbfa1 in immature chick chondrocytes decreased cell proliferation, induced glycosaminoglycan production, raised ALP activity and elevated type X collagen and MMP13 expression, and caused extensive cartilage-matrix mineralization [7]. Further, forced expression of the dominant negative form of Cbfa1 (DN-Cbfa1) suppressed type X collagen expression [12]. These results confirm that Cbfa1 is a positive inducer of chondrocyte maturation (Fig. 1) [7,13].

Chondrocyte Maturation and Endochondral Ossification is Accelerated in Cbfa1 Transgenic Mice

To investigate the function of Cbfa1 in chondrocytes during skeletal development, we generated two kinds of Cbfa1 transgenic mice under the control of promoter/enhancer elements of the α1 type II collagen gene (Col2a1) using type I Cbfa1 and type II Cbfa1 (Fig. 2) [12]. The Col2a1 promoter and enhancer fragments that were used mimicked the normal temporal and spatial expression of endogenous type II collagen, which is expressed in chondroprogenitor cells as well as chondrocytes. Evaluation of reporter activity indicated that mice overexpressing type I or type II Cbfa1 died at birth from respiratory failure. At embryonic day 18.5 (E18.5), they displayed severe dwarfism, with shortened limbs, domed skull, protruding tongue and shortened snout and mandible. Most skeletal elements, including the chondocranium, ribs, and vertebrae, showed massive mineralization with severely reduced cartilaginous portions; the thoracic cage was narrow and bell-shaped. The overall phenotypes of type I and type II Cbfa1 transgenic mice were generally similar, although type II Cbfa1 transgenic mice seemed to develop these severe abnormalities slightly earlier than type I Cbaf1 animals.

In E18.5 Cbfa1 transgenic mice, abnormal chondrocyte maturation and endochondral ossification was observed in nasal septum cartilage, presumptive intervertebral regions, and thyroid, cricoid, and tracheal cartilages. Further, the distal portion of Meckel's cartilage, which is normally replaced with fibrous tissue, displayed endochondral ossification. All of these findings indicate that chondrocyte maturation had been greatly enhanced in Cbfa1 transgenic mice. In E15.5 Cbfa1 transgenic mice, humerus, radius, and ulna were fused and the growth plate was disorganized. The fused joint regions contained enlarged chondrocytes expressing type X collagen and a small number of type II collagen-positive cells. In E18.5 Cbfa1 transgenic mice, the fused forelimb skeletal elements displayed gross mineralization. Therefore, activation of Cbfa1 signaling in immature chondrocytes promoted hypertrophy and precocious endochondral ossification (Fig. 1).

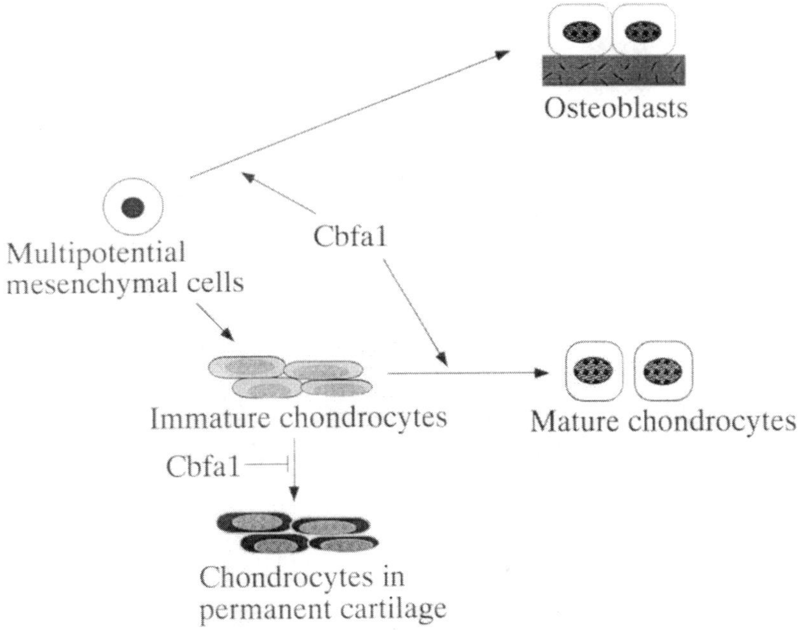

Figure 1. Cbfa1 is a fundamental transcription factor for both osteoblast differentiation and chondrocyte maturation. Cbfa1 induces both osteoblast differentiation and chondrocyte maturation and inhibits chondrocytes from acquiring permanent phenotype.

Chondrocyte Maturation and Endochondral Ossification is Inhibited in DN-Cbfa1 Transgenic Mice

To generate DN-Cbfa1 transgenic mice, we used two cDNAs encoding the runt domain only, or the runt domain with the N-terminal domain of type I Cbfa1 (Fig. 2) [12]. Both truncated forms of Cbfa1 exhibited strong dominant negative activity and were able to block the function of both type I and type II Cbfa1 in reporter assays using OSE elements. Transgenic mice overexpressing either form of DN-Cbfa1 also died soon after birth from respiratory failure, and displayed dwarfism with shortened limbs, just like Cbfa1-deficient mice. Most skeletal elements, including ribs, occipital bone, the base of the skull, and vertebral bodies, were still uncalcified cartilaginous tissues, although most of the skull and clavicles, which are formed by intramembranous ossification, were clearly mineralized.

In DN-Cbfa1 transgenic mice at E15.5, humerus, radius, and ulna remained entirely cartilaginous and were composed of immature chondrocytes expressing type II collagen, but not type X collagen. The formation of bone collars, which are formed by intramembranous ossification, was not observed. In E18.5 DN-Cbfa1 transgenic mice, bone collars of the humerus, radius, and ulna were apparent, but vascular invasion into the cartilage was rarely observed, indicating a severe delay or blockage of endochondral ossification. The inhibition of endochondral ossification in DN-Cbfa1 transgenic mice was linked to retarded chondrocyte maturation, since type II collagen was expressed but in most of the diaphyses, PTH/PTHrPR, Ihh, and type X collagen were not expressed. Transgenic mice overexpressing either form of DN-Cbfa1 showed similar phenotypes in all histological analyses.

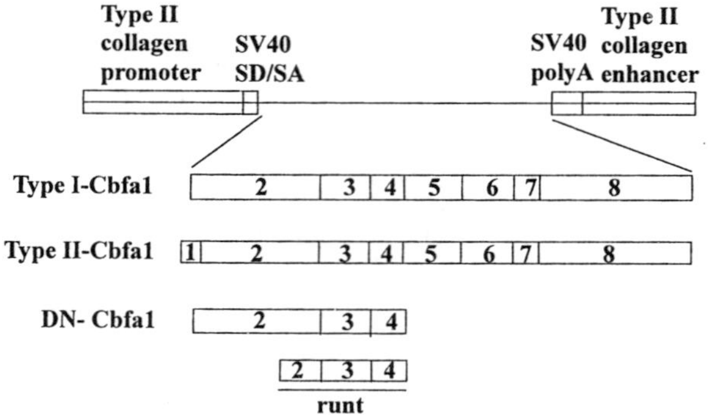

Figure 2. Diagrams of the DNA constructs used to generate Cbfa1 and DN-Cbfa1 transgenic mice. cDNAs of two Cbfa1 isoforms with different N-terminus and two DN-Cbfa1 cDNAs, which encode runt domain only or runt domain with N-terminal domain of type I Cbfa1.

Cbfa1 Plays an Important Role in the Specification of Cartilage Phenotype

Most skeletal elements in Cbfa1 transgenic mice were fused, except manus and pedis. Shoulder, elbow, hip, and knee joints were absent, and the vertebral bodies and arches were fused. Therefore, we examined the expression of growth and differentiation factor-5 (GDF-5) during limb development. In E12.5 Cbfa1 transgenic mice, humerus, radius, and ulna were already fused and GDF-5 expression was absent in the elbow joint. This result indicated that the transfers interrupted joint development in the elbow at an early stage, probably by affecting factors that determine the presumptive joint regions. In the shoulder, joint formation was observed and GDF-5 expression was detected at E12.5. However, the joint was fused at E15.5, indicating that Cbfa1 inhibited chondrocytes from acquiring the characteristics of articular cartilage.

We also examined tenascin expression, which is expressed in chondrocytes once cartilage tissue appears, but as cartilage development progresses it becomes limited to articular chondrocytes. In wild-type mice at E15.5, several layers of chondrocytes at the edge of the epiphysis expressed tenascin, whereas chondrocytes in the presumptive joint region of Cbfa1 transgenic mice did not. This result suggested that the cells had lost their permanent phenotype. In contrast, many chondrocytes in the DN-Cbfa1 transgenic mice expressed tenascin. Thus, in the DN-Cbfa1 transgenic mice most regions which contained developing cartilaginous elements retained the expression of a marker for permanent cartilage. The same regulation of tenascin gene expression by Cbfa1 was observed in vitro. Retrovirally forced expression of Cbfa1 decreased tenascin expression, while DN- Cbfa1 sustained it. These data indicate that Cbfa1 plays an important role in the specification of the cartilage phenotype (Fig. 1).

Conclusion

Cbfa1 is required for chondrocyte maturation. Further, Cbfa1 plays an important role in the specification of the cartilage phenotype. For these reasons, the temporal and spatial expression of Cbfa1 in chondrocytes is required for the formation of joints, permanent cartilages, and endochondral bones.

References

[1] Komori T, Kishimoto T 1998 Cbfa1 in bone development. Curr. Opin. in Genet. Dev. **8:**494-499.

[2] Yamaguchi A, Komori T, Suda T 2000 Regulation of osteoblast differentiation mediated by bone morphogenetic proteins, hedgehogs, and Cbfa1. Endocrine Rev. **21:**393-411.

[3] Ogawa E, Maruyama M, Kagoshima H, Inuzuka M, Lu J, Satake M, Shigesada K, Ito Y 1993 PEBP2/PEA2 represents a family of transcription factors homologous to the products of the Drosophila runt gene and the human AML1 gene. Proc Natl Acad Sci U S A. **90:**6859-6863.

[4] Mundlos S, Otto F, Mundlos C, Mulliken JB, Aylsworth AS, Albright S, Lindhout D, Cole WG, Henn, W, Knoll JHM, Owen MJ, Mertelsmann R, Zabel BU,. Olsen BR 1997 Mutations involving the transcription factor CBFA1 cause Cleidocranial Dysplasia. Cell **89:**773-779.

[5] Stewart M, Terry A, Hu M, O'Hara M, Blyth K, Baxter E, Cameron E, Onions DE, Neil JC 1997 Proviral insertions induce the expression of bone-specific isoforms of PEBP2alphaA (CBFA1): evidence for a new myc collaborating oncogene. Proc Natl Acad Sci U S A. **94:**8646-8651.

[6] Thirunavukkarasu K, Mahajan M, Mclarren KW, Stifani S, Karsenty G 1998 Two domains unique to osteoblast-specific transcription factor Osf2/Cbfa1 contribute to its transactivation function and its inability to heterodimerize with Cbfbeta. Mol. Cell Biol.**18:**4197.

[7] Enomoto H, Enomoto-Iwamoto M, Iwamoto M, Nomura S, Himeno M, Kitamura Y, Kishimoto T, Komori T 2000 Cbfa1 is a positive regulatory factor in chondrocyte maturation. J Biol Chem **275:**8695-8702.

[8] Komori T, Yagi H, Nomura S, Yamaguchi A, Sasaki K, Deguchi K, Shimizu Y, Bronson RT, Gao YH, Inada M, Sato M, Okamoto R, Kitamura Y, Yoshiki S, Kishimoto T 1997 Targeted disruption of Cbfa1 results in a complete lack of bone formation owing to maturational arrest of osteoblasts. Cell **89:**755-764.

[9] Otto F, Thornell AP, Crompton T, Denzel A, Gilmour KC, Rosewell IR, Stamp GWH, Beddington RSP, Mundlos S, Olsen BR, Selby PB, Owen MJ 1997 Cbfa1, a candidate gene for cleidocranial dysplasia syndrome, is essential for osteoblast differentiation and bone development. Cell **89:**765-771.

[10] Inada M, Yasui T, Nomura S, Miyake S, Deguchi K, Himeno M, Sato M, Yamagiwa H, Kimura T, Yasui N, Ochi T, Endo N, Kitamura Y, Kishimoto T, Komori T 1999 Maturational disturbance of chondrocytes in Cbfa1-deficient mice. Dev Dyn **214:**279-290.

[11] Kim IS, Otto F, Zabel B, Mundlos S 1999 Regulation of chondrocyte differentiation by cbfa1. Mech Dev **80:**159-170.

[12] Ueta C, Iwamoto M, Kanatani N, Yoshida C, Liu Y, Enomoto-Iwamoto M, Ohmori T, Enomoto H, Nakata K, Takada K, Kurisu K, Komori T 2001 Skeletal malformations caused by overexpression of Cbfa1 or its dominant negative form in chondrocytes. J Cell Biol **153:**87-100.

[13] Komori T 2000 A fundamental transcription factor for bone and cartilage. Biochem. Biophys. Res. Commun. **276:** 813-816.

Cell Maturation Specific Regulation of the PKC Signaling Pathway by 1α,25-$(OH)_2D_3$ and 24R,25-$(OH)_2D_3$ in Growth Plate Chondrocytes

Zvi Schwartz[1,2], Victor L. Sylvia[1], David D. Dean[1], and Barbara D. Boyan[1]
[1]*Departments of Orthopaedics, Periodontics, and Biochemistry, The University of Texas Health Science Center at San Antonio, San Antonio, Texas, USA; and* [2]*Department of Periodontics, Hebrew University Hadassah Faculty of Dental Medicine, Jerusalem, Israel*

Abstract. 1α,25$(OH)_2D_3$ and 24R,25$(OH)_2D_3$ regulate rat costochondral growth plate chondrocytes in a cell maturation-dependent manner. 1α,25$(OH)_2D_3$ primarily affects cells in the prehypertrophic and upper hypertrophic zones (GC, growth zone), whereas 24R,25$(OH)_2D_3$ primarily affects cells in the resting zone (RC). The cell-specific responses are mediated by protein kinase C (PKC). Only 1α,25$(OH)_2D_3$ regulates PKC in GC cells, and only 24R,25$(OH)_2D_3$ regulates PKC in RC cells. The effect of 1α,25$(OH)_2D_3$ is rapid, maximal by 9 minutes, and does not require protein synthesis, but the effect of 24R,25$(OH)_2D_3$ is maximal at 90 minutes and involves both gene expression and protein synthesis. Moreover, the direct effects of the metabolites on PKC in isolated plasma membranes and matrix vesicles is cell-specific. These observations suggest that separate mechanisms are involved. This paper reviews studies that support this hypothesis. In both GC and RC cells, regulation of PKC is membrane-mediated. Antibody 99 (Ab99), a polyclonal antisera generated to a 1,25$(OH)_2D_3$-binding protein in the basal lateral membrane of chick intestinal epithelium blocked the effect of 1α,25$(OH)_2D_3$, but not of 24R,25$(OH)_2D_3$, suggesting separate membrane receptors (mVDR). This was supported by Scatchard analysis of $[^3H]$-1,25$(OH)_2D_3$ and $[^3H]$-24,25$(OH)_2D_3$ binding. The receptors are not the classical nuclear VDR, since analogues of 1,25$(OH)_2D_3$ with low affinity for the nuclear VDR also increase PKC in a cell-specific manner, and isolated matrix vesicles lack immunoreactive nuclear VDR, but exhibit an Ab99-positive band at M_r 65,000. In both GC and RC cells and in their isolated plasma membranes, the PKC isoform that was affected was PKCα, whereas in matrix vesicles, the sensitive isoform was PKCζ. The effect of 1α,25$(OH)_2D_3$ on PKCα in GC cells required Gq-dependent, phosphatidylinositol-specific phospholipase C (PI-PLC), which generated inositol-1,4,5-trisphosphate and diacylglycerol (DAG), causing translocation of PKC to the plasma membrane. In addition, 1α,25$(OH)_2D_3$ stimulated cytosolic PLA_2, generating arachidonic acid (AA), which activated PKC. Metabolism of AA via cyclooxygenase-1 (Cox-1) resulted in prostaglandin production, including PGE_1 and PGE_2. PGE_2 acted on its EP1 receptor to further increase PKC activity. DAG mediated the effect of 24R,25$(OH)_2D_3$ in RC cells, but it was produced through the action of PLD2, and PKC translocation did not occur. Moreover, PLA_2 activity decreased, resulting in decreased production of AA and PGE_2. 1α,25$(OH)_2D_3$-dependent decreases in matrix vesicle PKCζ were mediated by Gq, but did not involve PLC or DAG. In contrast, the 24R,25$(OH)_2D_3$-dependent decrease in PKCζ was mediated by PLA_2, since the PLA_2 inhibitor quinacrine caused a synergistic decrease. These results show that 1α,25$(OH)_2D_3$ and 24R,25$(OH)_2D_3$ both exert their target cell specific effects through the PKC signaling pathway, but different receptors and different mechanisms are involved. Moreover, they regulate matrix vesicle PKC activity differently than in the cell.

Introduction

Steroid hormones regulate bone growth and development in a complex manner involving traditional nuclear receptor-mediated mechanisms. The process of endochondral ossification is sensitive to a number of steroid hormones including the seco-steroids 1α,25-$(OH)_2D_3$ and 24R,25-$(OH)_2D_3$. These vitamin D metabolites exert their effects on the cells in the growth plate in distinctly different ways (see Boyan et al. [1] for a review). 1α,25-$(OH)_2D_3$ is required for deposition of mineral in the extracellular matrix, in part through its action on Ca^{++} ion transport, and in part by regulating chondrocyte proliferation and hypertrophy. 24R,25-$(OH)_2D_3$ also plays a role by modulating the differentiation and maturation of resting zone cells. Studies using rat costochondral growth plate chondrocyte cultures indicate that 1α,25-$(OH)_2D_3$ primarily targets cells in the prehypertrophic and upper hypertrophic cell zones (growth zone), whereas 24R,25-$(OH)_2D_3$ targets cells in the resting zone. These observations are supported by in vivo studies examining rat tibia epiphyseal cartilage. [2]

Nuclear receptors for 1α,25-$(OH)_2D_3$ (1,25-nVDR) have been found in cartilage cells by a number of laboratories. [3-5] However, even though autoradiography indicates the existence of specific binding of radiolabeled 24R,25-$(OH)_2D_3$ in the growth plate, [6] no one has yet shown the presence of nuclear receptors for this vitamin D metabolite. As a result, the mechanisms by which 24R,25-$(OH)_2D_3$ might exert its effect on resting zone cells remain unclear.

We, and others, have identified a new class of receptors for steroid hormones like the vitamin D metabolites (see Nemere and Farach-Carson [7] for a review) that may account for their differential effects in growth plate cells. These receptors are present on the membranes and initiate rapid responses that themselves can elicit a genomic response through membrane receptor-mediated signal transduction pathways. In addition, they may modulate the activity of the nuclear receptor itself.

Physiological Significance of Membrane Receptors for Steroid Hormones

The role of membrane receptors for steroid hormones is of particular significance in bone growth and development since chondrocytes in the endochondral lineage have the goal of mineralizing their extracellular matrix, a process that involves extracellular membrane organelles called matrix vesicles. By controlling the composition of matrix vesicles through genomic mechanisms, the cell can ensure that appropriate machinery is in place in the extracellular matrix for controlling matrix maturation as well as growth factor activation. The cell can then secrete agents, like steroid hormones, into the matrix to interact directly with the matrix vesicle membrane receptors, thereby modulating the activity of that machinery. This implies that chondrocytes have the ability to produce their own steroid hormones, which is the case. Growth plate chondrocytes possess hydroxylase activity [8-10] and can synthesize both 1,25-$(OH)_2D_3$ and 24,25-$(OH)_2D_3$ in a regulated manner.[8,9,11]

In order to understand the mechanisms by which 1α,25-$(OH)_2D_3$ and 24R,25-$(OH)_2D_3$ exert their effects in the growth plate, we developed a cell culture model that has enabled us to compare chondrocytes at two distinct states of endochondral maturation. [12,13] The resting zone and growth zone are removed by sharp dissection from the costochondral cartilages of 125-g Sprague Dawley rats. The cells are isolated by enzymatic digestion of the cartilage matrix and grown in monolayer culture in Dulbecco's modified Eagle medium (DMEM) containing 10% fetal bovine serum (FBS), 1% antibiotics, and 50 μg/ml ascorbic acid. Because of the necessity to expand the cells in culture in order to isolate sufficient matrix vesicles for a typical set of experiments, we determined whether subpassaging the cultures would result in a loss of phenotype as has been reported for articular cartilage cells. Numerous studies examining the response of the cells to $1\alpha,25\text{-}(OH)_2D_3$ and $24R,25\text{-}(OH)_2D_3$, as well as to other regulatory factors, have shown that fourth passage resting zone and growth zone chondrocytes continue to express their differential phenotype. [1,14,15] Based on these studies, we now routinely use fourth passage cultures for most experiments.

Membrane Actions of $1\alpha,25\text{-}(OH)_2D_3$ and $24R,25\text{-}(OH)_2D_3$

$1\alpha,25\text{-}(OH)_2D_3$ exerts part of its effect on costochondral cartilage cells through a membrane receptor (1,25-mVDR) via a protein kinase C (PKC) -mediated signaling pathway. [16-18] The effect of $1\alpha,25(OH)_2D_3$ on PKC is seen in growth zone chondrocytes. In contrast, resting zone cells respond to $24R,25\text{-}(OH)_2D_3$ with an increase in PKC, [17] and recent evidence suggests that a 24,25-mVDR exists as well. [18] Whereas the effect of $1\alpha,25\text{-}(OH)_2D_3$ on PKC is rapid, reaching maximal activation within 9 minutes, the effect of $24R,25\text{-}(OH)_2D_3$ is comparatively slow, with maximal activation at 90 minutes. $1\alpha,25\text{-}(OH)_2D_3$ requires no new gene expression or protein synthesis to stimulate PKC whereas both gene expression and protein synthesis are involved in the response to $24R,25\text{-}(OH)_2D_3$. Sensitivity to each metabolite is cell-specific; $1\alpha,25\text{-}(OH)_2D_3$ does not increase PKC in resting zone cells and $24R,25\text{-}(OH)_2D_3$ does not increase PKC in growth zone cells.

These differences suggest that separate mechanisms are involved in the regulation of PKC by $1\alpha,25\text{-}(OH)_2D_3$ and $24R,25\text{-}(OH)_2D_3$ in their respective target cells. To test this, we investigated the contributions of a number of signaling pathways to the activation of PKC. In both cell types, the responsive isoform of PKC is PKCα, [19,20] which is sensitive to both Ca^{++} and phospholipid.[21] In addition, metabolites of membrane phospholipids such as diacylglycerol (DAG) and arachidonic acid have been shown to regulate PKCα activity directly. DAG binds to cytosolic PKC, resulting in translocation of the enzyme to the plasma membrane.[22] It can be produced through two pathways: either by the direct action of phospholipase C (PLC), or indirectly through the action of phospholipase D (PLD) and subsequent metabolism of phosphatidic acid. Arachidonic acid is produced through the action of phospholipase A_2 (PLA_2) and stimulates PKC by acting as a co-factor.[23]

Previous studies showed that $1\alpha,25\text{-}(OH)_2D_3$ and $24R,25\text{-}(OH)_2D_3$ regulate Ca^{++} flux and phospholipid metabolism in cell-specific ways. $1\alpha,25\text{-}(OH)_2D_3$ caused a rapid efflux of $^{45}Ca^{++}$ from growth zone cells and $24R,25\text{-}(OH)_2D_3$ caused a rapid influx of $^{45}Ca^{++}$ in cultures of resting zone cells.[24] Changes in phospholipid metabolism also occurred within minutes in response to treatment with the vitamin D metabolites. $1\alpha,25\text{-}(OH)_2D_3$ caused a rapid increase in the release of arachidonic acid by growth zone cells, while $24R,25\text{-}(OH)_2D_3$ caused a rapid but short decrease in arachidonic acid release. [25] Moreover, these effects were specific to the target cell.[26] Similarly, each metabolite caused a rapid change in the fluidity of the plasma membrane of its target cell,[27] suggesting that the composition of the membrane was modified.

In cultures treated for 24 hours with the vitamin D metabolites, $1\alpha,25(OH)_2D_3$ stimulated phospholipase A_2 (PLA_2) in matrix vesicles produced by growth zone cells, but $24R,25(OH)_2D_3$ inhibited PLA_2 in matrix vesicles produced by resting zone cells.[25] In addition, when isolated matrix vesicles were treated with the vitamin D metabolites directly, $1\alpha,25(OH)_2D_3$ stimulated PLA_2, and $24R,25(OH)_2D_3$ decreased PLA_2.[28] This effect was specific to matrix vesicles produced by the target cells. Finally, when purified PLA_2 was incubated directly with the metabolites, $1\alpha,25(OH)_2D_3$ increased activity, and $24R,25(OH)_2D_3$ decreased activity.[26]

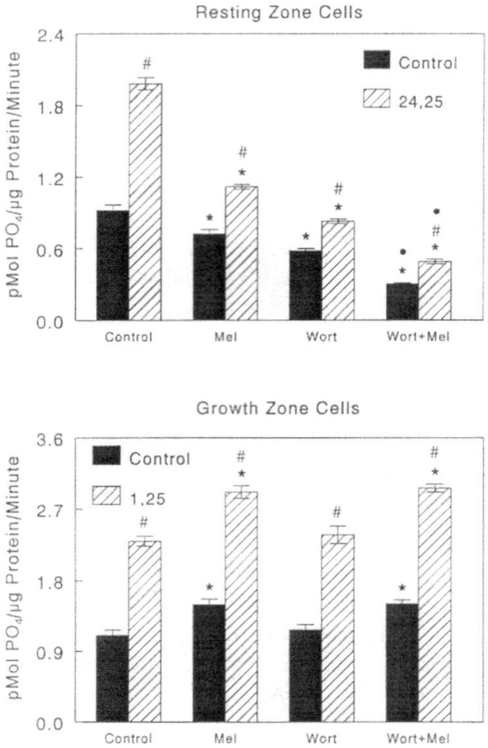

Figure 1. Comparative roles of phospholipase A_2 and phospholipase D in the activation of protein kinase C (PKC) by $24R,25(OH)_2D_3$ (24,25) and $1\alpha,25(OH)_2D_3$ (1,25) in rat costochondral growth plate chondrocytes. Resting zone cells (top panel) were treated with 10^{-7} M 24,25 for 90 min. Growth zone cells (bottom panel) were treated with 10^{-8} M 1,25 for 9 min. Phospholipase A_2 was activated with 0.3 µg/mL melittin. Phospholipase D was inhibited with 10 µM wortmannin. In addition, cells were treated with melittin and wortmannin together. Data shown are from a single representative experiment. Each variable was tested in six independent cultures in each experiment. Values are the mean ± SEM, N = 6. $^*P < 0.05$, treatment v. control; $^{\#}p < 0.05$, with 24,25 or 1,25 v. no vitamin D metabolite.

These observations suggested that PLA_2 might play a pivotal role in mediating the effects of the vitamin D metabolites on PKC. To test this, resting zone cells were incubated with $24R,25\text{-}(OH)_2D_3$ plus activators and inhibitors of PLA_2.[29] Activation of PLA_2 with melittin inhibited the basal level of PKC in control cultures and partially blocked the stimulatory effect of $24R,25\text{-}(OH)_2D_3$ on PKC (Fig. 1, top panel), whereas inhibition of PLA_2 with quinacrine stimulated PKC in control cultures and caused a further increase in PKC over that caused by the vitamin D metabolite (data not shown). When growth zone cells were treated with the PLA_2 activator, PKC in control cultures was increased, and the stimulatory effect of the vitamin D metabolite was enhanced (Fig. 1, bottom panel). In contrast, inhibition of PLA_2 with quinacrine decreased PKC in control cultures and decreased the stimulatory effect of $1\alpha,25\text{-}(OH)_2D_3$ on PKC (data not shown).

These cell-dependent, metabolite-specific effects could be reproduced by addition of arachidonic acid to the cultures. [29] Arachidonic acid may have multiple roles in the signaling pathway. Not only can it act directly on PKC, but also it can lead to new gene expression through interaction with PPAR in the nucleus.[30] In addition, arachidonic acid is a substrate for cyclooxygenase (Cox), leading to prostaglandin formation. $1\alpha,25\text{-}(OH)_2D_3$ stimulates PGE_1 and PGE_2 formation in growth zone cells whereas $24R,25\text{-}(OH)_2D_3$ inhibits production of these prostaglandins in resting zone cultures.[31] Moreover, exogenous PGE_2 stimulates PKC activity in growth zone cells and enhances the effect of $1\alpha,25\text{-}(OH)_2D_3$,[32] but it inhibits PKC activity in resting zone cells and reduces the stimulatory effect of $24R,25\text{-}(OH)_2D_3$.[33]

By comparing the effect of indomethacin, which is a general Cox inhibitor, with that of resveratrol, which blocks the constitutive Cox-1, and NS-398, which blocks the inducible Cox-2, we found that the vitamin D metabolites control the rate of PGE_2 production by regulating the rate of PLA_2 activity rather than the metabolism of arachidonic acid. [34] Once formed, PGE_2 signals through its EP1 receptor in both cell types to regulate PKC. [35] EP1 is a G-protein coupled receptor, [36] leading to cAMP production and activation of protein kinase A (PKA). In resting zone cells, this pathway is inhibitory with respect to PKC, whereas in growth zone cells, it is stimulatory. Since PLA_2 activity is decreased by $24R,25\text{-}(OH)_2D_3$ in the resting zone cells, the amount of PGE_2 ultimately produced is reduced and the inhibitory effect of the PKA signaling pathway is reduced, leading to a delayed increase in PKC.

$1\alpha,25\text{-}(OH)_2D_3$ caused a rapid increase in both inositol-1,4,5-tris-phosphate (IP3) and DAG in growth zone cells, [19] suggesting that PLC might also be involved in the mechanism of PKC activation. In addition, $1\alpha,25\text{-}(OH)_2D_3$ caused a rapid translocation of cytosolic PKC to the plasma membrane, [17] which is a consequence of DAG binding. In contrast, $24R,25\text{-}(OH)_2D_3$ had no effect on IP3 production and DAG production was delayed, with peak levels at 90 minutes, the same time at which maximal increases in PKC were observed. Moreover, $24R,25\text{-}(OH)_2D_3$ did not cause translocation of PKC to the plasma membrane. Instead, increases in PKC activity were associated with the cytosolic form of the enzyme. This argued against PLC and suggested that phospholipase D (PLD) might be responsible for the increase in DAG.

To test these hypotheses, the growth plate chondrocytes were treated with U73122 to inhibit phosphatidylinositol specific PLC (PI-PLC).[37] U73122 had no effect on $24R,25\text{-}(OH)_2D_3$-dependent stimulation of PKC in resting zone cells but it blocked the stimulatory effect of $1\alpha,25\text{-}(OH)_2D_3$ on PKC in growth zone cells (Table 1). Inhibition of phosphatidylcholine specific PLC (PC-PLC) with D609 had no effect on PKC activity in either cell type. [19,38] This indicated that PI-PLC mediates the effect of $1\alpha,25\text{-}(OH)_2D_3$ on PKC in growth zone cells and that PLC is not involved in the mechanism of $24R,25\text{-}(OH)_2D_3$ action in resting zone cells.

Both growth zone chondrocytes and resting zone chondrocytes possess PLD activity and express mRNAs for PLD1a and PLD1b, as well as PLD2, but enzyme activity and mRNA levels are significantly greater in the less mature cells. [39] Inhibition of PLD with wortmannin (Fig. 1, top panel) or EDS reduced the effect of $24R,25\text{-}(OH)_2D_3$ on PKC in resting zone cells but it had no effect on PKC activity in growth zone cells in response to $1\alpha,25\text{-}(OH)_2D_3$ (Fig. 1, bottom panel). [38] The effect of $24,25\text{-}(OH)_2D_3$ on PLD activity was G-protein insensitive, indicating that the form of the enzyme responsible was PLD2.

Table 1: Role of phospholipase C in the regulation of protein kinase C (PKC) by 24R,25$(OH)_2D_3$ and 1α,25$(OH)_2D_3$ in rat costochondral growth plate chondrocytes.

		PKC Specific Activity (pmol Pi/μg protein/min)	
Cell Type	**± U73122**	**Control**	**Vitamin D Metabolite**
Resting Zone	-	0.99 ± 0.05	2.07 ± 0.05#
	+	0.90 ± 0.07	2.17 ± 0.09#
Growth Zone	-	1.04 ± 0.14	2.96 ± 0.23#*
	+	0.93 ± 0.08	1.03 ± 0.08

Resting zone chondrocytes were treated with 10^{-7} M 24R,25$(OH)_2D_3$ for 90 minutes, and growth zone chondrocytes were treated with 10^{-8} M 1α,25$(OH)_2D_3$ for 9 minutes. Phospholipase C activity was inhibited with 10 μM U73122. Data are from a single representative experiment. Each variable was tested in six independent cultures in each experiment. Values are the mean ± SEM, N = 6. $^{\#}p < 0.05$, treatment with vitamin D metabolite vs. control; $^{*}p < 0.05$, U73122 vs. no U73122.

Mechanism of 1α,25-$(OH)_2D_3$ Action in Growth Zone Cells

The signaling pathways involved in the activation of PKC by 1α,25-$(OH)_2D_3$ in growth zone cells are summarized in Fig. 2. Activation of the 1,25-mVDR causes a rapid increase in PLC activity leading to an increase in IP3 and release of intracellular calcium stores, as well as to an increase in DAG. The DAG then binds to PKCα, resulting in translocation to the membrane and activation of the enzyme. 1α,25-$(OH)_2D_3$ also causes a rapid increase in cytosolic PLA_2 leading to increased release of arachidonic acid (AA), which can also stimulate PKC activity directly. Arachidonic acid also modulates gene expression through PPAR receptors. In addition, it serves as a substrate for Cox-1, resulting in prostaglandin production. PGE_2 acts via its EP-1 receptor to increase cAMP production and activate PKA. This G-protein dependent mechanism results in activation of Gq, [40] which can then activate PLC-beta. Both the PKC and the PKA pathways can result in phosphorylation of the ERK family of MAP kinases, [41] providing a mechanism for regulating gene expression.[42]

Mechanism of 24R,25-$(OH)_2D_3$ Action

The mechanism by which 24R,25-$(OH)_2D_3$ stimulates PKC activity in resting zone cells is summarized in Fig. 3. 24R,25-$(OH)_2D_3$ activates PKC through two pathways, one is nongenomic and one requires new gene expression and protein synthesis. The direct action of 24R,25-$(OH)_2D_3$ occurs through the activation of PLD, and the PLD2 isoform is responsible. PLD2 catalyzes the formation of DAG through a two-step reaction. The released DAG binds to PKC and activates the enzyme but the peak effect is not seen until 90 minutes whereas 1α,25$(OH)_2D_3$ exerts its effect within 9 minutes. 24R,25-$(OH)_2D_3$ does not cause translocation of PKC. There are other differences as well. PLA_2 activity and prostaglandin production are reduced and arachidonic acid inhibits the effect of the

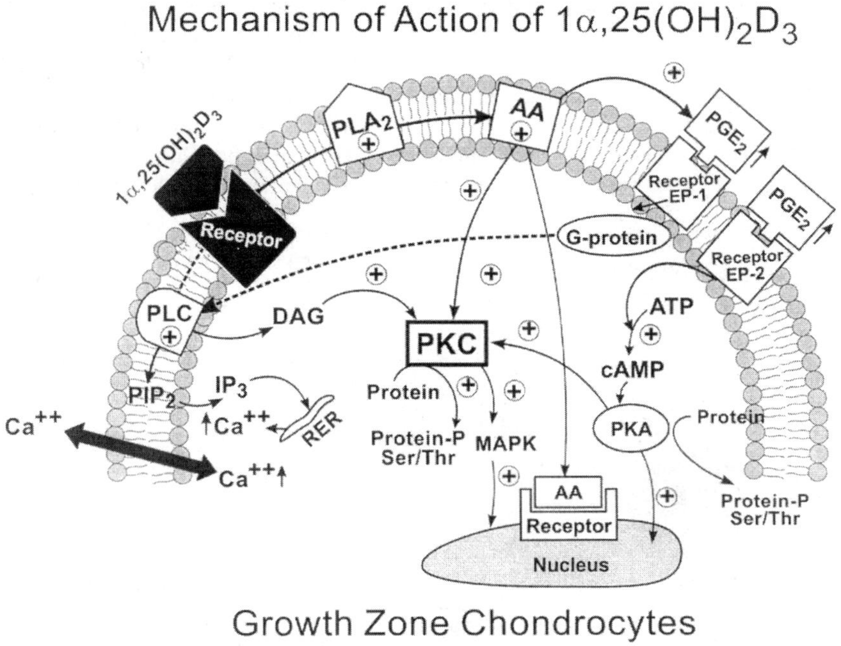

Figure 2. Proposed signal transduction pathways involved in the activation of PKC by 1α,25(OH)$_2$D$_3$ in growth zone chondrocytes. See text for additional details. PLA$_2$ (phospholipase A$_2$); AA (arachidonic acid); PGE$_2$ (prostaglandin E$_2$); DAG (diacylglycerol); PLC (phospholipase C); PKC (protein kinase C); PKA (protein kinase A); MAPK (mitogen-activated protein kinase); IP$_3$ (inositol trisphosphate); PIP$_2$ (phosphoinositol bisphosphate); cAMP (cyclic adenosine monophosphate); ATP (adenosine triphosphosphate); RER (rough endoplasmic reticulum).

hormone, as does prostaglandin. Once PKC activity is increased by 24R,25-(OH)$_2$D$_3$, the MAP kinase pathway is activated as well, potentially leading to new gene expression.

The 24,25-mVDR may account, at least in part, for the specific effects of 24R,25-(OH)$_2$D$_3$ on the physiology of resting zone cells in the absence of a 24,25-nVDR. Not only does 24R,25-(OH)$_2$D$_3$ modulate the differentiation of these cells in culture but it also induces the cells to acquire a growth zone cell phenotype with respect to sensitivity to 1α,25(OH)$_2$D$_3$.[43]

Differential Regulation of Cellular PKC and Matrix Vesicle PKC

1α,25(OH)$_2$D$_3$ stimulates PKC in growth zone chondrocytes and 24R,25-(OH)$_2$D$_3$ stimulates PKC in resting zone chondrocytes. The effects of these vitamin D metabolites on matrix vesicle PKC are also cell maturation specific. When cultures of cells are treated with either metabolite for 24 hours, 1α,25(OH)$_2$D$_3$ stimulates PKC activity in matrix vesicles produced by growth zone cells and 24R,25-(OH)$_2$D$_3$ increases PKC in matrix vesicles produced by resting zone cells. This is due to new matrix vesicle formation. [44] In contrast, when matrix vesicles are incubated directly with the hormones, 1α,25(OH)$_2$D$_3$

decreases PKC activity in matrix vesicles isolated from growth zone cell cultures, and 24R,25-$(OH)_2D_3$ decreases PKC activity in matrix vesicles isolated from resting zone chondrocytes. This direct effect is on PKCζ, an isoform of PKC that does not depend on either Ca^{++} or lipid for activity. The differential distribution of PKC isoforms between the plasma membrane, where PKCα predominates and is stimulated by 1α,25$(OH)_2D_3$ or 24R,25-$(OH)_2D_3$ depending on the cell type, and matrix vesicles produced by the cells, where PKCζ predominates and is inhibited, [20] explains in part how the cell can use secreted vitamin D metabolites to modulate activities of the matrix vesicles in the matrix.

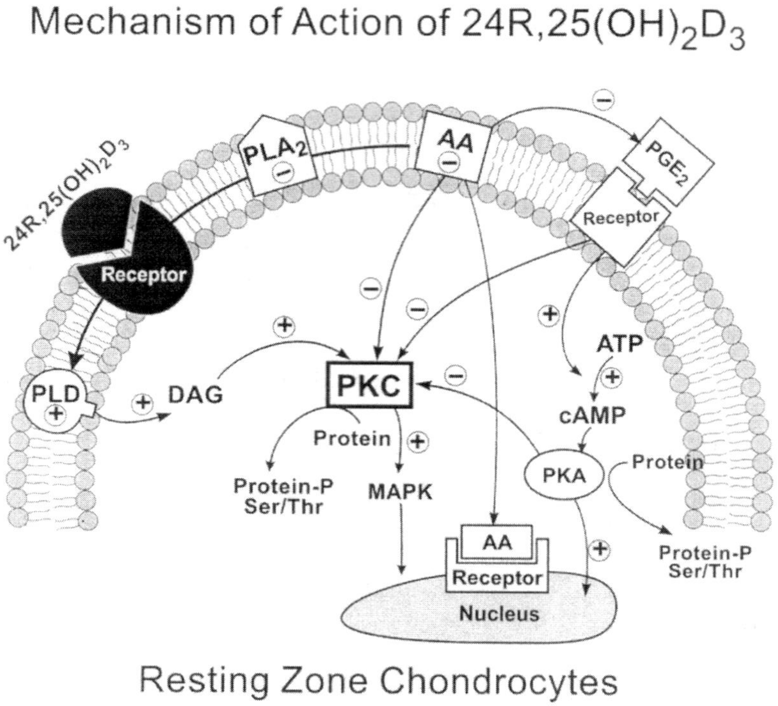

Figure 3. Proposed signal transduction pathways involved in the activation of PKC by 24R,25$(OH)_2D_3$ in resting zone chondrocytes. See text for additional details. PLA_2 (phospholipase A_2); AA (arachidonic acid); PGE_2 (prostaglandin E_2); DAG (diacylglycerol); PLD (phospholipase D); PKC (protein kinase C); PKA (protein kinase A); cAMP (cyclic adenosine monophosphate); ATP (adenosine triphosphosphate); MAPK (mitogen-activated protein kinase).

This ability may be critical to how the cell interacts with its extracellular matrix. Storage of latent TGF-β1 in the matrix is regulated by 1α,25$(OH)_2D_3$ and 24R,25-$(OH)_2D_3$ in a cell maturation-dependent manner through control of the synthesis of latent TGF-β1 binding protein (LTBP1).[45] Release of the large latent TGF-β1 complex and activation of small latent TGF-β1 are regulated by 1α,25$(OH)_2D_3$ through the direct action of the hormone on matrix vesicle membranes, resulting in the release of stromelysin-1 (MMP-3).[46,47]

Summary

It is increasingly evident that there is a relationship between the membrane mediated effects of steroid hormones and their nuclear receptor counterparts. For some responses, pathways initiated by membrane receptors may be sufficient in themselves. Other responses may require only traditional nuclear receptor regulation. For those events that involve both pathways, the interaction may be direct, via phosphorylation of the receptor, or indirect through phosphorylation of regulatory factors other than the nuclear receptor.

Acknowledgements

Many students and a stalwart staff have made these studies possible. The authors also thank our collaborators: Ilka Nemere, Tony Norman, Lynda Bonewald, Adele Boskey, and Gary Posner. Our studies have been supported by grants from NIDCR (DE05937 and DE08603).

References

[1] Boyan BD, Sylvia VL, Dean DD, Del Toro F, Schwartz Z 2001 Differential regulation of growth plate chondrocytes by 1α,25-$(OH)_2D_3$ and 24R,25-$(OH)_2D_3$ involves cell maturation specific membrane receptor activated phospholipid metabolism. Crit Rev Oral Biol Med, in press.

[2] Dean DD, Schwartz Z, Muniz OE, Carreno MR, Maeda S, Howell DS, Boyan BD 2001 Effect of 1α,25-$(OH)_2D_3$ and 24R,25-$(OH)_2D_3$ on metalloproteinase activity and cell maturation in growth plate cartilage in vivo. Endocrine **14:**311-323.

[3] Corvol MT, Dumontier MF, Garabedian M, Rappaport R 1978 Vitamin D and cartilage. II. Biological activity of 25-hydroxycholecalciferol and 24,25- and 1,25-dihydroxycholecalciferol in cultured growth plate chondrocytes. Endocrinology **102:**1269-1274.

[4] Balmain N, Hauchecorne M, Pike JW, Cuisinier-Gleizes P, Matlieu H 1993 Distribution and subcellular immunolocalization of 1,25-dihydroxyvitamin D_3 receptors in rat epiphyseal cartilage. Cell Mol Biol **39:**339-350.

[5] Klaus G, Von Eichel BV, May T, Hugel U, Mayer H, Ritz E, Mehls O 1994 Synergistic effects of parathyroid hormone and 1,25-dihydroxyvitamin D_3 on proliferation and vitamin D receptor expression of rat growth cartilage cells. Endocrinology **135:**1307-1315.

[6] Fine N, Binderman I, Somjen D, Earon Y, Edelstein S, Weisman Y 1985 Autoradiographic localization of 24R,25-dihydroxyvitamin D_3 in epiphyseal cartilage. Bone **6:**99-104.

[7] Nemere I, Farach-Carson MC 1998 Membrane receptors for steroid hormones: A case for specific cell surface binding sites for vitamin D metabolites and estrogens. Biochem Biophys Res Comm **248:**443-449.

[8] Pedrozo HA, Schwartz Z, Gomez R, Dean DD, Boyan BD 1996 1α- and 24R-hydroxylase activities are regulated by vitamin D_3 metabolites and TGF-β in chondrocytes *in vitro* (abstract #T483). J Bone & Min Res **11(Suppl 1):**S420.

[9] Schwartz Z, Pedrozo HA, Sylvia VL, Gomez R, Dean DD, Boyan BD 2001 1α,25-$(OH)_2D_3$ regulates 25-hydroxyvitamin D_3 24R-hydroxylase activity in growth zone costochondral growth plate chondrocytes via protein kinase C. Calcif Tissue Int, in press.

[10] Pedrozo HA, Boyan BD, Mazock J, Dean DD, Gomez R, Schwartz Z 1999 TGF-β1 regulates 25-hydroxyvitamin D_3 1α- and 24-hydroxylase activity in cultured growth plate chondrocytes in a maturation-dependent manner. Calcif Tissue Int **64:**50-56.

[11] Schwartz Z, Brooks BP, Swain LD, Del Toro F, Norman AW, Boyan BD 1992 Production of 1,25-dihydroxyvitamin D_3 and 24,25-dihydroxyvitamin D_3 by growth zone and resting zone chondrocytes is dependent on cell maturation and is regulated by hormones and growth factors. Endocrinology **130:**2495-2504.

[12] Boyan BD, Schwartz Z, Carnes DL, Jr., Ramirez V 1988 The effects of vitamin D metabolites on the plasma and matrix vesicle membranes of growth and resting cartilage cells *in vitro*. Endocrinology **122:**2851-2860.

[13] Boyan BD, Schwartz Z, Swain LD, Carnes DL, Jr., Zislis T 1988 Differential expression of phenotype by resting zone and growth region costochondral chondrocytes *in vitro*. Bone **9:**185-194.
[14] Boyan BD, Dean DD, Sylvia VL, Schwartz Z 1997 Cartilage and vitamin D: Genomic and nongenomic regulation by 1,25-$(OH)_2D_3$ and 24,25-$(OH)_2D_3$. In: Feldman D, Glorieux FH, Pike JW (eds) Vitamin D, Academic Press, San Diego, CA, pp. 395-421.
[15] Boyan BD, Dean DD, Sylvia VL, Schwartz Z 1998 Role of vitamin D: Genomic and nongenomic regulation of cartilage by 1,25-$(OH)_2D_3$ and 24,25-$(OH)_2D_3$. In: Buckwalter JA, Ehrlich MG, Sandell LJ, Trippel SB (eds) Skeletal Growth and Development: Clinical Issues and Basic Science Advances, American Academy of Orthopaedic Surgeons, Chicago,IL, pp. 333-359.
[16] Nemere I, Schwartz Z, Pedrozo H, Sylvia VL, Dean DD, Boyan BD 1998 Identification of a membrane receptor for 1,25-dihydroxy vitamin D_3 which mediates rapid activation of protein kinase C. J Bone Miner Res **13:**1353-1359.
[17] Sylvia VL, Schwartz Z, Schuman L, Morgan RT, Mackey S, Gomez R, Boyan BD 1993 Maturation-dependent regulation of protein kinase C activity by vitamin D_3 metabolites in chondrocyte cultures. J Cell Physiol **157:**271-278.
[18] Pedrozo HA, Schwartz Z, Rimes S, Sylvia VL, Nemere I, Posner GH, Dean DD, Boyan BD 1999 Physiological importance of the 1,25-$(OH)_2D_3$ membrane receptor and evidence for a membrane receptor specific for 24,25-$(OH)_2D_3$. J Bone Miner Res **14:**856-867.
[19] Sylvia VL, Schwartz Z, Curry DB, Chang Z, Dean DD, Boyan BD 1998 1,25-$(OH)_2D_3$ regulates protein kinase C activity through two phospholipid-independent pathways involving phospholipase A_2 and phospholipase C in growth zone chondrocytes. J Bone Miner Res **13:**559-569.
[20] Sylvia VL, Schwartz Z, Ellis EB, Helm SH, Gomez R, Dean DD, Boyan BD 1996 Nongenomic regulation of protein kinase C isoforms by the vitamin D metabolites 1α,25-$(OH)_2D_3$ and 24R,25-$(OH)_2D_3$. J Cell Physiol **167:**380-393.
[21] Newton AC 1995 Protein Kinase C: Structure, function and regulation. J Biol Chem **270:**28495-28498.
[22] Lapetina EG, Reep B, Ganong BR, Bell RM 1985 Exogenous sn-1,2-diacylglycerols containing saturated fatty acids function as bioregulators of protein kinase C in human platelets. J Biol Chem **260:**1358-1361.
[23] Nishizuka Y 1995 Protein Kinase C and lipid signaling for sustained cellular responses. FASEB J **9:**484-496.
[24] Langston GG, Swain LD, Schwartz Z, Del Toro F, Gomez R, Boyan BD 1990 Effect of 1,25$(OH)_2D_3$ and 24,25$(OH)_2D_3$ on calcium ion fluxes in costochondral chondrocyte cultures. Calcif Tissue Int **47:**230-236.
[25] Schwartz Z, Boyan BD 1988 The effects of vitamin D metabolites on phospholipase A_2 activity of growth zone and resting zone cartilage cells *in vitro*. Endocrinology **122:**2191-2198.
[26] Swain LD, Schwartz Z, Boyan BD 1992 1,25-$(OH)_2D_3$ and 24,25-$(OH)_2D_3$ regulation of arachidonic acid turnover in chondrocyte cultures is cell maturation-specific and may involve direct effects on phospholipase A_2. Biochim Biophys Acta **1136:**45-51.
[27] Swain LD, Schwartz Z, Caulfield K, Brooks BP, Boyan BD 1993 Nongenomic regulation of chondrocyte membrane fluidity by 1,25-$(OH)_2D_3$ and 24,25-$(OH)_2D_3$ is dependent on cell maturation. Bone **14:**609-617.
[28] Schwartz Z, Schlader DL, Swain LD, Boyan BD 1988 Direct effects of 1,25-dihydroxyvitamin D_3 and 24,25- dihydroxyvitamin D_3 on growth zone and resting zone chondrocyte membrane alkaline phosphatase and phospholipase-A_2 specific activities. Endocrinology **123:**2878-2884.
[29] Boyan BD, Sylvia VL, Curry D, Chang Z, Dean DD, Schwartz Z 1998 Arachidonic acid is an autocoid mediator of the differential action of 1,25-$(OH)_2D_3$ and 24,25-$(OH)_2D_3$ on growth plate chondrocytes. J Cell Physiol **176:**516-524.
[30] Bocos C, Gottlicher M, Gearing K, Banner C, Enmark E, Teboul M, Crickmore A, Gustafsson J 1995 Fatty acid activation of peroxisome proliferator-activated receptor (PPAR). J Steroid Biochem Molec Biol **53:**467-473.
[31] Schwartz Z, Swain LD, Kelly DW, Brooks BP, Boyan BD 1992 Regulation of prostaglandin E_2 production by vitamin D metabolites in growth zone and resting zone chondrocyte cultures is dependent on cell maturation. Bone **13:**395-401.
[32] Schwartz Z, Gilley RM, Sylvia VL, Dean DD, Boyan BD 1998 The effect of prostaglandin E_2 on costochondral chondrocyte differentiation is mediated by cAMP and protein kinase C. Endocrinology **139:**1825-1834.
[33] Helm SH, Sylvia VL, Harmon T, Dean DD, Boyan BD, Schwartz Z 1996 24,25-$(OH)_2D_3$ regulates protein kinase C through two distinct phospholipid-dependent mechanisms. J Cell Physiol **169:**509-521.

[34] Schwartz Z, Sylvia VL, Del Toro F, Hardin RR, Dean DD, Boyan BD 2000 24R,25-$(OH)_2D_3$ mediates its membrane receptor-dependent effects on protein kinase C and alkaline phosphatase via phospholipase A_2 and cyclooxygenase-1 (Cox-1) but not Cox-2 in growth plate chondrocytes. J Cell Physiol **182:**390-401.

[35] Del Toro F, Jr., Sylvia VL, Schubkegel SR, Campos R, Dean DD, Boyan BD, Schwartz Z 2000 Characterization of PGE_2 receptors (EP) and their role in 24,25-$(OH)_2D_3$-mediated effects on resting zone chondrocytes. J Cell Physiol **182:**196-208.

[36] Negishi M, Sugimoto Y, Ichikawa A 1995 Molecular mechanisms of diverse actions of prostanoid receptors. Biochim Biophys Acta **1259:**109-120.

[37] Bleasdale J, Bundy GL, Bunting S, Fitzpatrick FA, Huff RM, Sun FF, Pike JE 1989 Inhibition of phospholipase C-dependent processes by U73,122. Adv Prostag Thrombox Leuk Res **19:**590-593.

[38] Schwartz Z, Sylvia VL, Luna MH, DeVeau P, Whetstone R, Dean DD, Boyan BD 2001 The effect of 24R,25-$(OH)_2D_3$ on protein kinase C activity in chondrocytes is mediated by phospholipase D whereas the effect of 1α,25-$(OH)_2D_3$ is mediated by phospholipase C. Steroids, in press.

[39] Sylvia VL, Schwartz Z, Del Toro F, DeVeau P, Whetstone R, Dean DD, Boyan BD 2001 24R,25-$(OH)_2D_3$ regulates phospholipase D2 (PLD2) activity of costochondral chondrocytes in a metabolite specific and cell maturation dependent manner. Biochim Biophys Acta **1499:**209-221.

[40] Gilman AG 1987 G proteins: transducers of receptor-generated signals. Ann Rev Biochem **56:**615-649.

[41] Madden K, Sheu YJ, Baetz K, Andrews B, Snyder M 1997 SBF cell cycle regulator as a target of the yeast PKC-MAP kinase pathway. Science **275:**1781-1784.

[42] Cobb MH 1999 MAP kinase pathways. Prog in Biophys Molec Biol **71:**479-500.

[43] Schwartz Z, Dean DD, Walton JK, Brooks BP, Boyan BD 1995 Treatment of resting zone chondrocytes with 24,25-dihydroxyvitamin D_3 [24,25-$(OH)_2D_3$] induces differentiation into a 1,25-$(OH)_2D_3$-responsive phenotype characteristic of growth zone chondrocytes. Endocrinology **136:**402-411.

[44] Sylvia VL, Schwartz Z, Holmes SC, Dean DD, Boyan BD 1997 24,25-$(OH)_2D_3$ regulation of matrix vesicle protein kinase C occurs both during biosynthesis and in the extracellular matrix. Calcif Tissue Int **61:**313-321.

[45] Pedrozo HA, Schwartz Z, Mokeyev T, Ornoy A, Xin-Sheng W, Bonewald LF, Dean DD, Boyan BD 1999 Vitamin D_3 metabolites regulate LTBP1 and latent TGF-β1 expression and latent TGF-β1 incorporation in the extracellular matrix of chondrocytes. J Cell Biochem **72:**151-165.

[46] Maeda S, Dean DD, Schwartz Z, Boyan BD 2001 Activation of latent transforming growth factor-β1 by stromelysin-1 in extracts of growth plate chondrocyte-derived matrix vesicles. J Bone Miner Res **16:**1281-1290.

[47] Maeda S, Dean DD, Gomez R, Schwartz Z, Boyan BD 2001 The first stage of transforming growth factor β1 activation is release of the large latent complex from the extracellular matrix of growth plate chondrocytes by matrix vesicle stromelysin-1 (MMP-3). Calcif Tissue Int, submitted.

The Growth Plate
I.M. Shapiro et al. (Eds.)
IOS Press, 2002

Regulation of Chondrogenesis and Cartilage Maturation *In Vitro*: Role of TGF-β1, Thyroid Hormone, and Wnt Signaling

Maria Alice Mello, A. Cevik Tufan, Kathleen M. Daumer, Bruna Pucci, Toulouse Lafond, David J. Hall, and Rocky S. Tuan
Department of Orthopaedic Surgery, Thomas Jefferson University
Philadelphia, PA 19107

Abstract. Endochondral skeletal development involves the commitment and differentiation of mesenchymal cells into chondrocytes, and the subsequent, progressive proliferation, hypertrophy, and mineralization of the growth cartilage that lead to invasive vascularization and the replacement of the cartilage anlage by newly formed bone. Regulation of the maturation program of cartilage is therefore essential for proper skeletal formation. We have used two in vitro model systems to investigate the mechanisms regulating chondrocyte maturation. In the first system, chick embryonic limb bud mesenchymal cells are isolated at Hamburger-Hamilton Stage 23/24 prior to overt chondrogenesis, and plated as high cell density micromass cultures. Within 3-4 days, these cells differentiate into chondrocytes that, upon long-term culture up to 28 days, undergo progressive proliferation, hypertrophy, and mineralization. This maturation process is accompanied by cessation of cellular proliferation and the onset of programmed cell death (apoptosis), characterized by internucleosomal DNA fragmentation, appearance of apoptotic bodies, and TUNEL reactivity. Using this system, we have evaluated the effect of transforming growth factor-ß1 (TGF-ß1) and the thyroid hormone, triiodothyronine (T3), on chondrocyte hypertrophy. T3 stimulates chondrocyte hypertrophy and apoptosis in a dose and culture time-dependent manner. Interestingly, in the presence of both factors, TGF-ß1 inhibits the hypertrophy/apoptosis-stimulating effect of T3. These observations strongly suggest that T3 and TGF-ß1, which are both found in the growth cartilage in vivo, act to regulate cartilage maturation by modulating chondrocyte apoptosis. Using this system, we have recently examined whether members of the Wnt family of signaling molecules that are expressed in the embryonic limb and regulate mesenchymal chondrogenesis also influence cartilage maturation. Cultures are transfected prior to plating with replication-competent retroviral expression constructs of chicken Wnt-5a and Wnt-7a, as well as Chfz-1 and Chfz-7, two putative Wnt receptors of the Frizzled gene family. Mis-expression of Wnt-7a results in severe inhibition of chondrogenesis, resulting in the formation of a fibroblastic cell mass, completely devoid of cartilage phenotype, by culture Day 28. Wnt-5a mis-expression elicits a slight retardation of maturation during the "pre-hypertrophic stage", but the cultures appear to recover in their maturation program by Day 28. Interestingly, mis-expression of Chfz-7 also inhibits chondrogenesis and retards chondrocyte maturation and hypertrophy. These observations strongly suggest that Wnt functions in cartilage involve both mesenchymal chondrogenesis and chondrocyte maturation and hypertrophy. In the second experimental system, chondrocytes isolated from the upper and lower sternum of the chick embryo, which differ in their ability to mature and undergo hypertrophy, are examined and compared in terms of the expression profiles of genes specific for the hypertrophic and apoptotic programs, to analyze the nature of the cross-talk between these two pathways in the regulation of cartilage maturation.

Introduction

Endochondral skeletal development involves the commitment and differentiation of mesenchymal cells into chondrocytes, and the subsequent, progressive proliferation, hypertrophy, and mineralization of the growth cartilage that lead to invasive vascularization and the replacement of the cartilage anlage by newly formed bone. This process occurs during embryonic skeletal development, postnatal longitudinal bone growth, as well as fracture healing. The progress of chondrocyte maturation and hypertrophy is necessary for cartilage calcification and for the eventual replacement of the cartilage model by bone. Regulation of the maturation program of cartilage is therefore essential for proper skeletal formation. Impairment in cartilage maturation thus leads to skeletal developmental defects, osteochondrodysplasias, as well as delayed union or nonunion of fracture. Although the underlying mechanism of endochondral ossification remains to be completely understood, a number of growth factors, hormones, cytokines, and cell signaling molecules have been implicated as functionally involved in the regulation of endochondral ossification.[1] In this study, we have used two in vitro model systems to investigate the mechanisms regulating chondrocyte maturation. Specifically, we have focused on the role of transforming growth factor-ß1 (TGF-ß1), the thyroid hormone triiodothyronine (T3), and members of the Wnt signaling pathway in chondrocyte maturation and hypertrophy.

Clinically, the effects of thyroid hormone deficiency in skeletal development are indicated by significant decrease in longitudinal bone growth, causing dwarfism in individuals with congenital hypothyroidism, a condition called cretinism, as well as short stature in individuals with juvenile hypothyroidism. In patients with the congenital form of the disease the formation of secondary ossification centers is delayed and some centers are abnormal.[2] The most common hip pathologic condition in adolescents is slipped capital femoral epiphysis (SCFE) [3] which refers to a physical separation between the metaphysis and the epiphysis. The etiology of SCFE is not known, but has been correlated to endocrine conditions [4] such as hypothyroidism, hypopituitarism and hyperparathyroidism, and to metabolic conditions such as renal osteodystrophy, and others. It is generally believed that the cartilage growth plate of these individuals lacks the normal mechanical strengh, and that the slipping begins gradually and slowly, such that a minor trauma can cause the acute slipping between the metaphysis and the epiphysis.[5] Direct support for the functional involvement of thyroid hormone in cartilage maturation comes from the in vitro studies by Ballock et al [6,7] showing that in pellet cultures of chondrocytes, treatment with thyroid hormone promoted the formation of columnar organization of the cells characteristic of the growth plate.

As for TGF-ß1, it has been shown that it is expressed in the growth plate cartilage, with an expression profile that correlates inversely with chondrocyte maturation.[8] *In vitro* and *in vivo* studies have shown that TGF-ß1 treatment stimulates chondrogenesis, chondrocyte proliferation, and prevents cartilage hypertrophy.[9] It is thus possible that the cartilage maturation and hypertrophy program, which involves a balance between chondrocyte proliferation and programmed cell death or apoptosis, is regulated in part by the coordinated actions of TGF-ß1 and T3. Testing this hypothesis is one of the objectives of this study.

Wnts comprise a large family of highly conserved, secreted, cysteine-rich glycoproteins that function as extracellular signaling factors.[10-16] There are currently at least 16 Wnt family members in vertebrates (Wnt-Homepage: http://www.stanford.edu/~rnusse/wntwindow.html). Their name comes from fusing the name of the *Drosophila* segment polarity gene wingless (*wg*) with the name of one of its vertebrate homologs, *integrated.* Wnts share significant homology with *wg* and serve not only as secreted autocrine or paracrine factors primarily to control germ layer specification, and patterning during central nervous system, kidney, mammary gland and limb development in vertebrates, but also perform a wide variety of inductive and regulatory functions in oncogenic transformations.[10-16]

The expression of various Wnts has been identified in developing limb bud [15,17-19], and it has been shown that some Wnts may act in a chondro-inhibitory fashion when introduced during limb development.[19-24] It has been postulated that the dual consequences of Wnt signaling on cell adhesion and/or gene expression through β-catenin provide at least two potential mechanisms by which this key pathway can function in the regulation of limb chondrogenesis.[24] However, the specific targets and mechanism of Wnt regulation in chondrogenesis have not yet been fully understood. Furthermore, how the specificity of a Wnt ligand is determined during the process of limb chondrogenesis is not known.

Recent identification of the vertebrate homologues of *Drosophila* Frizzled gene as a large family of candidate transmembrane receptors for Wnt ligands, gives new insights into the complexity of the Wnt signaling. The Frizzled family of tissue polarity genes are first identified as the receptor for *Drosophila wg* gene.[25,26] So far there have been more than 10 vertebrate Frizzled genes identified (Wnt Homepage), some of which have been shown to be expressed in developing vertebrate limb.[18,26-29] An intriguing possibility is that distinct classes of Wnts might have different affinities for specific Frizzled receptors. Thus, members of the Frizzled family of genes may mediate distinct Wnt signaling pathways to regulate divergent processes during limb chondrogenesis. Testing these functional relationships is another objective of the present study.

The hypertrophic program in endochondral ossification involves the programmed cell death of chondrocytes, and thus represents a balance of proliferative and hypertrophic chondrocytes [1]; in addition, apoptosis of hypertrophic chondrocytes produces apoptotic bodies that participate in matrix mineralization. Several lines of evidence indicate that hypertrophy is regulated by the coordinated action of parathyroid hormone-related peptide (PTHrP) and Indian hedgehog (Ihh). PTHrP regulates the number of proliferating chondrocytes that enter the hypertrophic program. PTH/PTHrP receptor is a member of a G-protein coupled family [30] and acts through the cAMP/protein kinase A (PKA) and the phospholipase C/protein kinase C (PKC) signaling pathways.[31] Ihh is a member of the Hedgehog (Hh) gene family, which encodes a group of secreted factors that act as signals for embryonic patterning in many organisms, via a transmembrane receptor, Patched (Ptc), and a transcription factor, Gli.[32] In the developing skeleton, Ihh expression is localized to a zone of postmitotic, prehypertrophic chondrocytes immediately adjacent to the zone of proliferating chondrocytes. A link between Ihh and PTHrP has been established. Ihh and PTHrP regulate chondocyte differentiation through a negative feedback mechanism: production of Ihh by prehypertrophic chondrocytes induces PTHrP expression in the perichondrial cells. PTHrP inhibits additional proliferating chondrocytes to enter the hypertrophic program. When chondrocytes mature fully into hypertrophic chondrocytes, they stop expressing Ihh. At this point PTHrP expression in the perichondrium is attenuated and other proliferating chondrocytes can initiate their maturation.[33]

Less clear is the cell death mechanism in hypertrophic chondrocytes. Chondrocyte cell death occurs very fast (over 45 minutes in the 4-5 h life span of a hypertrophic chondrocyte) [34] and is accompanied by internucleosomal DNA fragmentation. Terminal chondrocytes display the characteristic morphology of apoptotic cells, such as chromatin and cellular condensation.[35] When hypertrophic chondrocytes enter the terminal phase of their life as fully functioning cells, local environmental conditions provide termination signals. In cartilage, several types of signals could serve to trigger programmed cell death. First, a change in chondrocyte metabolic state could be a signal. Immediately following hypertrophy, there is a switch in oxidative metabolism and a substantial shift in the reductive reserve [36], probably related to local hypoxia. Correlated with these metabolic events is an elevation in cytosolic calcium and possibly an elevation in nonmitochondrial oxidative metabolism.[37] Secondly, the local generation of growth factors, such as tumor necrosis factor (TNF) and transforming growth factor-ß (TGF-ß) may signal growth arrest prior to apoptosis.[38] Moreover, extracellular phosphate ions cause apoptosis of terminally differentiated chondrocytes [39], leading to the hypothesis that apoptosis in hypertrophic chondrocytes may be linked to the mineralization of their extracellular matrix.

Some of the molecular factors involved in hypertrophic chondrocyte apoptosis have been studied. Bcl-2 is the founding member of a family of proteins involved in apoptosis.[40] Bcl-2 is a homodimeric, anti-apoptotic factor present within organelle membranes, such as mitochondria, Golgi and endoplasmatic reticulum. Bax, another bcl-2 family member, heterodimerizes with Bcl-2 and when overexpressed, counters the anti-apoptotic effect of bcl-2, causing accelerated cell death. The ratio of Bcl-2 to Bax determines cell fate.[41] The spatial and temporal distribution of Bcl-2 in the developing embryo suggests that this protein regulates cell death during development. Bcl-2 is widely expressed in the developing limb bud, and Bcl-2 knock out mice have accelerated endochondral ossification, presence of short ears and short and deformed limbs.[42] This finding suggests a functional role for Bcl-2 in chondrocyte maturation. In fact, in Bcl-2 knock out mice there is a marked reduction in growth plate thickness, and shortening of the proliferative zone. Bcl-2 is expressed in chondrocytes and its expression in the developing growth plate decreases from the proliferating to hypertrophic chondrocytes, consistent with a role in regulating apoptosis. Bax shows an inverse pattern of expression, again consistent with its pro-apoptotic role. It has been suggested that Bcl-2 and PTHrP participate in the same regulatory pathways during skeletal development [43], because they have similar expression patterns in space and in time [44], and because Bcl-2 knock out mice have a similar phenotype to the PTHrP knockout mice. It is likely that PTHrP exerts its effect on skeletal development by pathways that induce Bcl-2 expression.

Other candidate regulators of chondrocyte apoptosis have been studied. Activation of the Fas/FasL system can trigger chondrocyte apoptosis in normal articular cartilage and Fas antigen is expressed in 30% of articular chondrocytes.[45] Another important candidate of apoptosis in chondrocytes is p63, a member of the p53 tumor suppressor family of apoptosis regulating factors. p63 null-mice have defects in limb and craniofacial development [46], most likely caused by abnormal ectodermal-mesenchymal signaling, but also possibly by abnormal maturation and apoptosis of cartilage. In fact p63 could play a regulating role in apoptosis similar to that of p53.

Presently little information is available regarding the mechanistic pathways regulating apoptosis in hypertrophic chondrocytes, and how and if apoptosis is connected to hypertrophy. We are interested in analyzing the mechanisms that regulate apoptosis in hypertrophic chondrocytes by identifying the induction signals that cause the cells to die, specifically the involvement of pro and anti-apoptotic regulator, Fas/FasL, Bcl-2 family members, and p53 family members. Information on the relationship between hypertrophy and apoptosis will permit us to determine if these two processes occur as parallel phenomena or if they are co-regulated.

We have used two in vitro model systems to investigate the mechanisms regulating chondrocyte maturation. In the first system, chick embryonic limb bud mesenchymal cells are isolated at Hamburger-Hamilton Stage 23/24 prior to overt chondrogenesis, and plated as high cell density micromass cultures. Within 3-4 days, these cells differentiate into chondrocytes that, upon long-term culture up to 28 days, undergo progressive proliferation, hypertrophy, and mineralization. Genes corresponding to members of the Wnt signaling pathway, specifically Wnt and Frizzled members, are transduced as retroviral constructs into these cultures, and their effects on chondrogenesis and chondrocyte maturation analyzed. In the second experimental system, chondrocytes isolated from the upper and lower sternum of the chick embryo, which differ in their ability to mature and undergo hypertrophy, are examined and compared in terms of the expression profiles of genes specific for the hypertrophic and apoptotic programs.

Materials and Methods

Long-term Culture of Chick Embryonic Limb Mesenchymal Cells
Limb buds were harvested from Hamburger-Hamilton Stage 23-24 chick embryos, enzymatically dissociated, and cells isolated and placed into high-density micromass cultures as described previously.[47,48] Long-term maintenance of the micromass cultures was carried out as described previously by Mello and Tuan.[49] Electroporation-mediated transfection of RCAS retroviral constructs of chicken Wnts (Wnt-5a and Wnt-7a) and Frizzled's (Chfz-1 and Chfz-7) was carried out using the procedure of Delise et al [50] as reported recently.[24] Wnt-5a and Wnt-7a represent two Wnt genes that are expressed in the limb bud mesenchyme and in the dorsal limb ectoderm, respectively [51] and Chfz-1 and Chfz-7 are two members of the Frizzled family of putative Wnt-receptors that have been shown to be expressed in the limb bud in distinct patterns.[18]

Culture of Chick Sternal Chondrocytes
Embryonic sterna were obtained from Day 14-15 chick embryos. Cells were obtained from the upper sternum, which normally undergo hypertrophy and mineralization in vivo, and placed into culture as described previously.[52] Briefly, freshly isolated chondrocytes were cultured in DMEM containing 4 Units/ml hyaluronidase and 10% NuSerum on non-tissue culture dishes at 37°C. After 5 days, the non-attached chondrocytes were replated in the same medium in the presence of on tissue culture plastic dishes (2.5×10^6 cells per 100 mm dish). After 3 days from the plating, 10 µg/ml ascorbic acid was added, and this day was considered as Day 0. At day 1, the concentration of ascorbic acid was increased to 50 µg/ml. When confluence was reached (Day 4), 35 nM retinoic acid and 10 mM ß-glycerophosphate were added. Medium was changed every other day and plates were protected from the light. The entire maturation program took place in 10 days.

Histochemistry, Immunhistochemistry, Western Blot, Northern Blot, BrdU Labeling, TUNEL Staining, and Reverse Transcription-Polymerase Chain Reaction (RT-PCR)
All analyses were carried out essentially as described previously.[24,49,52-54]

Results

Effect of TGF-ß1 and T3 on Chondrocyte Maturation in Long-Term Limb Mesenchyme Micromass Cultures
In the long-term micromass cultures, cells were maintained in serum-containing medium until Day 3, when chondrogenesis was clearly established, and then switched to a serum-free, ITS-supplemented medium. After 24 hours in this medium, TGF-ß1 (100 pM) and T3 (10 nM) were added either singly or in combination, and the cultures observed at specific time points up to Day 28.

Morphologically, by Day 21, only control cultures and cultures treated with T3 showed columnar organization of the chondrocytes. Quantitation of BrdU-labeled cells (as percent of total cells) localized within cartilage nodules of significant size revealed the following profiles during the course of culture. At day 7 no statistically significant difference was observed among control and the treated groups. At Day 14 the overall rate of proliferation decreased, and the cultures treated with only TGF-ß1 and with TGF-ß1 and T3 showed statistically significant increase compared to the control. At Day 21, the rate of proliferation of the cells treated with TGF-ß1 alone increased compared to Day 14; the cells treated with T3 alone stopped proliferating, and almost no BrdU-positive cells were seen in the nodules, whereas the number of positive cells in cultures treated with TGF-ß1 and T3 in combination decreased in relation to Day 14. By Day 28 there was a large variation in the level of BrdU incorporation between different nodules of the same culture, making it impossible to obtain a consistent estimate of cell number for subsequent analysis.

The average cell size was significantly larger in the cultures treated with T3 only or with both T3 and TGF-ß1, compared to the untreated control and the TGF-ß1 treated cultures, and the maximal effect was seen at Day 21. Collagen type X expression was detected by immunohistochemistry in all cultures, from Day 7 until Day 28. By Day 7, collagen type X was detected around and within the cells. By Day 14, control cultures and cultures treated with T3 or with T3 and TGF-ß1 in combination showed obvious matrix deposition of collagen type X; however, in the cultures treated with TGF-ß1 alone, collagen type X wad detected only around the cells. At day 21 all the cultures showed matrix deposition of collagen type X. The enzymatic activity of alkaline phosphatase, another functional marker of chondrocyte maturation, was signfiicantly higher in the T3 treated cultures by Day 21.

TUNEL staining to detect apoptosis revealed that cultures treated with T3 or with both T3 and TGF-ß1 contained significantly higher number of apoptotic cells, beginning on Day 14. By Day 21, over 95% of the cells in T3-treated cultures were apoptotic. Electorphoretic analysis of cellular DNA also revealed significantly higher internucleosomal DNA degradation in the cultures treated with T3 or with T3 and TGF-ß1 in combination.

Effect of Mis-expression of Wnt and Frizzled Members on Chondrocyte Maturation and Hypertrophy

Our approach was to transfect freshly isolated Stage 23/24 limb mesenchymal cells, by means of electroporation, with the replication-competent retroviral constructs (RCAS) expressing Wnt-5a, Wnt-7a, Chfz-1, and Chfz-7, respectively (Fig. 1). Empty RCAS vector was used as control. These cells were then placed into long-term micromass cultures, and their chondrogenic differentiation and maturation program examined. As reported previously [24], we observed chondro-inhibitory effects of mis-expressed Wnt-7a in short-term (3-4 days) micromass cultures (Fig.2A.c), whereas Wnt-5a had no effect on the condensation and chondrogenic differentiation of chick limb mesenchymal cells *in vitro* during this time period (Fig. 2A.b), when compared to the RCAS-empty vector transfected control cultures (Fig. 2A.a). In an analogous manner, mis-expression of ChFz-7 (Fig. 2A.e) was also chondro-inhibitory compared to the Chfz-1 expressing cultures (Fig. 2A.d). In our previous study [24], we showed that Wnt-7a mis-expression inhibited the chondrogenic differentiation of chick limb mesenchymal cells by altering the expression and stabilization of N-cadherin and related activities, i.e., cell-cell adhesion. Compared to Wnt-7a, the chondro-inhibitory effect of Chfz-7 mis-expression, on the other hand, was only partial, since the mesenchymal cells localized at the central region of the culture, known to higher cell density compared to the peripheral region, appeared to be able to overcome the Chfz-7 effect and differentiated into chondrocytes (Fig. 2A.e).

For the long-term micromass cultures, beginning from culture Day 7 and beyond, mis-expression of Wnt-7a resulted in severe inhibition of chondrogenesis, resulting in the formation of a fibroblastic cell mass that stained poorly with Alcian blue (Fig. 2B.e and 2C.g) and was completely devoid of cells exhibiting chondrocyte phenotype (Fig. 2B.f, 2C.h and 2C.i). Furthermore, Northern blot analysis of mRNA expression of collagen type X, a marker for the maturation and hypertrophy of chondrocytes, showed that, by culture Day 21, collagen type X expression was in fact undetectable in Wnt-7a mis-expressing cultures (Fig. 2D). On the other hand, Wnt-5a mis-expression elicited a slight retardation of maturation during the "pre-hypertrophic stage" observed by Alcian blue staining on Day 7 (Fig. 2B.c), but the Alcian blue staining in these cultures appeared to recover by Day 14 (Fig. 2C.d) and beyond (data not shown). However, Wnt-5a mis-expressing cultures exhibited a poor hypertrophic phenotype on Day 14 (Fig. 2C.e and 2C.f) and the expression of collagen type X was detectable but severely delayed even on culture Day 21 (Fig. 2D).

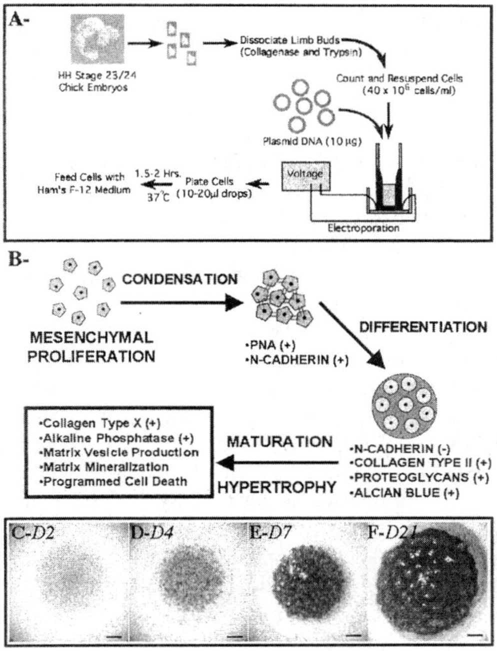

Figure 1. Schematic of long-term micromass culture of Stage 23/24 chick embryonic limb bud cells and their differentiation and maturation in vitro. (A) Procedure for micromass culture, also illustrating the electroporation step used for DNA transfection. (B) Schematic representation of the sequential events taking place in a long-term culture of limb mesenchyme micromass, involving cellular proliferation, condensation, chondrogenic differentiation, chondrocyte proliferation, maturation, hypertrophy, and matrix mineralization, accompanied by specific cellular and molecular markers. (C-F) Whole-mount morphology of long-term limb mesenchyme culture, stained with Alcian blue, showing the appearance of cartilaginous nodules by Days 3-4, the extensive matrix production during the chondrocyte proliferative phase (Day 7), and the final mature and hypertrophic cartilage mass (Day 21). All images are at same magnification (Bar = 1 mm).

Chfz-1 mis-expression in chick limb mesenchymal micromass cultures did not appear to influence any aspect of the maturation and hypertrophy phases of the chondrogenic program (Fig. 2B.g and h, 2C.j-l and 2D), whereas Chfz-7 mis-expression showed a slight delay of this process when compared to the RCAS-empty vector control cultures. Some areas of the Chfz-7 mis-expressing cultures showed reduced Alcian blue staining, whereas other areas appeared similar to the controls (Fig. 2B.i). All Chfz-expressing cultures did recover in terms of Alcian blue staining by Day 14 (Fig. 2C.m). The morphology of the hypertrophic chondrocytes in Chfz-7 cultures showed a combination of normal and synthetically less active cells on Day 14 (Fig. 2C.n and o), and collagen type X expression in these cultures was slightly delayed when examined on Day 21 (Fig. 2D).

Relationship Between Hypertrophy and Apoptosis in Sternal Chondrocytes In vitro

The goal of our study was to analyze the regulation of apoptosis in hypertrophic chondrocytes by identifying the induction signals that cause the cells to die, specifically the involvement of pro- and anti-apoptotic regulator, Fas/FasL, Bcl-2 family members, and p53 family members. We also wanted to analyze the relationship between hypertrophy and apoptosis to determine if these two processes occur as parallel phenomena or if they are co-regulated. Day 14-15 chick embryonic cephalic or upper sternal chondrocytes (USC), representing pre-hypertrophic chondrocytes capable of recapitulating the chondrocyte maturation program, were placed into culture *in vitro*. Non-hypertrophic chondrocytes (LSC) isolated from the caudal or lower third of the chick embryonic sterna, which were unable to undergo maturation *in vitro*, were used as a negative control.

Chondrocyte phenotype was initially assessed on the basis of matrix Alcian blue staining, and chondrocyte maturation on the basis of alkaline phosphatase histochemistry and collagen type X expression (Fig. 3A). In addition, cellular apoptosis was characterized by TUNEL staining (Fig. 3A). The time-dependent increase in alkaline phopshatase histochemical staining, collagen type X expression, and the number of TUNEL-positive cells confirmed that the cultured USC were capable of undergoing differentiation *in vitro* from pre-hypertrophic chondrocytes to apoptotic cells (the negative control corresponded to USC at the beginning of culture). Expression of the molecular markers of differentiation and maturation was also detected. Early cartilage phenotype was evident in terms of collagen type II immuno-staining (Fig. 3A), and collagen type II and aggrecan mRNA detection by RT-PCR analysis (Fig. 3B). Hypertrophic maturation of the chondrocytes was analyzed by monitoring collagen type X expression by immunohistochemistry (Fig. 3A) and by RT-PCR of mRNA level (Fig. 3B). Chondrocyte hypertrophy was apparent from the beginning of the *in vitro* culture, and by Day 10, many of the cells were lost from the culture as a result of extensive apoptosis, as detected by TUNEL staining. In contrast, LSC chondrocytes, while capable of expressing collagen type II and aggrecan, failed to undergo maturation and hypertrophy, as indicated by the lack of collagen type X expression.

The expression profile of various candidate regulatory genes of the hypertrophy and apoptotic programs in the USC and LSC cultures was analyzed by RT-PCR as a function of culture time (up to Day 10). RT-PCR was performed for those chicken genes whose sequence structures are available from GenBank. For the genes involved in hypertrophy, such as Cbfa-1 [55] and PTHrP, no changes were seen in their mRNA levels. On the other hand, Ihh was clearly induced in the USC cultures, consistent with cellular maturation from pre-hypertrophic to hypertrophic chondrocytes. It is noteworthy that, for the first time, we detected the induction of BMP-2 mRNA during chondroctye (USC) maturation *in vitro*. Interestingly, it was previously shown that BMP-2 treatment stimulated collagen type X expression in serum-free cultures of chondrocytes from the cephalic region of the sternum [56], although it was unlikely that BMP-2 acted alone. Our analysis also showed that there were no significant changes in the expression of the pro- or anti-apoptotic genes, Fas, Bcl-2, and p63, in association with chondrocyte maturation. Possibly, the changes could be at a protein level, these factors are known to be regulated primarily at the post-translational level.

We also tested the applicability of retroviral gene transduction as a means to study the function(s) of those genes whose expression was profoundly affected by chondrocyte maturation. Our initial results showed that the retroviral system of the replication incompetent pBABE [57], prepared as the 48-hour culture medium of the amphoteric Phoenix cells transfected with pBABE using calcium phosphate, was able to infect USC cultures. Our initial results showed that the infection efficiency was approximately 20%, on the basis of X-Gal staining of the recombinant ß-galactosidase activity in the infected cells. Current efforts are directed towards improving the efficiency of this gene transduction method for chondrocytes.

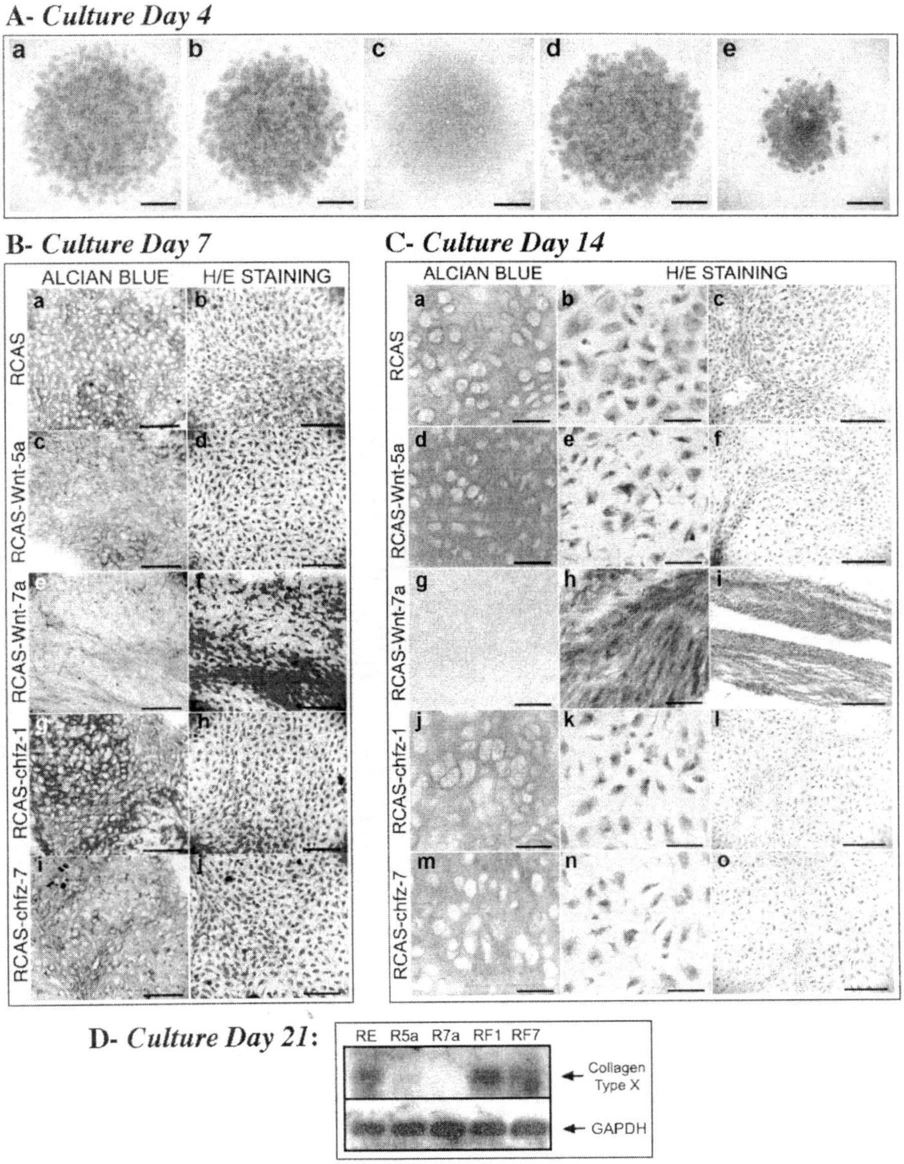

Figure 2. Effect of mis-expression of Wnt and Frizzled members on mesenchymal chondrogenesis and chondrocyte maturation in long-term culture of embryonic limb mesenchymal cells. (A) Alcian blue stained whole-mount cultures. a, control; b, Wnt-5a; c, Wnt-7a; d, Chfz-1; and e, Chfz-7. Bar = 1 mm. (B) Histology of Day 7 cultures stained with Alcian blue or hematoxylin/eosin (H/E). Bar = 50 μm. (C) Histology of Day 14 cultures stained with Alcian blue or H/E (c, f, i, l, and o are lower magnifications of b, e, h, k, and n, respectively). Bar = 100 μm for lower magnifications, and 25 μm for all higher magnifications. (D) Northern analysis of collagen type X mRNA expression in Day 21 cultures. RE, RCAS-empty; R5a, Wnt-5a; R7a, Wnt-7a; RF1, Chfz-1; and RF7, Chfz-7. See text for description.

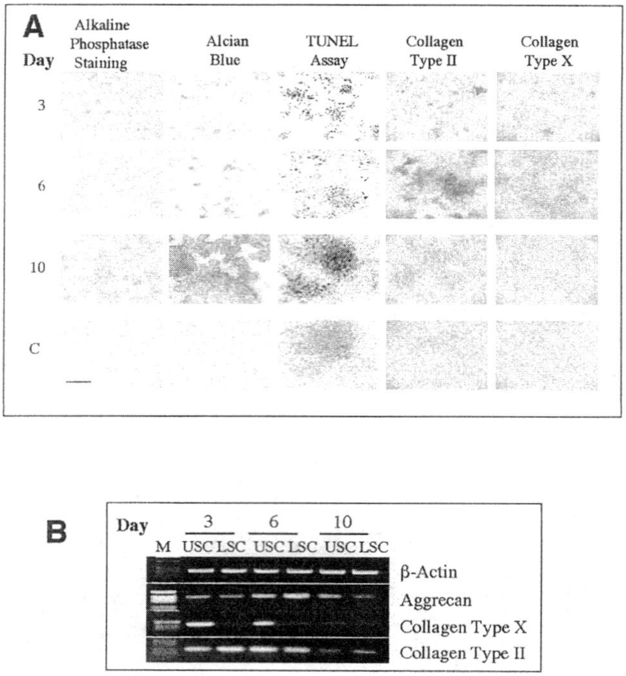

Figure 3. Maturation and hypertrophy of chick embryonic sternal chondrocytes in vitro. (A) Histology, alkaline phosphatase histochemistry, TUNEL labeling of apoptotic cells, and expression of collagen types II and X, on culture Days 3, 6, and 10. Controls (C) represent samples taken at the beginning of culture. See text for description. (B) Comparison of the upper (USC) and lower (LSC) sternal chondrocytes in terms of cartilage-specific gene expression analyzed by RT-PCR. While both USC and LSC express chondrogenic genes, only USC express collagen type X, characteristic of hypertrophic chondrocytes. ß-Actin was used as internal mRNA control.

Discussion

The results reported here, using two cell culture systems derived from the undifferentiated mesenchymal cells and from prehypertrophic sternal chondrocytes, have provided interesting insights on the regulation of chondrocyte hypertrophy and differentiation. First, as reported previously [49,58]; we have shown that the entire program of mesenchymal chondrogenesis and chondrocyte maturation, hypertrophy, and mineralization may be recapitulated, with high fidelity, in the high-density cultures of Stage 23/24 limb bud mesenchymal cells. This culture system thus represents a "complete" system as far as studying the life history of a chondrocyte. In comparison, other "complete" systems that used embryonic limb cells, such as that by Castagnola et al. [59] which used cells isolated from chick embryo tibae first plated in monolayer, then transferred to suspension culture, filtered, and re-plated on substrate-coated plates over the course of several weeks to promote their differentiation into the hypertrophic phenotype [59,60], are generally more labor-intensive.

Our results clearly indicate that both TGF-ß1 and the thyroid hormone, T3, are functionally involved in regulating chondrocyte maturation and hypertrophy. The parameters used to investigate the effects of TGF-ß1 and T3 on cartilage maturation include cell proliferation, cell hypertrophy, matrix calcification, and apoptosis. Our results showed that TGF-ß1 acts to stimulate cellular proliferation during the pre-hypertrophy and hypertrophy stages (as measured on Days 14 and 21], and that T3 inhibits cellular proliferation during the hypertrophy stage. The average cell size is also significantly increased with T3 treatment. It is noteworthy that collagen type X expression does not appear to affected by treatment by either TGF-ß1 or T3, while alkaline phosphatase activity is increased by T3, suggesting that these two generally accepted markers of chondrocyte hypertrophy are regulated by different mechanisms. Finally, T3 treatment also stimulates the overall level of apoptosis in the long-term limb mesenchyme micromass cultures, detected by both TUNEL staining as well as nuclear DNA ladder analysis, consistent with nuclear DNA fragmentation. In general, cultures co-treated with both TGF-ß1 and T3 exhibit characteristics intermediate between cultures treated singly with the two agents, suggesting that the two factors may share some common mechanistic pathways in influencing chondrocyte maturation and hypertrophy. Taken together, our observations strongly indicate that cartilage maturation represents a dynamic balance between cellular proliferation and apoptosis of the chondrocytes, and that TGF-ß1 and T3 influence cartilage maturation by modulating this balance.

Our results also clearly indicate the intriguing possibility of the functional involvement of Wnt signaling pathways in chondrocyte maturation. We [24] and others [21,22] have previously shown that Wnt members, for example Wnt-7a, act as chondro-inhibitory agents. Our recent study [24] suggests that this chondro-inhibitory effect of Wnt is mediated, at least in part, via the regulation of the expression and activity of the cell adhesion molecule, N-cadherin [48], as well as its interaction with ß-catenin, one of its cytosolic partners at the adherens junction. In addition, our study reported here is the first study to use long-term micromass culture of primary chick limb mesenchymal cells to investigate Wnt signaling during late events in chondrogenesis. The current understanding of Wnt involvement during post-differentiation events has been limited to in vivo studies [19,21,23,29,51] demonstrated the inhibitory effects of Wnt-1 and Wnt-7a misexpression using a similar micromass culture system; however, cells were maintained in culture for a total of only 4 days. Thus, the results from our system as reported here that concern the action of Wnt and Frizzled members are highly relevant in terms of understanding how Wnt signaling pathways function to pattern the forming skeletal anlage of the limb. We are currently analyzing the expression profiles of other markers of chondrocyte hypertrophy and apoptosis to further characterize the specific effects of Wnt and Frizzled in our system.

Using the cephalic sternal chondrocyte culture system, we have begun to map the coordinated expression of markers and candidate regulatory genes of the hypertrophy and apoptotic programs. While our data are still preliminary, we believe that once key changes in gene expression levels are identified, we will be able to use a retroviral gene transduction system to mis-express such candidate genes, and to follow the cellular and molecular consequences. The outcomes of such studies should provide important information on whether and how the hypertrophy program and apoptotic program of the maturing chondroctye are functionally linked, and on the nature of the cross-talk mechanisms.

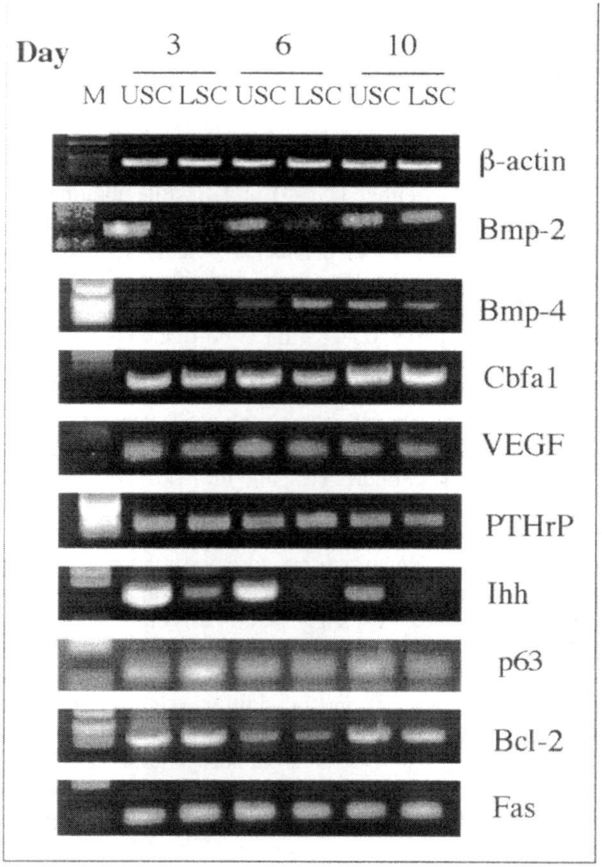

Figure 4. Expression of candidate regulatory genes of the hypertrophic and apoptosis pathways in cultures of sternal chondrocytes. USC, upper sternal chondrocytes; LSC, lower sternal chondrocytes. The genes examined by RT-PCR include: bone morphogenetic protein (BMP)-2 and –4, Cbfa1, vasculoendothelial growth factor (VEGF), parathyroid hormone related peptide (PTHrP), Indian hedgehog (Ihh), p63, Bcl-2, and Fas. ß-Actin was used as internal mRNA control. See text for description.

Acknowledgements

This work is supported in part by NIH grants (DE12864, AR39740, AR45181).

References

[1] Hickok NJ, Haas AR, and Tuan RS 1998 Regulation of chondrocyte differentiation and maturation. Micros Res Techn **43**: 174-190.

[2] Cummings RJ 1995 Metabolic and Endocrine Disorders. In: Canale ST, Beaty JH (eds.) Operative Pediatric Orthopeadics, 2nd ed, Mosby-Year, St. Louis, pp 383 - 397.

[3] Canale ST 1995 Fractures and Dislocations. In: Canale ST, Beaty JH (eds.) Operative Pediatric Orthopeadics, 2nd ed, Mosby-Year, St. Louis, pp 982- 983.

[4] Loder RT, Wittenberg B, and Silva G 1995 Slipped capital femoral epiphysis associated with endocrine disorders. J Pediatr Orthop **15**:349-356.

[5] Rockwood CAJr, Wilkins KE and Beaty JH 1996 Fractures in Children, 4th edition. Lippincott-Raven, Philadelphia- New York, pp 30 -31.

[6] Ballock R and Reddi AH 1994 Thyroxine Is the serum factor that regulates morphogenesis of columnar cartilage from isolated chondrocytes in chemically defined medium. J Cell Biol **126**:1311-1318.

[7] Ballock RT, Zhou X, Mink LM, Chen DH, and Mita BC 2001 Both retinoic acid and 1,25(OH)$_2$ vitamin D$_3$ inhibit thyroid hormone-induced terminal differentiaton of growth plate chondrocytes. J Orthop Res **19**: 43-49.

[8] O'Keefe RJ, Crabb ID, Puzas JE and Rosier RN 1994 Effects of transforming growth factor-•1 and fibroblast growth factor on DNA synthesis in growth plate chondrocytes are enhanced by insulin-like growth factor-I. J Orthop Res **12**:299- 310.

[9] Ballock RT, Heydemann A, Wakefield LM, Flanders KC, Roberts AB, Sporn MB 1993 TGF-beta 1 prevents hypertrophy of epiphyseal chondrocytes: Regulation of gene expression for cartilage matrix proteins and metalloproteases. Developmental Biology. **158**:414-429.

[10] Cadigan KM and Nusse R 1997 Wnt signaling: a common theme in animal development. Genes Dev **11**:3286-3305.

[11] Moon RT, Brown JD, Torres M 1997 WNTs modulate cell fate and behavior during vertebrate development. Trends Genet **13**:157-162.

[12] Brown JD and Moon RT 1998 Wnt signaling: why is everything so negative?. Curr Opin Cell Biol **10**:182-187.

[13] Dale TC 1998 Signal transduction by the Wnt family of ligands. Biochem J **329**:209-223.

[14] Giarre M, Semenov MV, Brown AM 1998 Wnt signaling stabilizes the dual-function protein beta-catenin in diverse cell types. Ann NY Acad of Sci **857**:43-55.

[15] Wodarz A and Nusse R 1998 Mechanisms of Wnt signaling in development. Annu Rev Cell Dev Biol **14**:59-88.

[16] Dierick H and Bejsovec A 1999 Cellular mechanisms of wingless/Wnt signal transduction. Curr Top Dev Biol **43**:153-90.

[17] Parr BA, Shea MJ, Vassileva G, McMahon AP 1993 Mouse Wnt genes exhibit discrete domains of expression in the early embryonic CNS and limb buds. Development **119**:247-261.

[18] Kengaku M, Twombly V, Tabin CJ 1997 Expression of Wnt and Frizzled genes during chick limb bud development. Cold Spring Harb Symp Quant Biol **62**:421-429.

[19] Kawakami Y, Wada N, Nishimatsu SI, Ishikawa T, Noji S, Nohno T 1999 Involvement of Wnt-5a in chondrogenic pattern formation in the chick limb bud. Dev Growth Differ **41**:29-40.

[20] Zakany J and Duboule D 1993 Correlation of expression of Wnt-1 in developing limbs with abnormalities in growth and skeletal patterning. Nature **362**: 546-549.

[21] Rudnicki JA and Brown AM 1997 Inhibition of chondrogenesis by Wnt gene expression in vivo and in vitro. Dev Biol **185**:104-118.

[22] Stott NS, Jiang TX, Chuong CM 1999 Successive formative stages of precartilaginous mesenchymal condensations in vitro: modulation of cell adhesion by Wnt-7A and BMP-2. J Cell Phys **180**:314-324.

[23] Hartmann C and Tabin CJ 2000 Dual roles of Wnt signaling during chondrogenesis in the chicken limb. Development **127**:3141-3159.

[24] Tufan AC and Tuan RS 2001 Wnt regulation of limb mesenchymal chondrogenesis is accompanied by altered N-cadherin-related functions. FASEB J 15:1436-1438.

[25] Vinson CR, Conover S, Adler PN 1989 A Drosophila tissue polarity locus encodes a protein containing seven potential transmembrane domains. Nature **338**:263-264.

[26] Park WJ, Liu J, Adler PN 1994 The frizzled gene of Drosophila encodes a membrane protein with an odd number of transmembrane domains. Mech Dev **45**:127-137.

[27] Wang Y, Macke JP, Abella BS, Andreasson K, Worley P, Gilbert DJ, Copeland NG, Jenkins NA, Nathans J 1996 A large family of putative transmembrane receptors homologous to the product of the Drosophila tissue polarity gene frizzled. J Biol Chem **271**:4468-4476.

[28] Nohno T, Kawakami Y, Wada N, Komaguchi C, Nishimatsu S 1999 Differential expression of the frizzled family involved in Wnt signaling during chick limb development. Cell Mol Biol **45**:653-659.

[29] Kawakami Y, Wada N, Nishimatsu S, Nohno T 2000 Involvement of frizzled-10 in Wnt-7a signaling during chick limb development. Dev Growth Differ **42**:561-569.

[30] Abou-Samra AB, Juppner H, Force T, Freeman MW, Kong XF, Schipani E 1992 Expression cloning of a common receptor for parathyroid hormone and parathyroid hormone-related peptide from rat osteoblast-like cells: a single receptor stimulates intracellular accumulation of both cAMP and inositol trisphospates and increases intracellular free calcium. Proc Natl Acad Sci USA **89**:2732-2736.

[31] Iida-Klein A, Varlotta V, and Hahn TJ 1989 Protein kinase C activity in UMR-106-01 cells: effects of parathyroid hormone and insuline. J Bone Miner Res **4**:767-774.

[32] Goodrich LV, Johnson RL, Milenkovic L, McMahon JA, and Scott MP 1996 Conservation of the hedge/patched signaling pathway from flies to mice: induction of a mouse patched gene by Hedgehog. Genes Dev **8**:277-289.

[33] Vortkamp A, Lee K, Lanske B, Segre GV, Kronenberg HM, and Tabin CJ 1996 Regulation of rate of cartilage differentiation by Indian hedgehog and PTH-related protein. Science **273**:613-622.

[34] Farnum CE, and Wilsman NJ 1989 Cellular turnover at the chondro-osseous junction of growth plate cartilage: analysis by serial sections at the light microscopical level. J Orthopaed Res **7**:654-666.

[35] Zenmyo M, Komiya S, Kawabata R, Sasaguri Y, Inoue A, and Morimatsu M 1996 Morphological and biochemical evidence for apoptosis in the terminal hypertrophic chondrocytes of the growth plate. J Pathol **180**:430-433.

[36] Mansfield K, Rajpurohit R, and Shapiro IM 1999 Extracellular phosphate ions cause apoptosis of terminally differentiated epiphyseal condrocytes. J Cell Phys **179**:276-286.

[37] Chuang LY, Hung WC, Chang CC, and Tsai JH 1994 Characterization of apoptosis induced by transforming growth factor-1 in human heptatoma cells. Anticancer Res **14**:147-152.

[38] Wallen-Ohman M, Lonnbro P, Schon A, and Borrebaeck CA 1993 Antibody-induced apoptosis in a human leukemia cell line is energy dependent: thermochemical analysis of cellular metabolism. Cancer Letters **75**:103-109.

[39] Mansfield K, Rajpurohit R, and Shapiro IM 1999 Extracellular phosphate ions cause apoptosis of terminally differentiated epiphyseal condrocytes. J Cell Phys **179**:276-286.

[40] Vaux DL, Cory S, and Adams JM 1988 Bcl-2 gene promotes haemopoietic cell survival and cooperates with c-myc to immortalize pre-B cells. Nature **335**:440-442.

[41] Sedlak TW, Oltvai, ZN, Yang E, Wang K, Boise LH, Thompson CB, and Korsmeyer SJ 1995 Multiple Bcl-2 family members demonstrate selective dimerization with Bax. Proc Natl Acad Sci USA **92**:7834-7838.

[42] Nakayama K, Nakayama KI, Negishi I, Kuida K, Sawa H, and Loh DY 1994 Target disruption of Bcl-2ab in mice: occurrence of gray hair, polycystic kidney disease, and lymphocytopenia. Proc Natl Acad Sci. USA **91**:3700-3704.

[43] Amling M, Neff L, Tanaka S, Inoue D, Kuida K, Weir E, Philbrick WM, Broadus AE, and Baron R 1997 Bcl-2 lies downstream of parathyroid hormone-related peptide in a signaling pathway that regulates chondrocyte maturation during skeletal development. J Cell Biol **136**:205-213.

[44] Lee JK, Deeds JD, and Segre GV 1995 Expression of parathyroid hormone–related peptide and its receptor messenger ribonucleic acids during fetal development of rats. Endocrinology **136**:453-463.

[45] Hashimoto S, Setareh M, Ochs R, and Lotz M 1997 Fas/Fas ligand expression and induction of apoptosis in chondrocytes. Arthritis Rheum **40**:1749-1755.

[46] Yang A, Schweitzer R, Sun D, Kaghad M, Walker N, Bronson RT, Tabin C, Sharpe A, Caput D, Crum C, and Mckeon F 1999 p63 is essential for regenerate proliferation in limb, craniofacial and epithelial development. Nature **398**:714-718.

[47] San Antonio JD and Tuan RS 1986 Chondrogenesis of limb mesenchyme *in vitro*: stimulation by cations. Devel Bio **115**:313- 324.

[48] Oberlender S and Tuan RS 1994 Expression and functional involvement of N-cadherin in embryonic limb chondrogenesis. Development **120**:177-197.

[49] Mello MA and Tuan RS 1999 High density cultures of embryonic limb bud mesenchymal cells: An in vitro model of endochondral skeletal development. In vitro Cell Develop Biol. **35**:262-269.

[50] DeLise AM and Tuan RS 2000 Electroporation Mediated DNA Transfection of Embryonic Chick Limb Mesenchymal Cells. In: Tuan RS and Lo CW (eds.) Developmental Biology Protocols, vol 3, Humana Press, Totowa, pp 377-382.

[51] Dealy CN, Roth A, Ferrari D, Brown AM, Kosher RA 1993 Wnt-5a and Wnt-7a are expressed in the developing chick limb bud in a manner suggesting roles in pattern formation along the proximodistal and dorsoventral axes. Mechanisms of Development **43**:175-186.

[52] Shapiro I, DeBolt K, Funanage VL, Smith SM, and Tuan RS 1992 Developmental regulation of creatine kinase activity in cells of the epiphyseal growth plate. J Bone Min Res **7**:493-500.

[53] DeLise AM, Stringa E, Woodward WA, Mello MA, and Tuan RS 2000 Embryonic Limb Mesenchyme Micromass Culture as an *In Vitro* Model for Chondrogenesis and Chondrocyte Maturation. In: Tuan RS and Lo CW (eds.) Developmental Biology Protocols, vol 3, Humana Press, Totowa, pp 359-375.

[54] Funanage V, Carango P, Shapiro I, Tokwoka T, and Tuan RS 1992 Creatine kinase activity is required for mineral deposition and matrix synthesis in endochondral growth cartilage. Bone Miner. **17**:228-236.

[55] Takeda S, Bonnamy JP, Owen MJ, Ducy P, and Karsenty G 2001 Continues expression of Cbfa1 in nonhypertrophic chondrocytes uncovers its ability to induce hypertrophic chondrocyte differentiation and partially rescues Cbfa1-deficient mice. Genes Dev **15**: 467-481.

[56] Volk SW, Luvalle P, Leask T, and Leboy PS 1998 A BMP responsive transcriptional region in the chicken type x collagen gene. J Bone Miner Res **13**:1521-1529.

[57] Morgenstern JP, and Land H 1990 Advanced mammalian gene transfer: high titer retroviral vectors with multiple drug selection a complementary helper-free packaging cell line. Nucleic Acids Res 18:3587-3596.

[58] Boskey AL, Stiner D, Doty SB, Binderman I, and Leboy P 1991 Studies of mineralization in tissue culture: optimal conditions for cartilage calcification. J Bone Miner Res **16**:11-36.
[59] Castagnola P, Moro G, Descalzi-Cancedda F, and Cancedda R 1986 Type X collagen synthesis during in vitro development of chick embryo tibial chondrocytes. J Cell Biol **102:**2310-2317.
[60] Descalzi-Cancedda F, Gentili C, Manduca P, and Cancedda R 1992 Hypertrophic chondrocytes undergo further differentiation in culture. J Cell Biol **117:**427-435.

The Growth Plate
I.M. Shapiro et al. (Eds.)
IOS Press, 2002

Local Production of Estradiol by Growth Plate Chondrocytes and its Gender-Specific Membrane Mediated Effects

V. L. Sylvia[1], D. Dombroski[1], I. Gay[1], D. D. Dean[1], Z. Schwartz[1,2,3], and B. D. Boyan[1,2,4]
Departments of [1]Orthopaedics, [2]Periodontics, and [4]Biochemistry, The University of Texas Health Science Center at San Antonio, San Antonio, TX 78229-3900; and [3]Hebrew University Hadassah School of Dental Medicine, Jerusalem, Israel 91010

Abstract. Growth plate chondrocytes from both male and female rats have the nuclear receptors for 17β-estradiol (E_2). However, cells from female rat costochondral growth plates respond to E_2 differently than cells from male rats. E_2 directly affects the fluidity of chondrocyte membranes derived from female, but not male, rats. In addition, E_2 activates PKC in a nongenomic manner in female cells, and chelerythrine, a specific inhibitor of PKC, inhibits E_2-dependent alkaline phosphatase activity in these cells, indicating that PKC is involved in the signal transduction mechanism. ERα and ERβ have been found in plasma membranes of E_2-sensitive cells, suggesting that E_2 mediates its effects through membrane receptor-mediated mechanisms in addition to nuclear receptor-mediated pathways. This paper reviews studies that examine this hypothesis. Fourth passage resting zone (RC) and growth zone (GC) chondrocytes from female and male rat costochondral cartilage were treated with 10^{-10} to 10^{-7} M E_2 in the presence of inhibitors and activators of enzymes involved in membrane-mediated signal transduction. To examine the role of classical estrogen receptors, cultures were also treated with the estrogen agonist diethylstilbesterol (DES) and the antagonist ICI 182780. To verify that the effects of E_2 on PKC were due to a membrane-mediated response, we examined the effects of E_2 that had been conjugated to bovine serum albumin (E_2-BSA) to prevent its uptake into the cell. To determine if the membrane receptor might function in an autocrine/paracrine manner, we examined whether costochondral chondrocytes themselves produce E_2 by examining aromatase expression and activity as a function of cell maturation state and gender. For these experiments, RC and GC cells from male rats were also used. E_2 and E_2-BSA elicited comparable effects. Both forms of E_2 stimulated PKC, but only in female cells. The phosphatidylinositol-specific phospholipase C (PLC) inhibitor U73122 blocked the increase in PKC activity. Inhibition of PC-PLC, PLD, and PLA_2 had no effect; also, activation of PLA_2 was without effect. ICI 182780 did not block the stimulatory effect of either E_2 or E_2-BSA and DES also did not alter their effects on PKC. The G-protein inhibitor GDPβS inhibited E_2-BSA stimulated PKC in both RC and GC cells. However, the G-protein activator GTPγS increased PKC in E_2-BSA treated GC cells only. Aromatase activity was present in male and female cells and was highest in female RC chondrocytes compared to male RC cells and male and female GC cells. Female RC cells also produced the highest levels of E_2. Moreover, regulation of E_2 production by 1α,25-$(OH)_2D_3$ was gender-dependent and was found in female cells.
The results of these studies show that the gender-specific effects of E_2 involve membrane-associated mechanisms that are independent of ERα/ERβ. The rapid nongenomic effect of E_2 and E_2-BSA on PKC is dependent on G-protein coupled PI-PLC. The membrane ERs may function in autocrine/paracrine regulation since RC and GC cells produce E_2 in a gender- and cell maturation-dependent and regulated manner.

Introduction

17β-Estradiol (E_2) regulates endochondral bone formation by direct effects on chondrocytes and indirectly through secretion of other hormones and local factors.[1-4] E_2 mediates its effects on cells, including chondrocytes,[5-7] through classical steroid hormone receptors.[8-10] Additionally, recent evidence suggests that some of the effects of estrogen are mediated by rapid, membrane-associated mechanisms. Using a rat costochondral model for studying agents which modulate chondrocyte differentiation and maturation along the endochondral developmental pathway,[11] we have shown that E_2 increases arachidonic acid turnover, phospholipase A_2 activity, and membrane fluidity in cultures of cartilage cells derived from female adolescent rats.[12] These effects were shown to be membrane-specific, gender-specific, and cell maturation-dependent and occurred in a nongenomic manner.

There is increasing support that membrane-mediated effects of steroid hormones proceed via pathways traditionally ascribed to peptide hormones, including protein kinase C (PKC)[13-16] and MAP kinase.[17] Recent data demonstrate that E_2 has the capacity to increase PKC activity exclusive of inducing new gene expression.[18] PKC appears to be involved in the biological response to E_2, since PKC inhibitors block the action of this steroid on DNA synthesis.[19,20] Studies using E_2-bovine serum albumin (BSA) conjugates, which remain extracellular,[21] suggest that it may not be necessary for E_2 to enter the cell to initiate membrane-associated mechanisms.[22,23] Moreover, the effect of E_2 on PKC is stereospecific,[18] indicating a receptor-mediated mechanism.

This evidence raises the possibility that receptors for E_2 exist on the plasma membrane. Studies in other laboratories support this hypothesis. Estrogen receptor alpha (ERα) has been found associated with plasma membranes of Chinese hamster ovary cells.[24] More recently, ERβ was also found to be membrane-associated in rat astrocytes.[25] It is possible that the membrane receptor responsible for PKC activation in chondrocytes is also ERα or ERβ, since membrane-associated ERα mediates some of the effects of E_2-BSA in osteoblasts.[26] However, none of these responses were reported to be gender-dependent, whereas the increase in PKC is, suggesting that a unique membrane receptor is involved.

The present paper reviews studies showing that (A) PKC mediates the effect of E_2 on chondrocyte proliferation, matrix synthesis, and differentiation; (B) the signaling pathways leading to increased PKC differ from those of other steroid hormones and do not involve traditional E_2-receptors; and (C) chondrocytes possess the ability to metabolize androgens to estrogen.

Materials and Methods

Reagents

17β-estradiol, 17β-estradiol-BSA, 17α-estradiol, diethylstilbesterol (DES), G-protein inhibitors (pertussis toxin and GDPβS), and quinacrine (phospholipase A_2 inhibitor) were purchased from Sigma Chemical Co. (St. Louis, MO). The estrogen receptor antagonist ICI 182780 was obtained from Tocris Cookson, Inc. (Ballwin, MO). Chelerythrine (PKC inhibitor) and U73122 (phospholipase C inhibitor) were purchased from Calbiochem (San Diego, CA). PKC assay reagents and DMEM were obtained from GIBCO-BRL (Gaithersburg, MD). The protein content of each sample was determined using the bicinchoninic acid (BCA) protein assay reagent obtained from Pierce Chemical Co.

(Rockford, IL). Gamma-[^{32}P]-ATP, [^{3}H]-thymidine and [^{35}S]-sulfate were obtained from NEN-DuPont (Boston, MA).

Role of PKC in Mediating the Physiological Response to E_2

The culture system used for these studies has been described in detail previously.[11] The chondrocytes were isolated from the resting zone (RC; reserve zone) and growth zone (GC; prehypertrophic/upper hypertrophic cell zones) of the costochondral junction of 125 g male and female Sprague-Dawley rats by enzymatic digestion and cultured in Dulbecco's modified Eagle medium (DMEM) containing 10% fetal bovine serum (FBS), antibiotics, and vitamin C.[18] Confluent, fourth passage cultures were used for all experiments.

To determine whether the physiological response of resting zone and growth zone chondrocytes to E_2 is mediated by PKC, the cells were treated with 10^{-10} to 10^{-7} M E_2, and PKC activity was inhibited by addition of (10 μM) chelerythrine to the cultures. DNA synthesis was estimated by measuring [^{3}H]-thymidine incorporation by quiescent resting zone and growth zone chondrocytes.[27,28] Proteoglycan synthesis was assessed by measuring [^{35}S]-sulfate incorporation by confluent cultures.[29,30] Alkaline phosphatase specific activity was measured in cell layer lysates as a function of release of *para*-nitrophenol from *para*-nitrophenylphosphate at pH 10.2.[28,31]

Mechanism of E_2-dependent PKC Activation

To determine the signaling pathways involved in E_2-dependent activation of PKC, confluent cultures in 24-well plates were treated for various time periods with E_2 in the absence or presence of various concentrations of inhibitors. Phospholipase C (PLC) involvement in E_2 action was assessed using U73122, an inhibitor of phosphatidylinositol (PI)-specific PLC,[32] and D609, an inhibitor of phosphatidylcholine (PC)-specific PLC.[33] To examine the role of phospholipase A_2 (PLA_2), quinacrine[34] was used to inhibit PLA_2 activity. Phospholipase D activity was measured using a modification of the method originally described by Brown et al.[35,36] in order to assess whether this enzyme is regulated by E_2. To assess the role of G-proteins, pertussis toxin (PTX, Gi inhibitor), and the non-hydrolyzable GDPβS (general G-protein inhibitor) were used.

E_2-BSA was used to test the hypothesis that E_2 exerts its effects on proliferation, matrix synthesis, and differentiation of growth plate chondrocytes via membrane-associated mechanisms that are mediated by the PKC signaling pathway in resting zone and growth zone chondrocyte cultures. In addition, cultures were treated with 17α-estradiol to examine the stereospecificity of the effect. Finally, to assess the role of the classical estrogen receptor in the effect of E_2 on PKC, female resting zone cells were treated with E_2 ± estrogen receptor agonist diethylstilbesterol (DES) or antagonist ICI 182780. Since E_2 activates PKC in resting zone from female rats as quickly as 3 minutes, reaches maximum activity at 90 minutes, and remains significantly higher than control even after 4 hours of treatment,[37] time points of 3, 9, 90 and 270 minutes were chosen for experiments examining time course. Dose-response experiments were done at 90 minutes. After the appropriate incubation period, cell layers were washed with phosphate buffered saline (PBS), and lysed in solubilization buffer. The cell layer lysates were assayed for protein content and PKC activity.

Local Production of E_2

Aromatase activity in resting and growth zone chondrocyte cultures from male and female rats was measured using the classical $[^3H]_2O$ assay as described by others in a number of publications.[38-40] 17β-Estradiol production was determined using a specific commercially available radioimmunoassay kit (Cat. #DSL-4800; Diagnostic Systems Laboratories, Inc., Webster, TX). Others have reported that E_2 production is regulated by $1\alpha,25(OH)_2D_3$.[41] To see if this was true for growth plate chondrocytes as well, growth zone cells were also treated with $1\alpha,25(OH)_2D_3$ and E_2 production measured.

Results

PKC activity was stimulated by E_2 in growth plate chondrocytes from female rats in a dose- and time-dependent manner.[18] In contrast, E_2 elicited only a slight increase in male cells. The results of the studies described here show that PKC mediates the physiological response of resting zone and growth zone cells from female rats to the hormone. In E_2-treated resting zone cells, the inhibitory effect of chelerythrine on $[^3H]$-thymidine incorporation was even greater than that of E_2. This was also the case for growth zone cells, but only at the lowest E_2 concentration examined. The stimulatory effect of E_2 on proteoglycan sulfation was blocked by PKC inhibition in both cell types. In addition, E_2 regulated alkaline phosphatase specific activity in female resting zone and growth zone cells via PKC. Whereas chelerythrine only partially blocked the effect of E_2 on alkaline phosphatase in resting zone cells, it completely blocked the effect in growth zone cells.

The E_2-dependent increase in PKC was mediated by a membrane-associated mechanism. PKC activity was increased in a time-dependent manner by E_2-BSA. The increase was noted by 9 minutes, but maximal effects were seen at 90 minutes in growth zone cells. The rate of increase in PKC activity between 9 and 30 minutes was greater in the more mature chondrocytes. In resting zone cells, maximal increases occurred later, and elevated activity was maintained longer. Treatment with E_2 together E_2-BSA failed to show additive or synergistic effects on PKC activity in both cell types.

The effects of E_2 were stereospecific. Only 17β-E_2 caused an increase in PKC or modulated cell response in a PKC-dependent manner. 17α-E_2 had no effect, suggesting a receptor-mediated mechanism. The effect of E_2 on PKC in growth plate chondrocytes did not involve classical nuclear E_2 receptors, however. PKC activity was not significantly changed in resting zone cells with either 10^{-9}-10^{-7} M diethylstilbesterol (DES) or ICI 182780, and neither compound affected E_2-stimulated PKC activity.

E_2 mediated its effects on PKC via PLC in the growth plate chondrocytes (Table 1). The PLC involved was phosphatidylinositol-specific; the PI-PLC inhibitor U73122 blocked the E_2 effect on PKC completely at all time points tested. In contrast, the PC-PLC inhibitor D609 was without effect. E_2 had no effect on PLD activity in either cell type, indicating that any DAG produced was via PI-PLC.[42] Regulation of PKC activity by E_2 also did not involve phospholipase A_2. Inhibition of phospholipase A_2 activity with quinacrine had no effect on PKC activity in E_2-treated cultures either at 9 minutes or at 270 minutes.[42]

Table 1. Effect of 17β-estradiol and the phosphatidylinositol specific-phospholipase C inhibitor U73122 in female resting zone (RC) chondrocytes.

	PKC Specific Activity (pmol PO_4/μg protein/min)		
Treatment	**9 Minutes**	**90 Minutes**	**270 Minutes**
Vehicle	0.89 ± 0.09	0.67 ± 0.03	0.56 ± 0.04
U73122	0.74 ± 0.04	0.62 ± 0.04	0.54 ± 0.04
E_2	1.39 ± 0.09*#	3.52 ± 0.25*#	1.84 ± 0.06*#
E_2 + U73122	0.74 ± 0.02•	0.73 ± 0.07•	0.74 ± 0.08•

Data represent the mean ± SEM of six cultures from one of three experiments, each with comparable results. *$P < 0.05$, E_2 vs. control; #$p < 0.05$, E_2 vs. U73122; •$p < 0.05$, E_2 ± U73122 vs. E_2 alone.

G-proteins mediated the effect of E_2 on PKC; however, neither G_i nor G_s was involved. The G_i inhibitor pertussis toxin had no effect on PKC activity in E_2-treated resting zone cells. Similarly, the G_s inhibitor cholera toxin had no effect. In contrast, PKC activity was reduced in cultures treated with E_2 in the presence of the general G-protein inhibitor GDPβS (Table 2), indicating that a G_q was responsible. With increasing concentrations of the inhibitor, PKC was reduced to levels below baseline whether or not E_2 was present.

Table 2. Effect of G-protein inhibitor, GDPβS, on PKC specific activity of female resting zone chondrocytes (RC) treated with 17β-estradiol.

	PKC Specific Activity (pmol PO_4/μg protein/min)		
Treatment	**9 Minutes**	**90 Minutes**	**270 Minutes**
Vehicle	0.88 ± 0.07	0.95 ± 0.07	0.74 ± 0.06
GDPβS	0.69 ± 0.06	0.75 ± 0.06	0.58 ± 0.05
E_2	1.02 ± 0.08*#	2.32 ± 0.15*#	0.88 ± 0.03*#
E_2 ± GDPβS	0.74 ± 0.05•	0.70 ± 0.05•	0.62 ± 0.04•

Data represent the mean ± SEM of six cultures from one of three experiments, all with comparable results. *$P < 0.05$, E_2 vs. control; #$p < 0.05$, E_2 vs. GDPβS; •$p < 0.05$, E_2 ± GDPβS vs. E_2 alone.

Local Production of E_2

Growth plate chondrocytes isolated from resting zone and growth zone costochondral cartilage of both male and female rats possessed aromatase activity.[43] Female resting zone cells had the highest level of aromatase activity, which was approximately two times greater than that of female growth zone cells. Male resting zone and growth zone cells also had significantly lower aromatase activity than female resting zone cells and comparable activity to female growth zone cells.

The costochondral cartilage cells secreted 17β-E_2 into their culture medium. All cultures synthesized 17β-E_2, however, female resting zone cells had the highest basal level, with a 2.5-fold greater production than any of the other cell types. Female growth zone cells produced the same amount of 17β-E_2 as the resting zone and growth zone cells from male rats.

1α,25$(OH)_2D_3$ regulated E_2 production by growth plate chondrocytes in a gender-dependent manner. Estrogen production by female chondrocytes was increased by 1α,25$(OH)_2D_3$ in a dose-dependent manner (Table 3). While the absolute amount of 17β-E_2 produced by female resting zone cells was greater, the stimulatory effect of 1α,25$(OH)_2D_3$ was greater in female growth zone cells. The response of male cells to 1α,25$(OH)_2D_3$ differed from that of female cells. Only at the highest concentrations of 1α,25$(OH)_2D_3$ was there an increase in 17β-E_2 production in the resting zone cell cultures, and this effect was relatively weak.

Table 3. Effect of 1α,25$(OH)_2D_3$ on 17β-estradiol production by growth plate resting zone (RC) chondrocytes from female rats.

	17β-Estradiol Production	
1α,25$(OH)_2D_3$ [M]	fMol E_2/Culture	fMol E_2/10^6 cells
0	658 ± 54	661 ± 68
10^{-10} M	716 ± 416	782 ± 66
10^{-9} M	842 ± 36*	1,044 ± 60*
10^{-8} M	914 ± 58*	1,267 ± 13*

Data represent the mean ± SEM of six cultures from one of two experiments, both with comparable results. *$P < 0.05$, 1α,25$(OH)_2D_3$ vs. vehicle.

Discussion

These studies have shown that E_2 exerts its physiological effects on chondrocytes from female costochondral cartilage via PKC-mediated mechanisms. Direct inhibition of PKC mimicked the effects of E_2 on proliferation and reduced E_2-dependent stimulation of alkaline phosphatase specific activity and proteoglycan sulfation, indicating that proliferation, differentiation, and matrix synthesis are all PKC-dependent. These effects were similar in resting zone and growth zone cells. The ability of chelerythrine to block the effects of E_2 on these biological processes emphasizes the importance of the PKC pathway in the action of E_2 in cartilage cells.

Regulation of PKC by the hormone is at the level of the plasma membrane. While it is receptor-activated, based on the stereospecificity of the response,[18] the classical E_2 receptors do not appear to play a role. Neither DES nor ICI 182780 stimulate PKC in the absence of E_2 or inhibit the response in the presence of E_2. This suggests that the mechanism of activation of PKC by E_2 is through a mechanism different than classic nuclear receptors, whether they are membrane-associated or not. This is supported by the fact that the effect of E_2 on PKC is gender-specific, whereas other effects of E_2 on osteoblasts appear to be gender-neutral.[26]

E_2 regulated PKC via a PLC-dependent mechanism, and the PI-PLC form of the enzyme is responsible. E_2 did not affect PLD activity, indicating that diacylglycerol produced via the PLD pathway did not contribute to PKC activation. Neither PC-PLC nor PLA_2 played a role. E_2 exerted its effect on PKC through a G-protein-dependent mechanism, however, involving G_q, but not G_i or G_s. This is based on the observation that the general G-protein inhibitor GDPβS inhibited the effect of E_2, whereas neither pertussis toxin nor cholera toxin did.

These results indicate that E_2 regulates PKCα through mechanisms that differ from those used by $1\alpha,25(OH)_2D_3$ and $24R,25(OH)_2D_3$ in the same cells (Table 4). The effect of $1\alpha,25(OH)_2D_3$ is limited to growth zone cells and involves PLA_2, as well as PI-PLC and G_q,[37] whereas the PLA_2 pathway is not involved in the action of E_2. Both E_2 and $1\alpha,25(OH)_2D_3$ induce translocation of cytosolic PKC to the plasma membrane[37,44] The effect of $24R,25(OH)_2D_3$ is limited to resting zone cells and involves only PLD.[45] Moreover, PKC translocation does not occur.

The fact that E_2 exerts a membrane effect suggests the possibility that the hormone may serve an autocrine regulatory role in cartilage. In support of this hypothesis, chondrocytes have been shown to possess aromatase activity and produce 17β-E_2 in a regulated manner.[43] Moreover, both aromatase activity and E_2 production were greatest in female resting zone cells. This may explain physiological differences in the response of growth plate resting zone and growth zone cartilage observed in organ culture[46] and in vivo,[2] where E_2 exerts its greatest effects in the resting zone of female tissues.

Taken together this study demonstrates that E_2 binds a membrane receptor, independent of the nuclear receptor, associated with a Gq protein activating phosphatidylinositol specific phospholipase C and subsequently increasing the activity of PKC. This stimulation by E_2 leads to cell differentiation and a decrease in proliferation with increased matrix production. More importantly, all of the physiologic effects mediated by E_2 via PKC are gender-dependent. Male chondrocytes fail to show this response to E_2. The importance of the gender-dependent increase in PKC may be related to the failure of $1\alpha,25(OH)_2D_3$ to increase local production of 17β-E_2 in male cells.

Table 4. Comparison of mechanisms involved in the stimulation of PKCα by 17β-estradiol, 1α,25$(OH)_2D_3$, and 24R,25$(OH)_2D_3$ in growth plate chondrocytes.

	Change in PKC Specific Activity		
	17β-E_2	**1α,25$(OH)_2D_3$**	**24R,25$(OH)_2D_3$**
Cell Specificity	RC > GC	GC	RC
Gender Specificity	F	M/F	M/F
Peak Effect	90 min	9 min	90 min
PLC	Yes	Yes	No
PLD	No	No	Yes
PLA_2	No	Yes (↑)	Yes (↓)
Genomic Component	Yes	No	Yes
Gq	Yes	Yes	No
Arachidonic Acid	No	Yes (↑)	Yes (↓)
PGE_2		Yes (↑)	No (↓)
Translocation	Yes	Yes	No

Acknowledgements

The authors thank Sandra Messier for her help in the preparation of the manuscript. The studies summarized in this paper were supported by the Susan G. Komen Foundation and PHS grants DE05937 and DE08603.

References

[1] Schwartz Z, Meincke J, Nasatzky E, Dean DD, Boyan BD 1995 Estrogen regulation of endochondral bone formation. In: Ornoy A (ed) Animal Models of Human Related Calcium Metabolic Disorders., CRC Press, Boca Raton, FL, pp. 149-164.

[2] Ornoy A, Giron S, Aner R, Goldstein M, Boyan BD, Schwartz Z 1994 Gender dependent effects of testosterone and 17 beta-estradiol on bone growth and modelling in young mice. Bone Miner **24:**43-58.

[3] Schwartz Z, Soskolne WA, Neubauer T, Goldstein M, Adi S, Ornoy A 1991 Direct and sex-specific enhancement of bone formation and calcification by sex steroids in fetal mice long bone in vitro (biochemical and morphometric study). Endocrinology **129:**1167-1174.

[4] Sheridan PJ, Aufdemorte TB, Holt GR, Gates GA 1985 Cartilage of the baboon contains estrogen receptors. Rheumatol Int **5:**279-281.

[5] Pinus H, Ornoy A, Patlas N, Yaffe P, Schwartz Z 1993 Specific beta estradiol binding in cartilage and serum from young mice and rats is age dependent. Connect Tissue Res **30:**85-98.

[6] Nasatzky E, Schwartz Z, Soskolne WA, Brooks BP, Dean DD, Boyan BD, Ornoy A 1994 Evidence for receptors specific for 17β -estradiol and testosterone in chondrocyte cultures. Connect Tissue Res **30:**277-294.

[7] Dayani N, Corvol MT, Robel P, Eychenne B, Moncharmont B, Tsagris L, Rappaport R 1988 Estrogen receptors in cultured rabbit articular chondrocytes: influence of age. J Steroid Biochem **31:**351-356.

[8] Means AR, O'Malley BW 1972 Mechanism of estrogen action: early transcriptional and translational events. Metab Clin & Experimen **21:**357-370.

[9] Jensen EV, DeSombre ER 1973 Estrogen-receptor interaction. Science **182:**126-134.

[10] Anstead GM, Carlson KE, Katzenellenbogen JA 1997 The estradiol pharmacophore: ligand structure-estrogen receptor binding affinity relationships and a model for the receptor binding site. Steroids **62:**268-303.

[11] Boyan BD, Schwartz Z, Swain LD, Carnes DL, Jr., Zislis T 1988 Differential expression of phenotype by resting zone and growth region costochondral chondrocytes *in vitro*. Bone **9:**185-194.

[12] Schwartz Z, Gates PA, Nasatzky E, Sylvia VL, Mendez J, Dean DD, Boyan BD 1996 Effect of 17β-estradiol on chondrocyte membrane fluidity and phospholipid metabolism is membrane-specific, sex-specific, and cell maturation-dependent. Biochim Biophys Acta **1282:**1-10.

[13] Wehling M 1997 Specific, nongenomic actions of steroid hormones. Ann Rev Physiol **59:**365-393.

[14] Pailler-Rodde I, Garcin H, Higueret P 1991 Effect of retinoids on protein kinase C activity and on the binding characteristics of the tri-iodothyronine nuclear receptor. J Endocrinol **128:**245-251.

[15] Morelli S, Boland R, de Boland AR 1996 1,25$(OH)_2$-vitamin D_3 stimulation of phospholipases C and D in muscle cells involves extracellular calcium and a pertussis-sensitive G protein. Molec & Cell Endocrinol **122:**207-211.

[16] Magda T, Lloyd V 1993 Protein kinase C activity and messenger RNA modulation by estrogen in normal and neoplastic rat pituitary tissue. Lab Invest **68:**472-480.

[17] Endoh H, Sasaki H, Maruyama K, Takeyama K, Waga I, Shimizu T, Kato S, Kawashima H 1997 Rapid activation of MAP kinase by estrogen in the bone cell line. Biochem Biophys Res Comm **235:**99-102.

[18] Sylvia VL, Hughes T, Dean DD, Boyan BD, Schwartz Z 1998 17β-Estradiol regulation of protein kinase C activity in chondrocytes is sex-dependent and involves nongenomic mechanisms. J Cell Physiol **176:**435-444.

[19] Fujimoto J, Hori M, Ichigo S, Morishita S, Tamaya T 1996 Estrogen induces expression of c-fos and c-jun via activation of protein kinase C in an endometrial cancer cell line and fibroblasts derived from human uterine endometrium. Gynecol Endocrinol **10:**109-118.

[20] Rajkumar K 1993 Effect of protein kinase C inhibitor on estradiol-induced deoxyribonucleic acid synthesis in rats. Steroids **58:**100-105.

[21] Zheng J, Ali A, Ramirez VD 1996 Steroids conjugated to bovine serum albumin as tools to demonstrate specific steroid neuronal membrane binding sites. J Psych Neurosci **21:**187-197.

[22] Stefano GB, Prevot V, Beauvillain J, Fimiani C, Welters I, Cadet P, Breton C, Pestel J, Salzet M, Bilfinger TV 1999 Estradiol coupling to human monocyte nitric oxide release is dependent on intracellular calcium transients: evidence for an estrogen surface receptor. J Immunol **163:**3758-3763.

[23] Benten WP, Lieberherr M, Giese G, Wunderlich F 1998 Estradiol binding to cell surface raises cytosolic free calcium in T cells. FEBS Letts **422:**349-353.

[24] Razandi M, Pedram A, Greene GL, Levin ER 1999 Cell membrane and nuclear estrogen receptors (ERs) originate from a single transcript: Studies of ERα and ERβ expressed in Chinese hamster ovary cells. Molec Endocrinol **13:**307-319.

[25] Hosli E, Ruhl W, Hosli L 2000 Histochemical and electrophysiological evidence for estrogen receptors on cultured astrocytes: colocalization with cholinergic receptors. Int J Dev Neurosci **18:**101-111.

[26] Kousteni S, Bellido T, Plotkin LI, O'Brien CA, Bodenner DL, Han L, Han K, DiGregorio GB, Katzenellenbogen JA, Katzenellenbogen BS, Roberson PK, Weinstein RS, Jilka RL, Manolagas SC 2001 Nongenotropic, sex-nonspecific signaling through the estrogen or androgen receptors: dissociation from transcriptional activity. Cell **104:**719-730.

[27] Schwartz Z, Schlader DL, Ramirez V, Kennedy MB, Boyan BD 1989 Effects of vitamin D metabolites on collagen production and cell proliferation of growth zone and resting zone cartilage cells in vitro. J Bone Miner Res **4:**199-207.

[28] Nasatzky E, Schwartz Z, Boyan BD, Soskolne WA, Ornoy A 1993 Sex-dependent effects of 17-β-estradiol on chondrocyte differentiation in culture. J Cell Physiol **154:**359-367.

[29] Nasatzky E, Schwartz Z, Soskolne WA, Brooks BP, Dean DD, Boyan BD, Ornoy A 1994 Sex steroid enhancement of matrix production by chondrocytes is sex and cell maturation specific. Endocrine J **2:**207-215.

[30] O'Keefe RJ, Puzas JE, Brand JS, Rosier RN 1988 Effects of transforming growth factor-beta on matrix synthesis by chick growth plate chondrocytes. Endocrinology **122:**2953-2961.

[31] Bretaudiere JP, Spillman T 1984 Alkaline phosphatases. In: Bergmeyer HU (ed) Methods of Enzymatic Analysis, 4. Verlag Chemica, Weinheim, Germany, pp. 75-92.

[32] Bleasdale J, Bundy GL, Bunting S, Fitzpatrick FA, Huff RM, Sun FF, Pike JE 1989 Inhibition of phospholipase C-dependent processes by U73,122. Adv Prostag Thrombox Leuk Res **19:**590-593.

[33] Muller-Decker K 1989 Interruption of TPA-induced signals by an antiviral and antitumoral xanthate compound: inhibition of a phospholipase C-type reaction. Biochem Biophys Res Comm **162:**198-205.

[34] Church D, Braconi S, Vallotton M, Lang U 1993 Protein kinase C-mediated phospholipase A_2 activation, platelet-activating factor generation and prostacyclin release in spontaneously beating rat cardiomyocytes. Biochem J **290:**477-482.

[35] Brown HA, Gutowski S, Moomaw CR, Slaughter C, Sternweis PC 1993 ADP-ribosylation factor, a small GTP-dependent regulatory protein, stimulates phospholipase D activity. Cell **75:**1137-1144.

[36] Sylvia VL, Schwartz Z, Del Toro F, DeVeau P, Whetstone R, Dean DD, Boyan BD 2001 24R,25-$(OH)_2D_3$ regulates phospholipase D2 (PLD2) activity of costochondral chondrocytes in a metabolite specific and cell maturation dependent manner. Biochim Biophys Acta **1499:**209-221.

[37] Sylvia VL, Schwartz Z, Curry DB, Chang Z, Dean DD, Boyan BD 1998 1,25-$(OH)_2D_3$ regulates protein kinase C activity through two phospholipid-independent pathways involving phospholipase A_2 and phospholipase C in growth zone chondrocytes. J Bone Miner Res **13:**559-569.

[38] Lephart ED, Simpson ER 1991 Assay of aromatase activity. In: Waterman MR, Johnson EF (eds) Methods in Enzymology, 206. Academic Press, p. 477.

[39] Noguchi T, Kitawaki J, Tamura T, Kim T, Kanno H, Yamamoto T, Okada H 1993 Relationship between aromatase activity and steroid receptor levels in ovarian tumors from postmenopausal women. J Steroid Biochem Molec Biol **44:**657-660.

[40] Vinggaard AM, Hnida C, Breinholt V, Larsen JC 2000 Screening of selected pesticides for inhibition of CYP19 aromatase activity in vitro. Toxicol in Vitro **14:**227-234.

[41] Hodgins MB, Murad S 1986 1,25-Dihydroxycholecalciferol stimulates conversion of androstenedione into oestrone by human skin fibroblasts in culture. J Endocrinology **110:**R1-R4.

[42] Sylvia VL, Boyan BD, Dean DD, Schwartz Z 2000 The membrane effects of 17β-estradiol on chondrocyte phenotypic expression are mediated by activation of protein kinase C through phospholipase C and G-proteins. J Steroid Biochem Molec Biol **73:**211-224.

[43] Sylvia VL, Gay I, Hardin R, Dean DD, Boyan BD, Schwartz Z 2001 Rat costochondral chondrocytes produce 17β-estradiol and regulate its production by $1\alpha,25(OH)_2D_3$. Bone, in press.

[44] Sylvia VL, Schwartz Z, Schuman L, Morgan RT, Mackey S, Gomez R, Boyan BD 1993 Maturation-dependent regulation of protein kinase C activity by vitamin D_3 metabolites in chondrocyte cultures. J Cell Physiol **157:**271-278.

[45] Sylvia VL, Schwartz Z, Del Toro F, DeVeau P, Whetstone R, Hardin RR, Dean DD, Boyan BD 2001 Regulation of phospholipase D (PLD) in growth plate chondrocytes by 24R,25$(OH)_2D_3$ is dependent on cell maturation state (resting zone cells) and is specific to the PLD2 isoform. Biochim Biophys Acta **1499:**209-221.

[46] Turnquist J, Ornoy A, Eini D, Schwartz Z 1992 Effects of 1 alpha(OH)-vitamin D3 and 24,25(OH)2-vitamin D3 on long bones of glucocorticoid-treated rats. Acta Anat (Basel) **145:**61-67.

Components of the Cartilage Extracellular Matrix Regulate Chondrocyte Apoptosis

Christopher S. Adams[1], Kyle D. Mansfield[2], Ramesh Rajpurohit[1], Hideharu Tachibana[1], Cristina M. Teixeira[1], and Irving M. Shapiro[1].

[1]Department of Orthopaedic Surgery, Thomas Jefferson University, Philadelphia, PA, 19107, [2]Abrahamson Cancer Center, School of Medicine, University of Pennsylvania, Philadelphia PA 19104-6002

Abstract. Ionic components of apatite and arginine-glycine-aspartic acid (RGD)-containing peptides of the cartilage extracellular matrix, induce chondrocyte apoptosis. Using a well-defined culture system to induce terminal differentiation, we show that medium supplemented with Ca and Pi, the ion pair caused rapid cell death; within 6 hours, almost all of the treated cells are apoptotic. We also noted that sensitivity to the apoptogen is dependent on chondrocyte maturation. RGD-containing peptides kill chondrocytes; in this case, cell death is sequence specific. Thus GRGDSP is a more effective apoptogen than RGDS. Use of caspase inhibitors and a fluorescent caspase substrate indicate that these apoptogens activate caspase-3. To elucidate the mechanism of apoptosis, we evaluated mitochondrial function in relationship to Reactive Oxygen Species (ROS) generation and thiol status. We noted that there is a maturation-dependent early loss of mitochondrial function, possibly leading to uncoupling of oxidative phosphorylation from electron transport. Two other events correlate with the change in mitochondrial function. First, there is a low level of ROS generation; ROS accumulation is markedly increased when apoptosis is activated. Second, there is a fall in the thiol reductive reserve. Since thiols serve to protect the cells from ROS, this change would confirm that mitochondria are involved in the apoptotic response. We argue that at the chondro-osseous junction, the combination of a high local ion concentration and peptide fragments, as well as an increased sensitivity of the cells to apoptogens activates the death process. Since both maturation-dependent intracellular metabolic events and extracellular changes conspire to activate the deletion process, we propose that chondrocyte apoptosis is both self- and environmentally-regulated.

"During the penetration of connective tissue in their [the chondrocytes] capsule, the vesicular cartilage cells are believed by most, but not all, investigators to perish"

Textbook of Histology,
Maximov and Bloom, p. 131, 1938,
Saunders Company, Philadelphia.

Introduction

All of the appendicular bones grow through the activities of cells contained within a specialized cartilage, the growth plate. In most mammals, this cartilage is transitory in that it is only present during early post-natal growth. At maturity, the cartilage disappears and there is a fusion of the primary and secondary ossification centers. While the residence time of the cartilaginous growth plate is measured in months or years, the life history of the chondrocytes that make up the growth plate is completed in days or even in hours.

A great deal is now known concerning chondrocyte function in terms of macromolecular synthesis, response to cytokines and growth factors, and the induction of cartilage calcification. What is poorly understood is the fate of the terminally differentiated chondrocyte and the mechanism of their removal from the growth plate. This lack of information may relate to the curious structure of the growth plate cartilage itself. It was assumed that cells in the most mature region of the plate were buried in a dense, calcified, non-porous matrix. In this hostile environment, starved of nutrients and removed from an adequate oxygen supply, chondrocytes were thought to undergo necrosis and perish [1-3]. Since those original observations, new experimental finding and use of culture systems that mimic *in vivo* conditions have been used to generate new insights that directly pertain to the mechanism of chondrocyte deletion.

Chondrocyte Transdifferentiation into Metaplastic Bone

After necrosis, a second possible fate of the terminally differentiated chondrocyte is its conversion into a bone cell to form "metaplastic bone" [4]. Central to this viewpoint is the idea that the differentiated hypertrophic cells of cartilage do not die, instead they are converted into osteoblasts [5-14]. Circumstantial evidence provides some support for this transdifferentiation process, in that markers of former cartilage cells are found among the cells of bone. While some of the initial evidence in support of metaplastic bone formation came from ultrastructural analysis [12,15] , more recently in vitro and organ culture studies provided evidence of transdifferentiation [16-23]. Building on the transdifferentiation concept, Roach and co-workers suggested that in the growth plate, there was "asymmetric" cell division; one daughter cell died while the other underwent transdifferentiation and was converted into an osteoblast [24]. While the metaplastic bone hypothesis requires further study, the case for transdifferentiation during callus formation in bone healing remains compelling [6,9,25,26].

Chondrocyte Apoptosis in the Growth Plate

The notion that cell death is a regulated physiological process prompted a re-review of the process of chondrocyte deletion in the growth plate. As promulgated by Kerr, Wyllie and their colleagues [27], apoptosis is a process of programmed cellular destruction, cell suicide. In contrast to necrosis, apoptosis is characterized by fragmentation of DNA and activation of a specific group of enzymes, caspases, some of which are regulated by oxidative events at the mitochondrial level [28]. A correlated change in membrane phospholipid assembly provides the signal for removal of the dead cell remnants without eliciting an inflammatory response [29].

Considerable evidence has been assembled in support of the notion that hypertrophic chondrocytes die through the activation of apoptosis. Apoptotic cells were originally identified in the growth plate on the basis of morphological criteria [30-34]. However, until the development of specialized techniques for epiphyseal chondrocyte fixation [35], analysis of intracellular structural changes was impossible at the light microscopic level. It was subsequently noted that the most mature cells in growth plate exhibited fragmented nuclei [31]. With the development of DNA end labeling techniques (TUNEL) [36], analysis of apoptotic rates in growth plates was possible [37-38]. Thus, evidence accumulated by end-labeling, morphologic, and ultramicroscopic procedures indicates that cells in the growth plate die by apoptosis.

Matrix Regulation of Chondrocyte Apoptosis

Much of the early work on chondrocyte apoptosis was related to cartilage degradation in arthritis. As such, many arthritis-related factors are known to induce chondrocyte apoptosis. These include TNF-alpha [39] and Fas ligand [40-41]. Other more generalized apoptogens, including nitric oxide (NO) [42], hydrogen peroxide [43], staurosporine [44], and serum withdrawal [45], also cause chondrocyte apoptosis. However, with the exception of NO, there is little reason to suspect that in the developing growth plate, these factors are present, and in sufficient concentration to induce apoptosis. For this reason, we have begun to examine the apoptogenic activities of intrinsic factors that may be present in the developing growth plate. We focused our attention on both inorganic and organic constituents of the extracellular matrix. We reasoned that at the chondro-osseous junction, linked to apatite deposition and dissolution, there would be a high inorganic ion flux. Likewise, matrix hydrolysis would result in liberation of peptide and glycan fragments. The objective of this report is to describe the effect of these matrix constituents on chondrocyte apoptosis. For all of these studies, culture systems were utilized which induce the chondrocyte to recapitulate many of the phenotypic changes displayed by cells of the epiphyseal growth plate.

Phosphate Induces Chondrocyte Apoptosis

Recent published work has demonstrated that Pi, one of the major components of the inorganic matrix of calcified cartilage, is an effective chondrocyte apoptogen [46]. An elevation in the medium Pi concentration from normal (~1 mM) to 3-5 mM activated apoptosis. Treatment with 5-7 mM Pi caused a dramatic increase in cell death over a 24 h period. How the Pi activated cell death was not clear. One promising avenue of inquiry was that Pi contributed to the loss of mitochondrial membrane potential of terminally differentiated chondrocytes [47]. Related to this observation, treatment with Pi increased the intracellular anion concentration, indicating that Pi uptake is required for chondrocyte apoptosis. Phosphate transport into chondrocytes is mediated by specific membrane co-transporters. Two Na-Pi transporters were identified in growth plate cartilage; when inhibited, cell killing was blocked [48]. However, the Km of the transporters for Pi was much lower than the medium Pi concentration. Thus, while the transporters are involved with the apoptotic process, apoptosis is not Na-Pi transporter-dependent.

Results

Calcium Promotes Pi-dependent Apoptosis

Given that Pi, a major component of hydroxyapatite, could serve as a specific skeletal cell apoptogen, we next investigated the impact of Ca^{2+} on Pi-dependent chondrocyte apoptosis [It may be recalled that Ca^{2+} and Pi are the major ion pair of apatite]. First, we examined the effect of Ca^{2+} by adding EDTA or EGTA, two effective chelators, to media containing an elevated concentration of Pi. Fig. 1A shows that cation chelation reduces the apoptotic effect of Pi. When additional Ca^{2+} (1 mM) is added to the media along with increased, but not apoptogenic levels of Pi (3 mM), cell death is achieved (Fig. 1B). Cell death is through the apoptotic pathway, as evidenced by TUNEL positive cells and ultrastructural evidence for apoptotic changes (Fig. 2). Phosphate transport inhibitors, phosphonoformic acid and alendronate, both inhibit the apoptotic effect, while pre-treatment with several Ca^{2+} channel inhibitors, including verapamil and nifedipine, as well as the intracellular Ca^{2+} chelators, BAPTA-AM and EGTA-AM, fail to inhibit ion-pair induced apoptosis. These results indicate that Ca^{2+} modulates the apoptotic effect of Pi at the level of the cell membrane. However, the mechanism by which these ions conspire to activate apoptosis is not as yet understood.

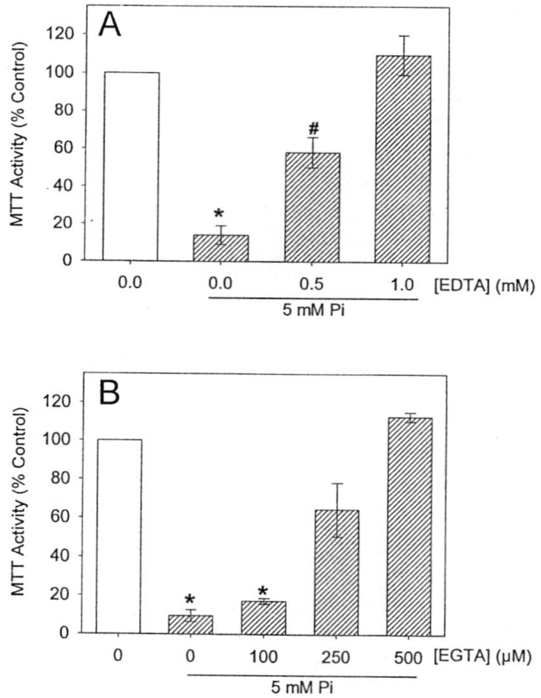

Figure 1. Effect of EDTA and EGTA on Pi induced cell death. Tibial chondrocytes were treated for 24 h with 1 or 5 mM Pi and either 0-1.0 mM EDTA (A) or 0-500 μM EGTA (B). Cell viability was assessed by the MTT assay. Note that while 5 mM Pi resulted in almost 90% cell death, EDTA and EDTA protected the cells from death in a dose dependent manner. * = significantly different from control; # = significantly different from control and 5 mM Pi; n = 4.

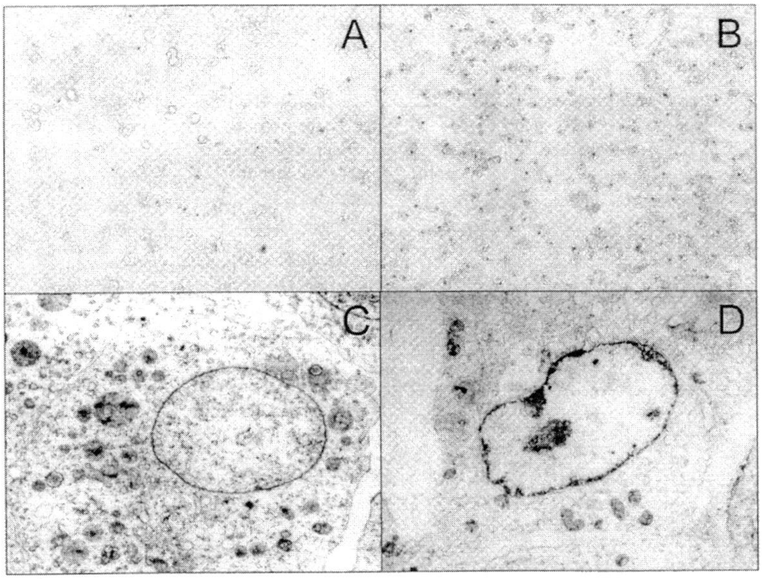

Figure 2. TUNEL and TEM analysis of tibial chondrocytes treated with Pi and Ca^{2+}. Cells were treated for 24 h with combinations of Pi and Ca^{2+} and the extent of apoptosis determined by the TUNEL assay. There was no apoptosis in the presence of 1 mM Pi and 1.8 mM Ca^{2+} (A). When the Pi concentration was raised to 3 mM, and the Ca^{2+} concentration to 2.8 mM, there was a dramatic increase in TUNEL positive cells (B) (Magnification:100X). Ultrastructurally, the control cells appear healthy (C), while ion-pair treated chondrocytes exhibited an apoptotic morphology including condensed chromatin and vacuolation (D) (Magnification: 6200X).

Matrix Peptide Fragments Induce Apoptosis

Having demonstrated that cell death can be caused by components of the inorganic extracellular matrix and that the apoptotic effect was accentuated by the synergy of two of those components, we next examined the organic matrix for possible apoptogens. We focused our search on cell attachment proteins. These proteins contain peptide domains that bind to membrane receptors. One such attachment motif is the arginine-glycine-aspartic acid (RGD) sequence. This sequence binds to the αV-β_3 integrin receptor on chondrocytes and bone cells. Directly related to this interaction, Buckley and co-workers [49] demonstrated that short peptides containing the RGD sequence were capable of inducing apoptosis in breast cancer cell lines and in lymphocytes. We sought to demonstrate that these peptides induced apoptosis in connective tissue cells. Treatment with a number of RGD-containing peptides induces apoptosis of growth plate chondrocytes (Fig. 3); as the cells remained attached to the underlying substrate, this effect is distinct from anoikis [50]. Interestingly, the RGD tripeptide itself did not induce chondrocyte apoptosis. In fact, while RGDS causes a small, but significant, increase in cell death, within the time period studied (24 h), the hexapeptides, GRGDSP and GRGDNP, are far better apoptogens. These results suggest that fragments of major proteins of the organic matrix of cartilage may exert an apoptotic effect on epiphyseal chondrocytes.

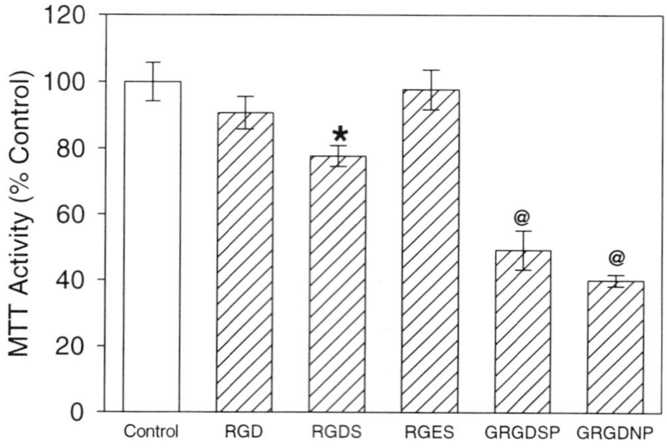

Figure 3. Effect of RGD-containing peptides on chondrocyte viability. Cells were treated with the following RGD-containing peptides (at a concentration of 5 mM): RGD, RGDS, RGES, GRGDSP, and GRGDNP. After 24 h, cell vitality was determined by the MTT assay. RGDS, GRGDSP and GRGDNP caused a significant elevation in chondrocyte death. Values shown represent the mean and SEM; the experiments were repeated 3-5 times.
* $p<0.05$ when compared to control. @ $p<0.05$ when compared to RGD, RGDS and RGES as well as control cells.

Mechanisms of Ion-Pair Mediated Apoptosis

Since we had isolated a number of agents that killed epiphyseal chondrocytes, we next examined the mechanism of apoptosis. Dealing first with the end result of apoptosis, it is known that endonucleases generate DNA fragments. When we pre-treated chondrocytes with the endonuclease inhibitor, aurotricarboxylic acid, we found that we could abrogate apoptosis (Fig. 4A). Since these endonucleases are activated by caspase-3, we used the specific inhibitor of this enzyme, DEVD-CHO, to explore the importance of caspases in ion-pair induced apoptosis. Fig. 4B shows that an increasing dose of DEVD-CHO blocks Pi-induced apoptosis. These results indicate that Pi induces apoptosis by first activating caspase-3 and subsequently enhancing endonuclease activity.

In a number of cell types, the generation of reactive oxygen species (ROS) is linked to the activation of apoptosis. Since chondrocyte mitochondria may be uncoupled and apoptosis is mediated through the caspase-3 system, we examined ROS generation in chondrocytes treated with the Ca^{2+} and Pi ion pair. ROS generation was detected using hydroethidine, a non-fluorescent molecule that is reduced to ethidium in the presence of ROS. Since ethidium is fluorescent and stains nuclear material, the presence of stained nuclei is indicative of ROS attack. We treated chondrocytes with 5 mM Pi and 2.8 mM Ca^{2+} to induce cell death and followed ROS generation with hydroethidine. Fig. 5 shows that after 75 minutes, the nuclei become maximally fluorescent. Over the following 90 minutes, the fluorescence decreases and returns to baseline. Results of this investigation indicate that an early event in the transduction of Ca^{2+} and Pi-dependent apoptosis is the generation of ROS.

To ascertain if caspase-3 activity is elevated in maturing cells, we treated chondrocytes with the fluorescent caspase-3 substrate, Phiphilux-G1D2 (OncoImmun, Gaithersburg MD). Fig. 6 shows that chondrocyte maturation is accompanied by upregulation of caspase-3 activity. While the initial fluorescence is quite low, if the cells are treated with 35 nM retinoic acid (RA) to induce terminal differentiation, there is a dramatic increase in substrate fluorescence. It is interesting to note that despite the high level of caspase-3 in these cells, they do not die. Thus, downstream caspase-3 inhibitors are probably functional.

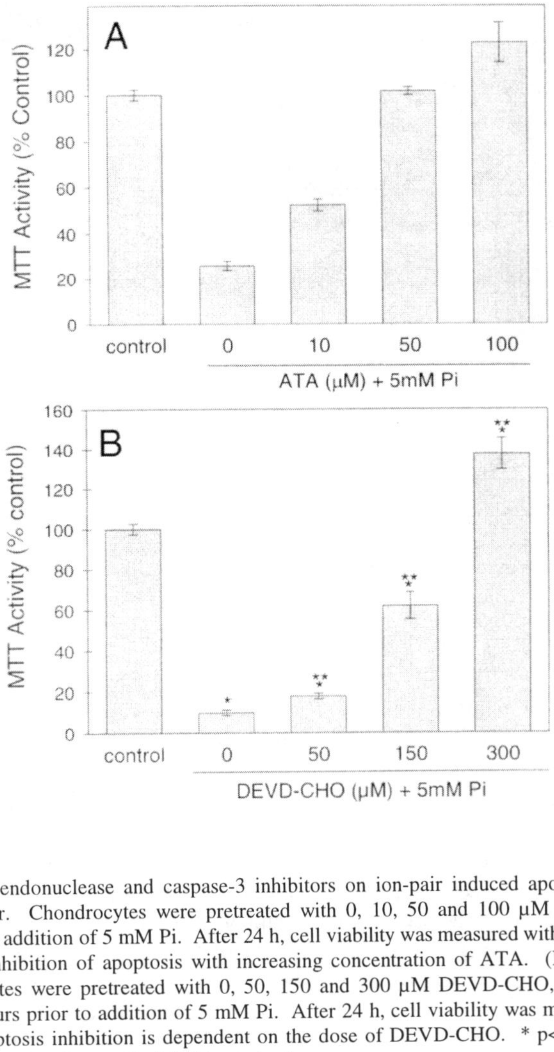

Figure 4. Effect of endonuclease and caspase-3 inhibitors on ion-pair induced apoptosis. (A) Effect of endonuclease inhibitor. Chondrocytes were pretreated with 0, 10, 50 and 100 μM aurotricarboxylic acid (ATA) for 2 h prior to addition of 5 mM Pi. After 24 h, cell viability was measured with an MTT assay. Note the dose-dependent inhibition of apoptosis with increasing concentration of ATA. (B) Effect of caspase-3 inhibitor. Chondrocytes were pretreated with 0, 50, 150 and 300 μM DEVD-CHO, a peptide inhibitor of caspase-3, for two hours prior to addition of 5 mM Pi. After 24 h, cell viability was measured with the MTT assay. Note that apoptosis inhibition is dependent on the dose of DEVD-CHO. * $p<0.05$ in comparison to control. ** $p<0.05$ in comparison to 5 mM Pi treated cells.

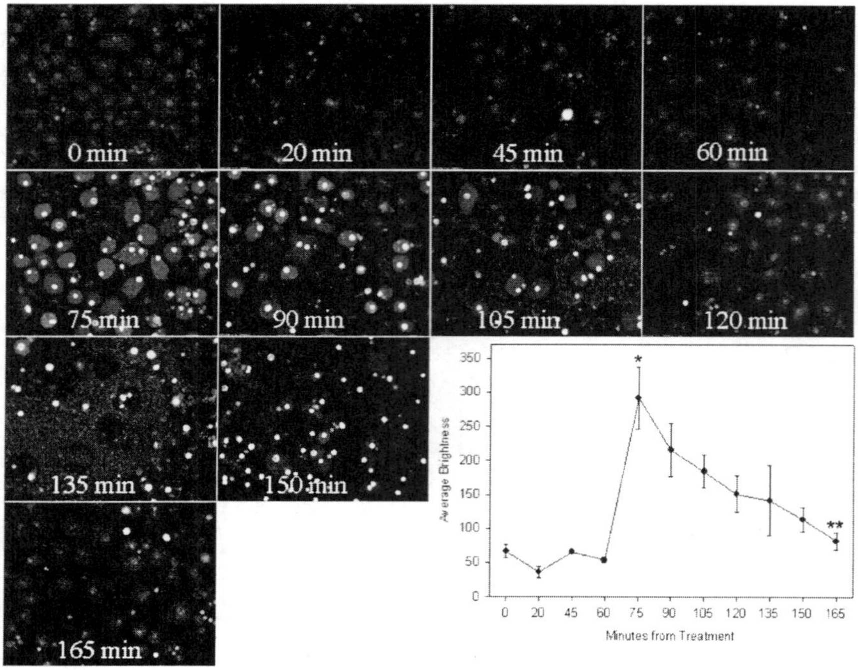

Figure 5. Reactive Oxygen Species (ROS) generation in the presence of Pi and Ca^{2+}. Tibial chondrocytes were treated with 3 mM Pi and 2.8 mM Ca^{2+} for 0 to 180 min after which they were stained with 5 μg/ml hydroethidine for 20 min, washed and analyzed by confocal microscopy. A representative experiment is shown. Maximum ROS generation was evident by 75 min after which there was a slow return to baseline levels. The inset graph shows average brightness across 3 separate fields for each time period. Note, the initial lag period after which there was a dramatic increase in nuclear fluorescence between 60 and 75 min, followed by a slow decay in the signal over the remaining 90 min. * = significantly different from 0 min; ** = significantly different from 75 min.

Chondrocyte Maturation, Mitochondrial Function and the Induction of Apoptosis

In vivo, when terminally differentiated, epiphyseal chondrocytes undergo apoptosis [37, 38]. To model these events *in vitro*, embryonic chondrocytes were treated with very low doses of RA. After 4-7 days of treatment, these cells evidence high alkaline phosphatase activity and an elevated level of type X collagen expression. Mitochondrial activity in the mature cells was then evaluated using the voltage sensitive probe rhodamine 123. Since the fluorescent yield of the dye is proportional to the transmembrane potential, functional mitochondria are highly fluorescent, whereas uncoupled mitochondria exhibit a low level of membrane fluorescence (Fig. 7). When initially isolated, the distribution of mitochondria in immature chondrocytes is heterogeneous, being concentrated around the nucleus and the plasma membrane (Fig. 7A). Following maturation with RA, the mitochondria become evenly distributed throughout the cytoplasm (Fig. 7B). Analysis of the confocal images indicate that there is a loss of fluorescence following treatment with the retinoid. To measure the change in mitochondrial fluorescence, the chondrocytes were stained with Rhodamine 123 and evaluated by flow cytometry (Fig. 7C-D). While both immature and terminally differentiated cells form a single well defined population (Fig. 7C), RA treatment causes the mean fluorescence to shift to the left (Fig. 7D). The decrease in mitochondrial membrane potential lends support to the observational studies and confirms that chondrocyte maturation is accompanied by a change in mitochondrial function.

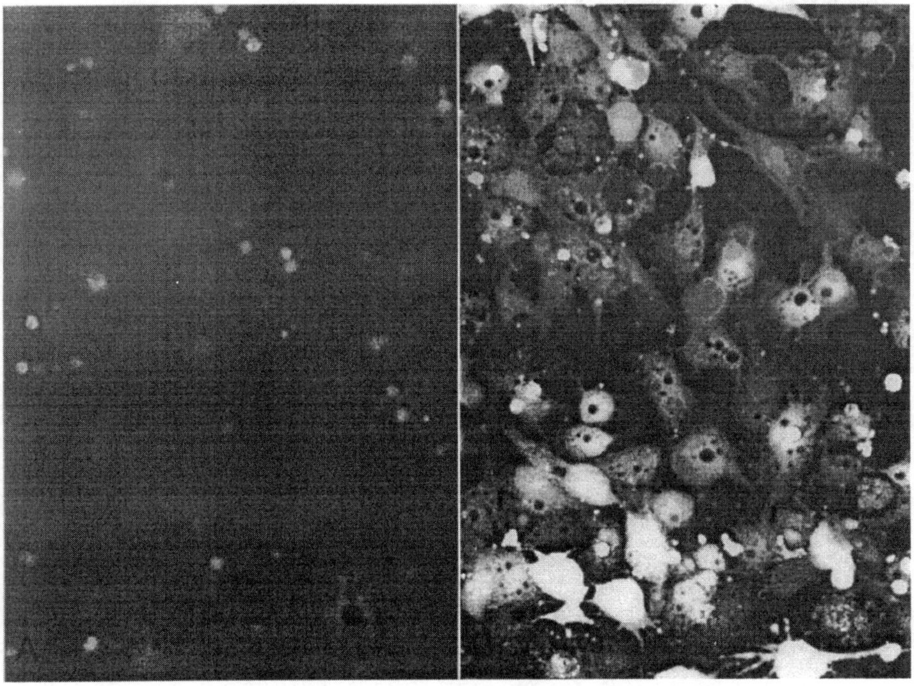

Figure 6. Maturational changes in intracellular caspase-3 levels. Tibial Chondrocytes were treated with 35 nM RA for 5 days and intracellular caspase-3 levels probed with the fluorescent caspase-3 substrate, Phiphilux-G1D2. Control cells show minimal fluorescence (A), while the RA-treated cells indicate a significant increase in active caspase-3 (B) (Magnification: 400X).

Since radical generation results in a loss of reductive reserve, we hypothesized that maturation would be characterized by a decrease in chondrocyte glutathione levels. We examined the thiol reductive capacity of the cell with CellTracker Green, a CMFDA-based thiol sensitive dye (Molecular Probes, Eugene, OR). Fig. 8A-C shows that maturation is accompanied by a decrease in the thiol reserve of the cell (Fig. 8A-C). In fact, while a small decrease in thiols is seen with 10 nM RA treatment, 35 nM RA causes almost complete loss of intracellular thiols.

Next, we addressed the question: Are mature growth plate chondrocytes more sensitive to apoptogens than immature cells? Cells were matured with RA for 10 days and exposed to 1, 3, 4, and 5 mM Pi (Fig. 9). The control cells are insensitive to any concentration of Pi as an apoptogen. However, if the cells are matured with RA, there is a significant increase in cell death in the presence of both 4 and 5 mM Pi. Thus, treatment of the cells with Pi for 24 hours resulted in killing of 60 – 70 % of all chondrocytes. While 5 mM Pi is an effective apoptogen in preconfluent chondrocytes, it is clear that confluence decreases sensitivity to the apoptogen. Relating this information to the growth plate, it is clear that maturation dependent regulators exist that modulate both down- and up-stream events in the apoptotic cascade.

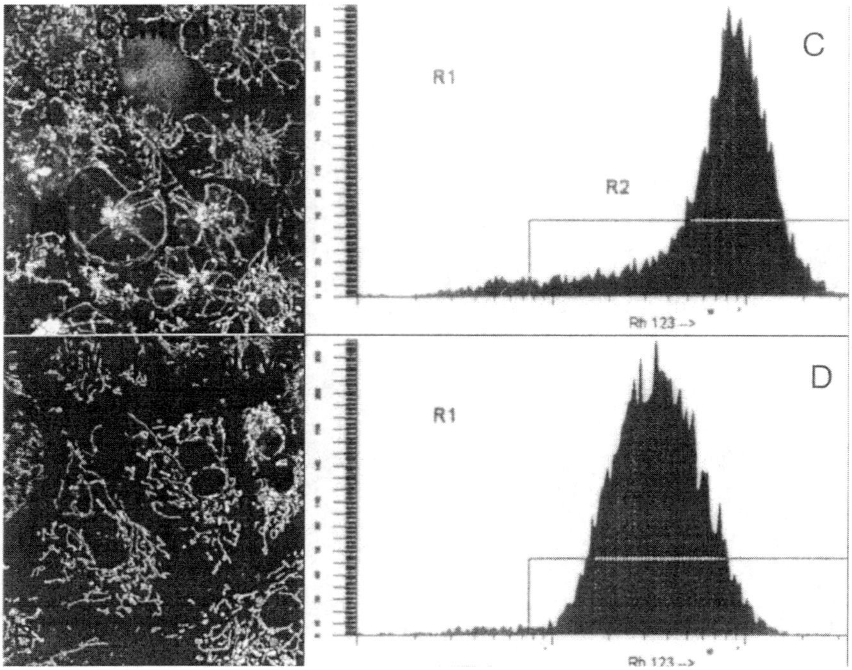

Figure 7. Maturation-dependent changes in chondrocyte mitochondria. Tibial chondrocytes were treated with 50 nM RA for 3 days and the mitochondria examined with the fluorescent probe, Rhodamine 123. Confocal microscopy demonstrated a characteristic distribution of mitochondria in control cells (A). Two distinct cellular mitochondrial morphologies were observed: one in which mitochondrial staining is punctuate and diffuse, the other where mitochondria are found primarily perinuclear and at the cell membrane. However, following treatment with RA, the distribution of mitochondria becomes homogenous (B). Mitochondria are now distributed throughout the cytoplasm in a string-like organization (Magnification: 600X). The magnitude of fluorescence is also altered by RA treatment. Rhodamine stained chondrocytes were analyzed by flow cytometry. Control cells demonstrate a normal distribution of fluorescence (C), while RA treatment shifts the curve to the left (D), indicating a loss of fluorescence and consequently a decrease in mitochondrial membrane potential.

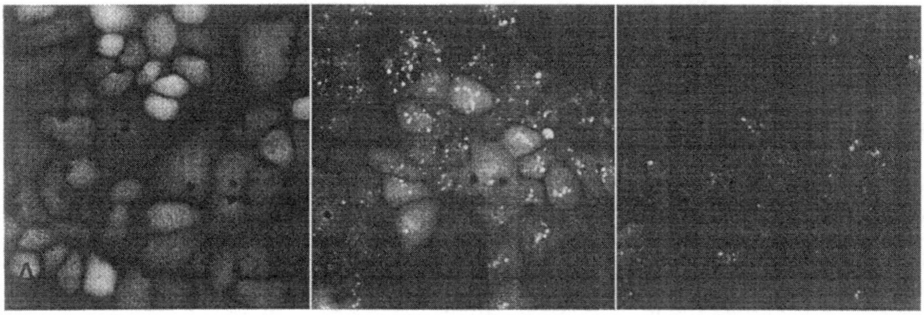

Figure 8. Maturational changes in intracellular thiol levels. Tibial chondrocytes were treated with 10 and 35 nM RA for 5 days and intracellular thiol levels evaluated with Cell Tracker Green. Control cells demonstrate stable, but variable levels of Cell Tracker Green fluorescence (A), while cells treated with 10 nM (B) and 35 nM (C) RA show a decrease in fluorescence, indicating a loss of thiol reserve in these matured cells (Magnification: 400X).

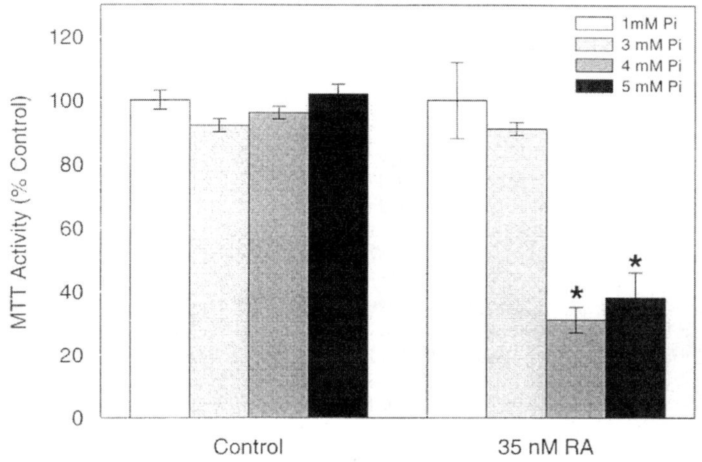

Figure 9X Maturational changes in sensitivity to Pi as an apoptogens. Tibial chondrocytes were treated with 35 nM RA for 10 days and subjected to 1, 3,4, and 5 mM Pi as an apoptogen. Note that both 4 and 5 mM Pi kill a significant number of cells in the RA treated cohort, while having no effect on control cells.

Discussion

We have started to dissect out some of the maturational events that pre-dispose the mature chondrocyte to apoptotic stimuli. Studies with voltage sensitive probes indicate that there is a maturation-dependent early loss of mitochondrial function, possibly leading to uncoupling of oxidative phosphorylation from electron transport. Two other events correlate with the change in mitochondrial function. First, there is a low level of ROS generation; ROS accumulation is markedly increased when apoptosis is activated. Second, there is a fall in the thiol reductive reserve. Since thiols serve to protect the cell from ROS, this change would further signal the role of disturbed mitochondrial function in the apoptotic response. Therefore, the possibility exists that one of the earliest events is the generation of radicals which lowers the reductive reserve of the cell while at the same time promoting caspase activation. As far as the executioners are concerned, we have firm evidence that both the caspase and endonuclease enzyme systems participate in this process. Thus, inhibition of these mediators blocks ion-pair induced apoptosis.

In concert with the change in mitochondrial function, the terminally differentiated cells are engulfed in a matrix that contains a huge amount of potential apoptogens: hydroxyapatite crystallites and the proteins of the organic matrix. At the chondro-osseous junction, these apoptogens are released from the matrix by resorbing cells, chondroclasts. The combination of a high local ion concentration and peptide fragments, as well as the increased sensitivity of the cells to apoptogens activates the death process. Since both maturation-dependent intracellular metabolic events and extracellular changes conspire to activate the deletion process, we propose that chondrocyte apoptosis is both self- and environmentally-regulated.

Acknowledgements

This work was supported by NIH Grants DE-13319, DE-10875, and DE-05748.

References

[1] Trueta,J. and Little,K. (1960) The vascular contribution to osteogenesis. II. Studies with the electron microscope. J.Bone Joint Surg., 42B:367-376.

[2] Young,R.W. (1962) Cell proliferation and specialization during endochondral osteogenesis in young rats. J.Cell Biol., 14:357-370.

[3] Anderson,C.E. and Parker,J. (1966) Invasion and resorption in enchondral ossification. An electron microscopic study. J.Bone Joint Surg., 48A:899-914.

[4] Haines,R.W. and Mohuiddin,A. (1968) Metaplastic bone. J. Anat., 103:527-538.

[5] Holtrop,M.E. (1966) The origin of bone cells in endochondral ossification. In: Third European Symposium on Calcified Tissues. H.Fleisch, H.J.J.Blackwood, and M.Owens, eds. Springer Verlag, Berlin, pp. 32-36.

[6] Holtrop,M.E. (1972) The ultrastructure of the epiphyseal plate II. The hypertrophic chondrocyte. Cal. Tiss. Res., 9:140-151.

[7] Crelin,E.S. and Koch,W.E. (1967) An autoradiographic study of chondrocyte transformation into chondroclasts and osteocytes during bone formation in vitro. Anat. Rec., 158:473-483.

[8] Lufti,A.M. (1971) The fate of chondrocytes during cartilage erosion in the growing tibia in the domestic fowl (*Gallus domesticus*). Acta Anat., 79:27-35.

[9] Shimomura,Y., Yoneda,T., and Suzuki,F. (1975) Osteogenesis by chondrocytes from growth cartilage of rat rib. Cal. Tiss.Res., 19:179-187.

[10] Kahn,A.J. and Simmons,D.J. (1977) Chondrocyte-to-osteocyte transformation in grafts of perichondrium-free epiphyseal cartilage. Clin.Orthop., 129:299-304.

[11] Weiss,A., von der Mark,K., and Silbermann,M. (1987) Fully differentiated chondrocytes alter their phenotype expression in vitro and transform into bone cells. In: Current Advances in Skeletogenesis 3. S.Hurwitz and J.Sela, eds. Heiliger Publishing, Jerusalem, pp. 40-43.

[12] Yoshioka,C. and Yagi,T. (1988) Electron microscopic observations on the fate of hypertrophic chondrocytes in condylar cartilage of rat mandible. J. Cran. Gen. Dev. Biol., 8:253-264.

[13] Moskalewski,S. and Malejczyk,J. (1989) Bone formation following intrarenal transplantation of isolated murine chondrocytes: Chondrocyte-bone cell transdifferentiation. Dev., 107:473-480.

[14] Galotto,M., Campanile,G., Robino,G., Candedda,F.D., Bianco,P., and Candedda,R. (1994) Hypertrophic chondrocytes undergo further differentiation to osteoblast-like cells and participate in the initial bone formation in developing chick embryo. J. Bone Min. Res., 9:1239-1249.

[15] Hunziker,E.B., Schenk,R.K., and Cruz-Orive,L.-M. (1987) Quantitation of chondrocyte performance in growth-plate cartilage during longitudinal bone growth. J. Bone Joint Surg., 69A:162-173.

[16] Strauss,P.G., Closs,E.I., Schmidt,J., and Erfle,V. (1990) Gene expression during osteogenic differentiation in mandibular condyles in vitro. J. Cell Biol., 110:1369-1378.

[17] Closs,E.I., Murray,A.B., Schmidt,J., Schon,A., Erfle,V., and Strauss,P.G. (1990) c-fos expression precedes osteogenic differentiation of cartilage cells in vitro. J. Cell Biol., 111:1313-1323.

[18] Cancedda,F.D., Gentili,C., Manduca,P., and Cancedda,R. (1992) Hypertrophic chondrocytes undergo further differentiation in culture. J. Cell Biol., 117:427-435.

[19] Gentili,C., Bianco,P., Neri,M., Malpeli,M., Campanile,G., Castagnola,P., Cancedda,R., and Cancedda,F.D. (1993) Cell proliferation, extracellular matrix mineralization, and ovotransferrin transient expression during in vitro differentiation of chick hypertrophic chondrocytes into osteoblast-like cells. J. Cell Biol., 122:703-712.

[20] Roach,H.I. (1992) Transdifferentiation of chondrocytes into cells capable of producing a mineralized bone matrix. Bone Min., 19:1-20.

[21] Roach, H.I., Erenpreisa,J., and Aigner,T. (1995) Osteogenic differentiation of hypertrophic chondrocytes involves asymmetric cell divisions and apoptosis. J. Cell Biol., 131:483-494.

[22] Thesingh, C.W., Groot,C.G., and Wassenaar,A.M. (1991) Transdifferentiation of hypertrophic chondrocytes into osteoblasts in murine fetal metatarsal bones induced by co-cultured cerebrum. Bone Min., 12:25-40.

[23] Groot, C.G., Thesingh,C.W., Wassenaar,A.M., and Scherft,J.P. (1994) Osteoblasts develop from isolated fetal mouse chondrocytes when co-cultured in high density with brain tissue. In Vitro Cell. Dev. Biol., 30A:547-554.

[24] Roach, H.I. and Erenpreisa,J. (1996) The phenotypic switch from chondrocytes to bone-forming cells involves asymmetric cell division and apoptosis. Conn. Tiss. Res., 35:85-91.

[25] Scammell, B.E. and Roach, H.I. (1996) A new role for the chondrocyte in fracture repair: Endochondral ossification includes direct bone formation by former chondrocytes. J Bone Min. Res., 11:737-745.

[26] Hughes,S.S., Hicks,D.G., O'Keefe,R.J., Hurwitz,S.R., Crabb,I.D., Krasinskas,A.M., Loveys,L., Puzas,J.E., and Rosier,R.N. (1995) Shared phenotypic expression of osteoblasts and chondrocytes in fracture callus. J. Bone Min. Res., 10:533-544.

[27] Kerr, J.F.R., Wyllie, A.H., and Currie,A.R. (1972) Apoptosis: a basic biological phenomenon with wide-ranging implications in tissue kinetics. Brit. J. Cancer, 26:239-257.

[28] Matsuyama, S., Llopis, J., Deveraux,Q.L., Tsien,R.Y., and Reed,J.C. (2000) Changes in intramitochondrial and cytosolic pH: early events that modulate caspase activation during apoptosis. Nature Cell Biol., 2:318-325.

[29] Koopman, G., Reutelingsperger,C.P., Kuijten,G.A., Keehnen,R.M., Pals,S.T., and van Oers, M.H. (1994) Annexin V for flow cytometric detection of phosphatidylserine expression on B cells undergoing apoptosis. Blood, 84:1415-1420.

[30] Hargest,T.E., Leach,R.M., and Gay,C.V. (1985) Avian tibial dyschondroplasia I. Ultrastructure. Amer. J. Pathol., 119:175-190.

[31] Farnum,C.E. and Wilsman,N.J. (1987) Morphologic stages of the terminal hypertrophic chondrocyte of growth plate cartilage. Anat. Rec., 219:221-232.

[32] Farnum,C.E. and Wilsman,N.J. (1989) Condensation of hypertrophic chondrocytes at the chondro-osseous junction of growth plate cartilage in Yucatan swine: Relationship to long bone growth. Amer. J. Anat., 186:346-358.

[33] Farnum,C.E. and Wilsman,N.J. (1989) Cellular turnover at the chondro-osseous junction of growth plate cartilage: Analysis by serial sections at the light microscopical level. J. Orthop. Res., 7:654-666.

[34] Lewinson,D. and Silbermann,M. (1992) Chondroclasts and endothelial cells collaborate in the process of cartilage resorption. Anat. Rec., 233:504-514.

[35] Hunziker,E.B., Herrmann,W., and Schenk,R.K. (1983) Ruthenium hexamine trichloride (RHT)-mediated interaction between plasmalemmal components and pericellular matrix proteoglycans is responsible for the preservation of chondrocytic plasma membranes in situ during cartilage fixation. J. Histochem. Cytochem., 31:717-727.

[36] Wijsman,J.H., Jonker,R.R., Keijzer,R., Van de Velde,C.J.H., Cornelisse,C.J., and Van Dierendonck,J.H. (1993) A new method to detect apoptosis in paraffin sections: In situ end-labeling of fragmented DNA. J. Histochem. Cytochem., 41:7-12.

[37] Bronckers,A.L.J.J., Goei,W., Luo,G., Karsenty,G., D'Souza,R.N., Lyaruu,D.M., and Burger,E.H. (1996) DNA fragmentation during bone formation in neonatal rodents assessed by transferase-mediated end labeling. J. Bone Min. Res., 11:1281-1291.

[38] Hatori,M., Klatte,K.J., Teixeira,C.C., and Shapiro,I.M. (1995) End labeling studies of fragmented DNA in the avian growth plate: evidence of apoptosis in terminally differentiated chondrocytes. J. Bone Min. Res., 10:1960-1968.

[39] Aizawa,T., Kon,T., Einhorn,T.A., and Gerstenfeld,L.C. (2001) Induction of apoptosis in chondrocytes by tumor necrosis factor-alpha. J. Orthop. Res., 19:785-796.

[40] Kuhn,K., Hashimoto,S., and Lotz,M. (2000) IL-1 beta protects human chondrocytes from CD95-induced apoptosis. J. Immunol., 164:2233-2239.

[41] Hashimoto,S., Setareh,M., Ochs,R.L., and Lotz,M. (1997) Fas/Fas ligand expression and induction of apoptosis in chondrocytes. Arth. Rheum., 40:1749-1755.

[42] Blanco,F.J., Ochs,R.L., Schwarz,H., and Lotz,M. (1995) Chondrocyte apoptosis induced by nitric oxide. Amer. J. Pathol., 146:75-85.

[43] Asada,S., Fukuda,K., Nishisaka,F., Matsukawa,M., and Hamanisi,C. (2001) Hydrogen peroxide induces apoptosis of chondrocytes; involvement of calcium ion and extracellular signal-regulated protein kinase. Inflamm. Res., 50:19-23.

[44] Gibson,G., Lin,D.-L., and Roque,M. (1997) Apoptosis of terminally differentiated chondrocytes in culture. Exp. Cell Res., 233:372-382.

[45] Feng,L., Precht,P., Balakir,R., and Horton,W.E., Jr. (1998) Evidence of a direct role for Bcl-2 in the regulation of articular chondrocyte apoptosis under the conditions of serum withdrawal and retinoic acid treatment. J. Cell. Biochem., 71:302-309.

[46] Mansfield,K., Rajpurohit,R., and Shapiro,I.M. (1999) Extracellular phosphate ions cause apoptosis of terminally differentiated epiphyseal chondrocytes. J. Cell. Physiol., 179:276-286.

[47] Rajpurohit,R., Mansfield,K., Ohyama,K., Ewert,D., and Shapiro,I.M. (1999) Chondrocyte death is linked to development of a mitochondrial membrane permeability transition in the growth plate. J. Cell. Physiol., 179:287-296.

[48] Mansfield,K., Teixeira,C.C., Adams,C.S., and Shapiro,I.M. (2001) Phosphate ions mediate chondrocyte apoptosis through a plasma membrane transporter mechanism. Bone, 28:1-8.

[49] Buckley,C.D., Pilling,D., Henriquez,N.V., Parsonage,G., Threlfall,K., Scheel-Toellner,D., Simmons,D.L., Akbar,A.N., Lord,J.M., and Salmon,M. (1999) RGD peptides induce apoptosis by direct caspase-3 activation. Nature, 397:534-539.

[50] Frisch,S.M. and Francis,H. (1994) Disruption of epithelial cell-matrix interactions induces apoptosis. J. Cell Biol., 124:619-626.

The Growth Plate
I.M. Shapiro et al. (Eds.)
IOS Press, 2002

The Release and Activation of TGF-β2 Associated with Chondrocyte Hypertrophy and Apoptosis

Gary Gibson, Xinli Wang and Maozhou Yang
Bone and Joint Center, Henry Ford Hospital, 2799 West Grand Blvd., Detroit, 48202
Corresponding author: Gary Gibson, Bone and Joint Center, Henry Ford Hospital, 2799 West Grand Blvd., Detroit, MI 48202.

Abstract. The final step in chondrocyte terminal differentiation is apoptosis at the interface with invading vascular cells. The transition from cartilage to bone at this site requires the coordinated and controlled removal of cartilage, stimulation of blood vessel invasion, recruitment of osteoblasts and stimulation of bone formation on cartilage remnants. Chondrocyte apoptosis is thus well positioned to play a central role in these processes. We have shown that hypertrophic chondrocytes cultured under serum free conditions released peak levels of TGF-β2 after approximately 7 days coincident with early stages of their apoptosis. TGF-β2 has the potential to mediate many of the processes critical to endochondral bone formation occurring in the vicinity of hypertrophic chondrocyte apoptosis. However, to perform these functions TGF-β2 must be first dissociated from inhibitory complexes. We show here that chondrocyte apoptosis is necessary for the dissociation of high molecular weight complexes of TGF-β2 associated with its activation.

Introduction

The formation of long bones from cartilaginous anlagen involves an ordered process of chondrocyte terminal differentiation that is the keystone of endochondral ossification. Chondrocyte terminal differentiation involves a series of steps that can be considered to begin with cell division progress through hypertrophy and end with apoptosis [1]. Hypertrophic differentiation, defined by cell expansion and the expression of type X collagen, is probably the most studied step in this process and emerging as a critical control point [2-5]. The expression of the hypertrophic phenotype is controlled both negatively and positively by a number of growth factors including PTHrP, FGF-2, TGF-β, BMPs and thyroxine [6-14]. Apoptosis of chondrocytes, the next step in chondrocyte terminal differentiation is restricted to the last row of hypertrophic chondrocytes in the growth plate [15-18]. This places chondrocyte apoptosis at the center of the transition from cartilage to bone and suggest it may play an important role in the local cellular events, including destruction of the cartilage matrix, attraction of invading blood vessel and stimulation of bone formation on cartilage remnants.

TGF-β plays an Essential and Multifunctional Role in Endochondral Differentiation

Endochondral bone formation is complex and requires the co-ordination of a number of signaling systems to generate the proper shape, length and structure of bones. In later development the lateral co-ordination of growth plate development and maintenance of the tight limits on growth plate thickness are critical to maintain normal skeletal growth. Recent studies suggest TGF-β2 fills an essential multifunctional role that may link diverse signaling pathways in this process. Mice carrying a dominant negative mutation of the TGF-β receptor had defective endochondral ossification with an apparently precocious differentiation of hypertrophic chondrocytes, an expanded zone of hypertrophic chondrocytes and defective vascular resorption [10]. These mice also developed severe articular cartilage damage that appeared to result from hypertrophic differentiation of articular chondrocytes and resembled osteoarthrosis. The pathology of these mice was, in part, explained by the recent demonstration that TGF-β and parathyroid hormone-related peptide (PTHrP) act in a common signal cascade that results in the inhibition of hypertrophic differentiation [19]. PTHrP and Indian hedgehog (Ihh), another extracellular signaling molecule form part of a negative feedback loop suggested to assure slow steady bone growth [20]. Though the precise details and the association with Ihh are not known, recent data suggests TGF-β acts upstream of PTHrP to regulate chondrocyte hypertrophic differentiation [21].

TGF-β was also shown to stimulate bone growth and mineralization independent of PTHrP. These effects were suggested to involve the FGF-2 signaling pathway [19] but are equally likely to involve the insulin-like growth factor (IGF) signaling pathway. A connection between the IGF signaling pathway, also critical to limb development, and TGF-β has been demonstrated to occur via interactions with the insulin-like growth factor binding proteins [22,23]. TGF-β has also been shown to exert a profound influence on processes associated with endochondral ossification that involve cells other than chondrocytes. TGF-β is a potent inducer of both bone deposition by osteoblasts and osteoclastic bone turnover [24] acting as a central component coupling bone formation to bone resorption [25]. TGF-β has an essential function in angiogenesis though its mode of action is not clearly understood [26]. As in other effects of TGF-β its actions on angiogenesis appear closely linked to interaction with soluble and matrix bound proteins such as betaglycan and endoglin [27,28]. Mutations in the gene for endoglin form the basis for the vascular disorder hereditary hemorrhagic telangiectasia [29,30].

The TGF-β2 Isotype Functions in Skeletal Development

Although the active forms of TGF-β1, TGF-β2 and TGF-β3 have very similar molecular structures and appear capable of stimulating widely overlapping effects in cell and organ culture they have distinct functions in vivo that appear to result from a combination of matrix or cell-specific expression and differences in protein interactions [31,32]. The development of TGF-β null mice has enabled demonstration of the critical involvement of TGF-β2 in endochondral bone growth. Of the mice carrying TGF-β null mutations only the TGF-β2 null mice had obvious skeletal defects; these included retarded growth of axial and appendicular bones [33,34].

Although TGF-β mRNA expression has been shown in a variety of cell types including growth cartilage chondrocytes and neighboring osteoblastic and haematopoietic cells a clear and prominent association with the last few hypertrophic chondrocytes of growth cartilage at the interface with invading vascular cells suggested this was the primary site of release of TGF-β in the vicinity of the transition from cartilage to bone [35]. In support of these observations cell culture studies [36-38] have demonstrated that substantial quantities of TGF-β2 are released by hypertrophic chondrocytes late in hypertrophic differentiation. TGF-β2 released by hypertrophic chondrocytes had an unusually high level of activation. Very few cell types or cell lines have been shown to release significant levels of active TGF-β [39].

Latency and Activation of TGF-β

TGF-β isotypes are synthesized as large inactive precursor proteins that are proteolytically processed in the Golgi into a mature TGF-β and an aminopropeptide, termed the latency associated peptide (LAP) that remain noncovalently linked. This complex, termed the small latent complex, associates covalently (via disulfide bridges) and noncovalently with a family of binding proteins (latent TGF-β binding proteins, LTBPs) and other multifunctional proteins via its LAP region (Fig. 1). Several studies suggest LTBPs mediate the attachment of TGF-β1 to extracellular matrix proteins, in many instances by transglutaminase generated cross-links [40]. Although the mature TGF-β1, TGF-β2 and TGF-β3 are highly homologous, the sequences corresponding to their LAP regions are more divergent [41]. Consequent differences in LAP interactions with LTBPs and other proteins may account, in part, for functional differences in these molecules. TGF-β2 and TGF-β3 have been shown capable of binding to some of the LTBPs [32], however the composition of their large latent complexes is not known. TGF-β within latent complexes is unable to bind to its receptors.

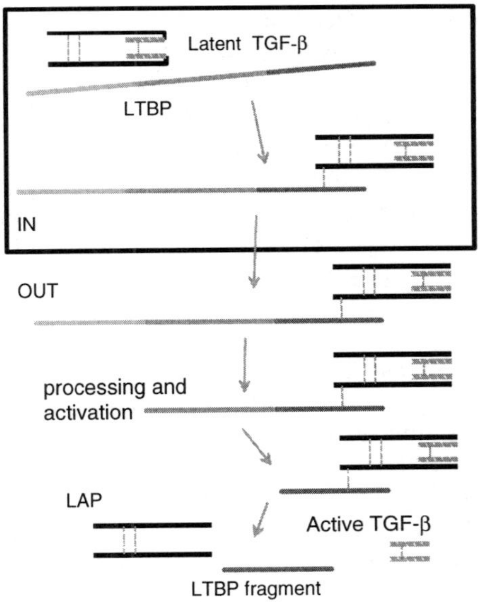

Figure 1. Model of the activation of latent TGF-β. A diagrammatic representation of the synthesis of proTGF-β, the cleavage of precursor latency associated peptide (LAP) and association with latent TGF-β binding protein (LTBP) inside the cell. Activation is depicted in association with the extracellular degradation of LTBP and processing and release of the LAP. This model is an oversimplification and does not include interactions of the TGF-β complex with a number of matrix and cell surface proteins that are involved in its latency and activation including thrombospondin, mannose 6 phosphate receptor and cell surface proteoglycans and endoglin. Dashed lines represent disulfide bonds.

One of the primary points of regulation of TGF-β activity is control of its conversion from the latent precursor to the biologically active form. Physical agents such as high and low pH, chaotropes, detergents or heat together with proteolysis are capable of activating latent TGF in vitro [40]. The mechanism of in vivo activation of latent TGF-β1 (the only isotype studied) remains to be clearly defined. At least three proteolytic cleavages have been suggested to be necessary to 1) release TGF-β1 from its association with the extracellular matrix, 2) release TGF-β1 from its complex with LTBP and LAP and 3) dissociate TGF-β1 from cell-surface proteoglycans. Although a number of candidate enzymes have been proposed to perform some or all these cleavages the proteinases necessary for activation remain to be characterized. Activation of latent TGF-β1 also appears to be mediated by association of mannose-6 phosphate oligosaccharide groups of LAP with the mannose-6 phosphate receptors on cell surfaces. Recent studies have also shown that thrombospondin 1 is an important activator of TGF-β1 and potentially other TGF-β isotypes [42].

Characterization of mutations causing Camurati-Engelmann (CED) disease or progressive diaphyseal dysplasia have recently demonstrated the essential role played by the LAP region of TGF-β in the control of TGF-β activity [43]. CED is an autosomal dominant disorder characterized by hyperostosis and sclerosis of the diaphysis of long bones. Missense mutations that result in amino acid substitutions in the LAP region of TGF-β1 appear to decrease the stability of latent complexes formed by TGF-β1. The resultant increased susceptibility to activation appears to result in increased TGF-β1 activity, increased synthesis and decreased degradation of bone in the diaphysis of long bones [43]. Although the mutation causes skeletal changes the onset of pathology appears relatively late in development (before age 30 [44]) consistent with the proposed association of TGF-β1 with bone remodeling rather than skeletogenesis.

In the studies described here we demonstrate that the release and activation of TGF-β2 from chondrocytes occurs in association with their hypertrophic differentiation and apoptosis respectively. Chondrocyte apoptosis is initiated early in hypertrophic differentiation but its morphological consequences are delayed in vivo until close interaction with invading vascular cells. Inhibitor studies suggested caspases or enzymes activated by the caspase proteinase cascade stimulate TGF-β2 activation.

Methods

Classification of Stages of Sterna Development

Embryonic chick sterna provide a convenient tissue from which to isolate chondrocytes in various stages of terminal differentiation [45]. The two presumptive primary ossification centers of the chick sterna become apparent as translucent areas on either side of the cephalic region of the sterna at 17 to 18 days of development. These contain an almost pure population of hypertrophic chondrocytes. At about 19d of development small areas of calcification are observed over the primary ossification center. By 20d of development extensive calcification of the sterna is observed. Sterna were classified according to the morphological appearance of the primary ossification center. This system enabled the isolation of sterna at distinct stages of differentiation and largely eliminated the variability due to different rates of embryo development arising from slight differences in incubator conditions or batch to batch variability. Chondrocytes were also isolated from the caudal region of sterna. This region remains cartilaginous for several months after hatching [46] and contains a population of smaller chondrocytes.

Chondrocyte Isolation and Culture

The presumptive primary ossification centers were carefully dissected from surrounding cartilage. The caudal regions were dissected from the caudal half of the sterna. Chondrocytes were isolated from these regions by overnight collagenase (1mg/ml) digestion in Dulbecco's modification of Eagle's medium (dme) containing fetal calf serum (10%) [46]. Isolated cells were lysed immediately for analysis of caspase activity, used for RNA isolation or cultured in alginate beads [46]. Hypertrophic chondrocytes isolated from 18d chick embryo sterna primary ossification centers were cultured in DME containing cysteine (1mM), pyruvate (1mM), and gentamicin (50 μg/ml) and ascorbate (50 μg/ml) (37°C, 5% CO_2) in alginate beads, as described previously, at a density of 10^7 cells/ml of alginate bead and $5X10^6$ cells per 2.5 cm culture well [46]. In some experiments the medium was supplemented with 0.1% serum or the caspase inhibitor zVADfmk (50 μM). Chondrocytes isolated from regions of 14d chick embryo sterna that would form the primary ossification centers were also cultured in the presence of thyroxine (100 ng/ml) conditions previously shown to initiate hypertrophic differentiation and result in apoptosis [46]. Cells were harvested at selected times by dissolution of alginate in 0.15M citrate, pH 7.5 at 37 °C for 10 min. and lysed for analysis of caspase activity. Conditioned culture medium was collected for analysis of TGF-β2 activity and structure. Cell viability was assessed by the MTT assay [47].

Assay of Caspase Activity

Caspase activity was measured in chondrocyte lysates generated by freeze/thawing $3x10^6$ cells in 100 ul of cell lysis buffer containing PIPES (50mM), pH 7.0, KCl (50mM), $MgCl_2$ (2mM) and inhibitors of non-specific proteinase activity (phenylmethylsulfonyl fluoride (1mM), leupeptin (1μg/ml), pepstatin A (1μg/ml) and antipain (50μg/ml)) [48]. A 50 μl sample of cell lysate was incubated with 15 μM benzyloxycarbonyl-Asp-Glu-Val-Asp-7-amino-4-trifluoromethylcoumarin (zDEVD-afc) or benzyloxycarbonyl-Tyr-Val-Ala-Asp-7-amino-4-trifluoromethylcoumarin (zYVAD-afc), in 1.5 ml of HEPES (0.1M), pH 7.4, sucrose (1%), CHAPS (0.1%), DTT (2mM) at 37° C. Cleavage of these substrates releases the fluorescent moiety, 7-amino-4-trifluoromethylcoumarin (afc) allowing quantitative analysis of caspase activities. Fluorescence was measured using 400 nm excitation and 505 nm emission at multiple time intervals (initially every 5 min.) up to 120 min. When activity was present a linear increase in fluorescence was observed during the initial time period (usually 30 min. depending on activity). Caspase activity was calculated from this linear region and expressed as nM of substrate digested (calculated from fluorescence of a known concentration of standard afc solution) /min/10^6 chondrocytes.

Caspase activity was further characterized by inhibition with peptide aldehyde inhibitors zDEVD-CHO, zVEID-CHO and zYVAD-CHO (Enzyme System Products, CA). Chondrocytes ($2x10^7$) from the primary ossification center of 18 and 20d chick embryos were lysed and aliquots equivalent to $3x10^6$ cells incubated with a series of concentrations of caspase inhibitors for 5 min. at 37° C in substrate buffer. Substrate (zYVAD-afc or zDEVD-afc) was then added and the release of fluorescence monitored as described above. The concentration of inhibitor required to reduce caspase activity by 50% was used as an indication of the type of caspase present.

Quantitative Real Time RT-PCR

Quantitative with real time RT-PCR was used to quantify relative levels of collagen types II and X mRNAs. RNA was extracted from chondrocyte pellets (5 x 10^6 cells) using Trizol reagent (Life Technologies, Rockville, MD) and purified by chloroform-phenol separation and precipitation with isopropanol as described by the suppliers. Reverse transcription reactions were conducted using 1 to 5 μg total RNA and a superscript II reverse transcriptase kit (Life Technologies, Rockville, MD) as described by the supplier.

One-tenth (2 μl) of each RT reaction was amplified by Lightcycler PCR (Roche Molecular Biochemicals, Indianapolis, IN) using for type II collagen sense GCA GAG AAC ATC AAC GGC GGT and antisense CAG GCG CGA GGT CTT CTG CGA and for type X collagen sense AAG GGG CCA CCA CAC TTT CTA and antisense TTC TCC AGG CTT CCC TAT CCC primers. (Roche Molecular Biochemicals, Indianapolis, IN).

ELISA Assay for TGF-β2

A sandwich ELISA assay was used to measure TGF-β2 (R&D, Minneapolis, MN). Latent TGF-β2 was activated prior to assay by acidification of samples with HCl to 0.17 M and incubation for 10 min. This was followed by neutralization by addition of NaOH to 0.14M and HEPES to 0.07M. The assay employs a monoclonal antibody specific for TGF-β2 coated onto a microplate. Standards and samples were pippetted into the wells and TGF -β2 present bound. After washing an enzyme linked polyclonal antibody specific for TGF-β2 was added to the wells and detected by addition of substrate.

Expression of the LAP Region of Chick TGF-β2

As a source of antigen the region of the chick TGF- β2 gene equivalent to the LAP part of the molecule, was isolated, characterized and expressed in E-coli. A recombinant protein was purified for antibody production by affinity chromatography. The cDNA sequence covering the LAP region of chick TGF- β2 was amplified with upstream primer cgg atc cCT GTC TAC CTG CAG CAC CCT and downstream primer cca agc tta ACT GGG CTG TTG CGA CTC. The product was purified and cleaved with Bam HI and Hind III and cloned in frame into pET28a vector (Novagen, Madison, WI). The sequence identity of the cDNA clones was confirmed using BigDye-labeled terminator cycle sequencing. The confirmed construct was introduced into BLR(DE3)pLysS E-coli host (Novagen, Madison, WI) for expression .

PolyHis tagged LAP was isolated from the E-coli pellet and purified using immobilized metal affinity chromatography on Ni-NTA agarose (Qiagen, Valencia, CA). After washing the column LAP was eluted with urea, (8M), NaH2PO4 (0.1M), Tris (0.01M) pH 4.5 and further purified using SDS PAGE. Approximately 2 mg pure LAP was used for production of rabbit polyclonal antibodies (Covance Research Products, Denver, PA). The polyclonal antisera generated had a high titer (1:30,000) against the antigen, recombinant chick TGF-β2 LAP, as determined by ELISA. Western blot analysis of crude lysates of E-coli expressing TGF-β2 LAP contained a single band consistent with the size of the expected TGF-β2 LAP and its disulfide dimer.

Immunoblot Analysis

Culture medium from chondrocyte cultures was subjected to SDS PAGE run under reducing and nonreducing conditions. Proteins were detected by Western blot using TGF-β2 LAP specific antibodies, TGF-β2 specific antiserum (sc 90 , Santa Cruz Biotechnology, CA) or preimmune serum with chemiluminescent detection. In some experiments samples of culture medium were subjected to 2 dimensional electrophoresis prior to immunoblot analysis. Samples were electrophoresed under non-reducing conditions in the first dimension and under reducing conditions (2% mercaptoethanol in sample buffer) in the second dimension.

Results

Chondrocyte Caspase Activity

Caspase activities assayed with fluorescent peptide substrates were demonstrated in hypertrophic chondrocytes. Substrate specificity indicated two types of caspase with peak activities at distinct stages of hypertrophic chondrocyte terminal differentiation. At or slightly before the time of hypertrophic differentiation (as monitored by type X collagen expression, Fig. 2) chondrocytes contained high activity against a preferred substrate of the caspase I family, zYVAD-AFC. Activity against the preferred substrate of the caspase II family, zDEVD-AFC, was more prominent later in development and coincided with the start of calcification and resorption (Fig. 3a). Chondrocytes from the caudal region of sterna showed only very low caspase activity (Fig. 3b).

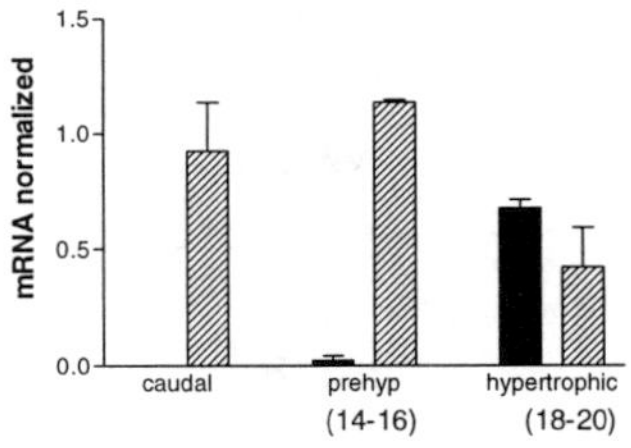

Figure 2. Collagen mRNA expression by chick embryo sterna chondrocytes. Chondrocytes were isolated from the caudal region or primary ossification center of sterna from chick embryos of selected age. Expression of collagen types II and X were measured using quantitative real time PCR. Bars represent the median level with standard deviation determined for four preparations of chondrocytes.

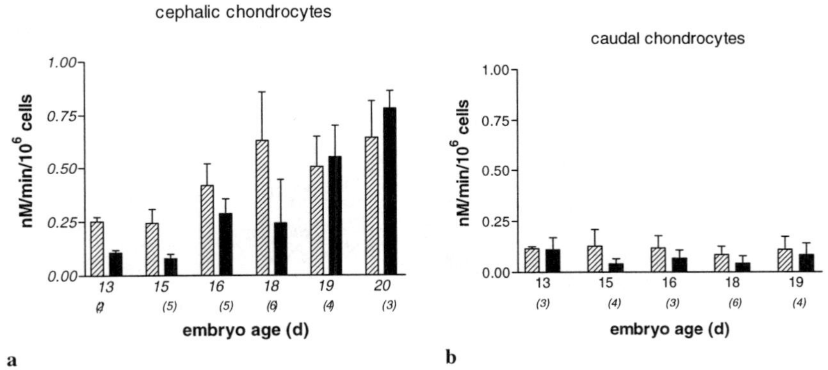

Figure 3. Caspase activity in lysates of a, chondrocytes isolated from the primary ossification center and b, chondrocytes isolated from the caudal region of chick embryo sterna at the indicated stages of embryonic development. Caspase activity was measured with peptide substrates zYVAD-afc (hatched bars) or zDEVD-afc (solid bars). Activities are expressed as nM product produced/min./ million cells. Bars represent the median activity and standard deviation. Numbers in parenthesis indicate the number of chondrocyte samples analyzed.

The inhibition profile determined using peptide aldehyde inhibitors confirmed the presence of caspase activity from at least two families. The inhibition profile of activity detected using the DEVD substrate, prominent at late stages of apoptosis, was consistent with caspase 3. The activity detected with the YVAD substrate, associated with earlier stages of chondrocyte hypertrophy cannot be easily categorized based on these studies (Table 1). It is, however, clear that multiple caspase activities are present, some being expressed well before the morphological signs of chondrocyte apoptosis are visible.

Table 1. Inhibition of caspase activity by peptide inhibitors.

	Inhibitor	Concentration at ½ maximal inhibition nM
d 20 sterna.		
zDEVD-AFC substrate.	DEVD-cho	0.1
	VEID-cho	200
	YVAD-cho	20000
d18 sterna		
zDEVD-AFC substrate	DEVD-cho	0.1
	VEID-cho	200
	YVAD-cho	10000
zYVAD-AFC substrate	DEVD-cho	2000
	VEID-cho	200
	YVAD-cho	7000

Chondrocytes induced to express the hypertrophic phenotype in culture exhibited a profile of caspase activity similar to that seen in chondrocytes that differentiated in vivo. Early in culture, coincident with type X collagen expression, chondrocytes contained peak levels of activity resembling the caspase I family. Later in culture coincident with the onset of apoptosis high levels of activity of the caspase II family were observed. TGF-β2 was released from hypertrophic chondrocytes in culture coincident with peak levels of caspase I-type activity detected in these cells (Fig. 4).

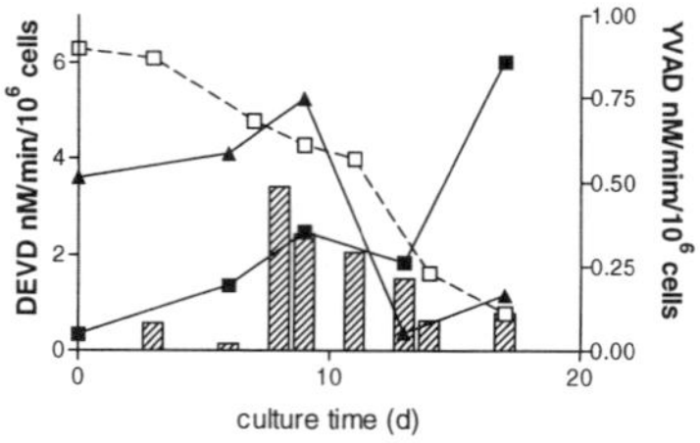

Figure 4. Caspase activity in cells in culture. Prehypertrophic chondrocytes were cultured in the presence of thyroxine. Viability was measured by MTT assay (dashed line open squares); TGF-β2 (ng/ml bars) by ELISA. Caspase activity determined with zYVAD-AFC (triangles) and zDEVD-AFC (solid squares) substrates is expressed as nM product produced/min./million cells. Note the different scales for DEVD and YVAD activity.

TGF-β2 Released by Hypertrophic Chondrocytes in Culture

We have shown previously that TGF-β2 is released from hypertrophic chondrocytes in culture. A high proportion of this was present in an active form [38]. We have recently observed that addition of ascorbate (50 μg/ml) to the culture medium of hypertrophic cells further increased the level of TGF-β2 released into the medium and in the absence of serum resulted in an increased activation of TGF-β2, reaching approximately 30% after 9 days culture (Fig. 5b). Addition of serum to these cultures even at very low concentrations (0.1%) inhibited the apoptosis of the chondrocytes (Fig. 5a) and stimulated TGF-β2 release (data not shown). Inhibition of chondrocyte apoptosis with low levels of serum also reduced the activation of TGF-β2 to levels of less than 5% (Fig. 5b). Addition of the cell permeable, broad-specificity, caspase inhibitor zVADfmk to the culture medium significantly delayed chondrocyte apoptosis (Fig. 5c). It did not influence the release of latent TGF-β2 (data not shown) but significantly reduced the activation of TGF-β2 (Fig. 5d).

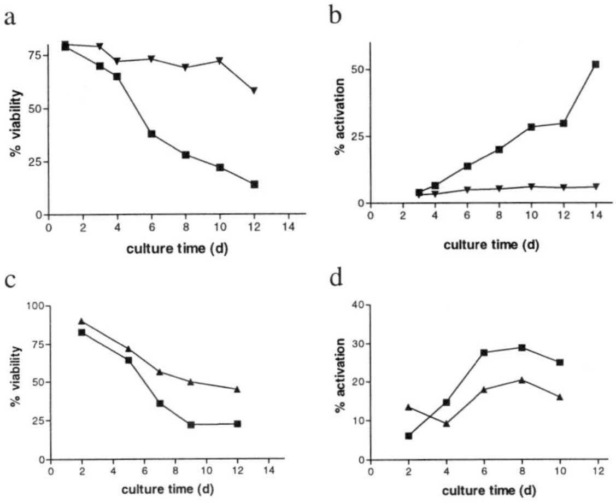

Figure 5. The affect of serum and zVADfmk on chondrocyte viability and the activation of TGF-β2 released into the culture medium. Hypertrophic chondrocytes isolated from 18d chick embryo sterna were cultured in DME (**squares**). Chondrocytes were also cultured in DME containing 0.1% serum (a) and (b) or DME containing zVADfmk (50 μM) (c) and (d) (**triangles**). Cell viability was determined by the MTT assay. TGF-β2 and its activation were determined by ELISA.

Preparation of Antibodies Recognizing the LAP Region of Chick TGF-β2

Antibodies recognizing the LAP region of chick TGF-β2 were generated to characterize the organization and intracellular processing of the TGF- β2 latent complex. As a source of antigen we isolated, and characterized the region of the chick TGF- β2 gene equivalent to the LAP part of the molecule, expressed it in E-coli and used the purified recombinant protein for antibody production. PolyHis tagged LAP was isolated from the E-coli pellet and purified using immobilized metal affinity chromatography followed by SDS PAGE. Polyclonal antisera generated against this protein had a high titer (1:30,000) and high specificity. Immunoblots of crude lysates of E-coli expressing TGF-β2 LAP contained a single band consistent with the size of the expected TGF-β2 LAP and its disulfide dimer (data not shown).

Changes in the TGF-β2 Complex Associated with Activation

Immunoblot analysis of hypertrophic chondrocyte-conditioned culture medium using the antibody raised against recombinant TGF-β2 LAP demonstrated multiple molecular forms of the TGF-β2 complex. A disulfide stabilized dimer of 80 kD was prominent. The 40 kD monomer was the only prominent band detected under reducing conditions. Additional bands of 92 and 105 kD and several of greater than 220 kD were clearly detectable by the LAP antibody in nonreduced samples (Fig. 6a, b and c). No bands were clearly visible in control gels run under nonreducing conditions (Fig. 6d).

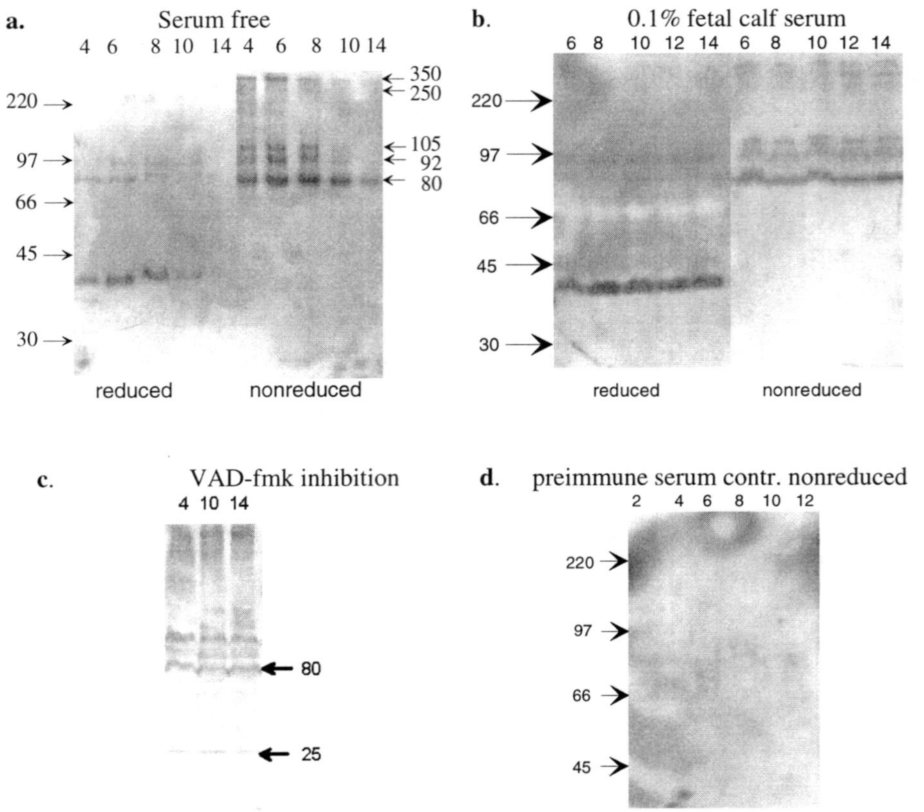

Figure 6. Western blot analysis of TGF-β2 in the culture medium of hypertrophic chondrocytes. Culture medium from hypertrophic chondrocyte cultures was subjected to immunoblot analysis using TGF-β2 LAP specific antibody. Samples of serum-free culture medium, A and D, medium containing 0.1% fetal calf serum, B and medium containing the caspase inhibitor zVAD-fmk (described in Fig. 5), C were subjected to SDS PAGE (7% A&B, 5% C & D acrylamide) run under reducing and nonreducing conditions. Proteins were detected by immunoblot using TGF-β2 LAP specific antibodies A, B, preimmune serum, D. or both TGF-β2 LAP-specific and TGF-β2-specific antibodies, C with chemiluminescent detection. The day of culture is at the top of images. The migration of molecular weight markers (kD) is shown by arrow on the left. The calculated molecular size of detected bands is shown on the right.

Western blot analysis of culture medium from hypertrophic chondrocytes after various times in culture revealed changes in the TGF-β2 complex that coincided with the activation of TGF-β2 as detected by ELISA (shown in Fig. 5). At early culture times where very low levels of activation were detected high molecular weight forms of LAP were prominent. With culture and the activation of TGF-β2 the 80 kD dimer predominated and higher molecular weight forms were lost. The 92 and 105 kD molecular species appeared at about 4 days culture under serum-free conditions coincident with the onset of apoptosis (Fig. 5a) but declined in significance with further culture. In some studies (not evident in Fig. 6) additional bands of approximately 120 kD were also visible. The inclusion of low concentrations of serum (0.1 % fcs) in the culture medium appeared to suppress degradation of higher molecular weight proteins as well as the 80 kD dimer. The higher molecular weight species (>200kD), the 92 and 105 kD species as well as the 80 kD dimer remained prominent throughout culture in serum-containing medium (Fig. 6b). Addition of the broad-specificity caspase inhibitor to the culture medium of hypertrophic chondrocytes generated a similar inhibition of the degradation of the higher molecular mass forms of the TGF-β2 complex.

Two-dimensional SDS PAGE using non-reducing condition in the first and reducing conditions in the second dimension showed that all the higher molecular species contained the mature 40 kD LAP. Some also contained disulfide bound mature TGF-β2 (Fig. 7).

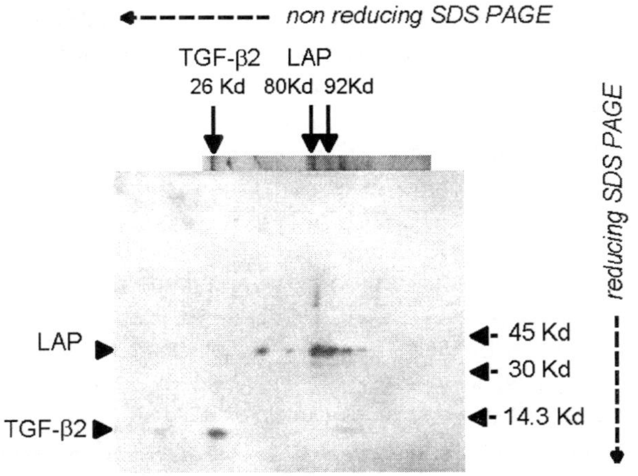

Figure 7. Two dimensional electrophoresis of TGF-β2 from the culture medium of hypertrophic chondrocytes. A sample of the culture medium from hypertrophic chondrocytes, similar to the 8 day sample shown in Fig. 6 a was subjected to SDS PAGE run under non reducing condition. A track was cut from the non-reducing gel and subjected to electrophoresis under reducing conditions (2% mercaptoethanol). The gel was then processed for immunoblotting using a mixture of two antibodies recognizing TGF-β2 and the LAP region of TGF-β2. The film generated by chemiluminescent development is shown. A duplicate track run under non-reducing conditions is shown at the top of the image. The migration position of molecular weight markers is shown at right and suggested TGF-β2 complex components at left.

Discussion

These studies confirm the release of large quantities of TGF-β2 by hypertrophic chondrocytes at about the time of their apoptosis in culture. The release of a substantial proportion of active TGF-β2 by hypertrophic chondrocytes as demonstrated in current studies is very unusual. Nearly all cell types studied to date release only latent forms of the TGF-β isotypes. The release of TGF-β2 from hypertrophic chondrocytes coincided with presence of high levels of caspase activity against a preferred substrate of the group I subfamily of caspases. Caspases of this type have been implicated in cytokine activation and release [49] however inhibition of caspase activity did not result in diminished TGF-β2 release suggesting this caspase activity was not necessary for TGF-β2 release.

It is, however, clear that multiple caspase activities are present, some present well before the morphological signs of chondrocyte apoptosis are visible. These studies suggest mechanisms operate in hypertrophic chondrocytes to prevent the destructive action of caspases and enable hypertrophic chondrocyte function. Studies of nematodes with mutations in the apoptotic pathway have demonstrated that the presence of active caspases does not necessarily result in immediate cell death. Cells containing active caspases were shown to function for an extended time in vivo [50]. Many factors influence caspase activity including compartmentation [51] and the action of inhibitors of apoptosis. High levels of Bcl-2 and IAP have been detected in prehypertrophic chondrocytes using quantitative pcr analysis in our laboratory (data not shown) and by immunohistochemistry [7]. A decline in these inhibitors observed later in chondrocyte hypertrophy (data not shown and [7]) might contribute to the initiation of morphological changes associated with chondrocyte apoptosis at the vascular interface or at later stages of culture.

Studies of the nature of the TGF-β2 latent complex showed the presence of bands on SDS PAGE consistent with the presence of the small and several large latent complexes. These appear similar in some aspects to the complexes shown formed by TGF-β1 [40]. The 80 kD band detected in nonreduced samples can be assumed to be the LAP dimer non-covalently associated with mature TGF-β2. At present we do not know the identity of the higher molecular weight components detected in nonreduced samples. Two dimensional electrophoresis suggested the proteins in the range of 90 to 120 kDa were similar to the small latent complex but unlike that seen with TGF-β1 appear linked with mature TGF-β2 and/or another peptide by disulfide bonds. We have detected four protein bands in this region. All generated the LAP monomer when electrophoresed under reducing conditions. In addition two of the bands were shown to contain TGF-β2 when run under reducing conditions. Model structures for these complexes are presented in Fig. 8.

Although samples studied in Fig. 7 did not contain significant quantities of large latent complexes studies of other samples (not shown) have demonstrated only a single band of LAP associated with these complexes. The size of the components observed suggested some bands may represent the a large latent complex homologous to that formed by TGF-β1, TGF-β1 LAP dimer and latent TGF-β binding protein (LTBP). Attempts to detect this protein using polyclonal antibodies to the human protein have not been successful (data not shown) although this may be due to the species specificity of the antibody.

Activation of TGF-β2 was associated with the loss of higher molecular species detected after electrophoresis under non-reducing conditions. This suggests these complexes are associated with maintaining TGF-β2 in a latent form. Release of the active mature TGF-β2 peptide appears to be associated with degradation or dissociation of these complexes. Chondrocyte apoptosis, though not essential for the release of TGF-β2, appeared to be tightly linked to TGF-β2 activation through the caspase proteinase cascade. Factors that inhibited apoptosis also inhibited the activation of TGF-β2. Inhibition of caspase activity using the caspase inhibitor zVAD-fmk delayed the loss of the higher molecular mass complexes and inhibited TGF-β2 activation. Members of the group I family of caspases including caspase 1 have been shown to activate a number of cytokines [52,53]. However, it appears unlikely that TGF-β2 activation is a direct result of caspase cleavage. TGF-β2 activation is presumed to be an extracellular process. Activation of TGF-β2 observed would thus appear to be a consequence of the caspase cascade via the activation of other as yet unidentified proteinases.

The studies suggest activation of growth factors like TGF-β2, involving the dissociation of latent complexes, can arise as a consequence of the process of apoptosis. In the case of hypertrophic chondrocyte apoptosis this would serve to localize the active growth factor to a region of morphological transition. As discussed previously TGF-β2 would be expected to influence many of the processes of endochondral ossification in this region. Regulation of its activity by processes associated with apoptosis would provide a signaling loop that has the potential to provide critical control of growth plate homeostasis.

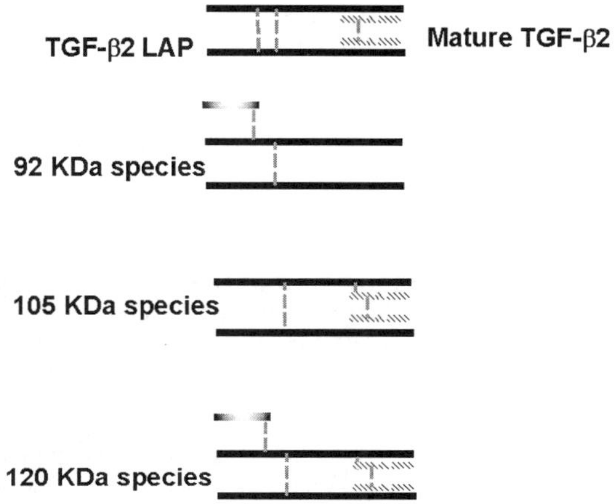

Figure 8. Model of TGF-β2 complex structures. Model structures of some of the TGF-β2 species detected by immunoblot analysis using antibodies that detect TGF-β2 LAP and mature TGF-β2 are depicted. The structures are based on published studies of TGF-β1 and immunoblot of 2 dimensional electrophoresis (Fig. 7). The small latent complex contains an 80 kDa LAP dimer as detected on SDS PAGE and the mature TGF-β2 dimer. We suggest the 92 kDa species contains a peptide of approximately 10 kDa (shaded bar) disulfide bonded to the LAP dimer. The 105 kDa species and 120 kDa contain mature TGF-β2 (hatched bars) disulfide bound to the LAP dimer. The 120 kDa species contains the LAP dimer disulfide bound to both the peptide and mature TGF-β2.

Acknowledgements

This study was supported by a grant from the NIH to G.G.

References

[1] Hunziker EB. Mechanism of longitudinal bone growth and its regulation by growth plate chondrocytes. Microsc Res Technique 1994;28(6):505-19.

[2] Gibson GJ, Bearman CH, Flint MH. The immunoperoxidase localization of type x collagen in chick cartilage and lung. Collagen Relat. Res 1986;6:163-84.

[3] Schmid TM, Linsenmayer TF. Developmental acquisition of type x collagen in embryonic tibiotarsus. DevBiol 1985;107:373-81.

[4] Bohme K, Winterhalter KH, Bruckner P. Terminal differentiation of chondrocytes in culture is a spontaneous process and is arrested by transforming growth factor-beta 2 and basic fibroblast growth factor in synergy. Exp Cell Res 1995;216(1):191-8.

[5] Szuts V, Mollers U, Bittner K, et al. Terminal differentiation of chondrocytes is arrested at distinct stages identified by their expression repertoire of marker genes. Matrix Biology 1998;17(6):435-48.

[6] Chung UI, Lanske B, Lee KC, Li E, Kronenberg H. The parathyroid hormone parathyroid hormone-related peptide receptor coordinates endochondral bone development by directly controlling chondrocyte differentiation. Proc Natl Acad Sci USA 1998;95(22):13030-5.

[7] Amling M, Neff L, Tanaka S, et al. Bcl-2 lies downstream of parathyroid hormone-related peptide in a signaling pathway that regulates chondrocyte maturation during skeletal development. J Cell Biol 1997;136(1):205-13.

[8] Bohme K, Conscienceegli M, Tschan T, Winterhalter KH, Bruckner P. Induction of Proliferation or Hypertrophy of Chondrocytes in Serum-Free Culture - The Role of Insulin-Like Growth Factor-I, Insulin, or Thyroxine. J Cell Biol 1992;116(4):1035-42.

[9] Mancilla EE, Deluca F, Uyeda JA, Czerwiec FS, Baron J. Effects of fibroblast growth factor-2 on longitudinal bone growth. Endocrinology 1998;139(6):2900-4.

[10] Serra R, Johnson M, Filvaroff EH, et al. Expression of a truncated, kinase-defective TGF-beta type II receptor in mouse skeletal tissue promotes terminal chondrocyte differentiation and osteoarthritis. J Cell Biol 1997;139(2):541-52.

[11] Enomotoiwamoto M, Iwamoto M, Mukudai Y, et al. Bone morphogenetic protein signaling is required for maintenance of differentiated phenotype, control of proliferation, and hypertrophy in chondrocytes. J Cell Biol 1998;140(2):409-18.

[12] Ito H, Akiyama H, Shigeno C, Nakamura T. Bone morphogenetic protein-6 and parathyroid hormone-related protein coordinately regulate the hypertrophic conversion in mouse clonal chondrogenic EC cells, ATDC5. Biochim Biophys Acta 1999;1451(2-3):263-70.

[13] Volk SW, Luvalle P, Leask T, Leboy PS. A BMP responsive transcriptional region in the chicken type X collagen gene. J Bone Miner Res 1998;13(10):1521-9.

[14] Ballock RT, Reddi AH. Thyroxine is the serum factor that regulates morphogenesis of columnar cartilage from isolated chondrocytes in chemically defined medium. J Cell Biol 1994;126(5):1311-8.

[15] Farnum CE, Wilsman NJ. Condensation of hypertrophic chondrocytes at the chondro-osseous junction of growth plate cartilage in Yucatan swine: relationahip to long bone growth. Am J Anat 1989;186:346-58 (p).

[16] Gibson GJ, Kohler WJ, Schaffler MB. Chondrocyte apoptosis in endochondral ossification of chick sterna. Develop Dynam 1995;203(4):468-76.

[17] Vu TH, Shipley JM, Bergers G, et al. MMP-9/gelatinase B is a key regulator of growth plate angiogenesis and apoptosis of hypertrophic chondrocytes. Cell 1998;93(3):411-22.

[18] Silvestrini G, Mocetti P, Ballanti P, Digrezia R, Bonucci E. In vivo incidence of apoptosis evaluated with the TdT FragEL (TM) DNA fragmentation detection kit in cartilage and bone cells of the rat tibia. Tissue Cell 1998;30(6):627-33.

[19] Serra R, Karaplis A, Sohn P. Parathyroid hormone-related peptide (PTHrP)-dependent and -independent effects of transforming growth factor beta (TGF-beta) on endochondral bone formation. J Cell Biol 1999;145(4):783-94.

[20] Vortkamp A, Lee K, Lanske B, Segre GV, Kronenberg HM, Tabin CJ. Regulation of rate of cartilage differentiation by Indian hedgehog and PTH-related protein. Science 1996;273(5275):613-22.

[21] Vortkamp A, Pathi S, Peretti GM, Caruso EM, Zaleske DJ, Tabin CJ. Recapitulation of signals regulating embryoinic bone formation during postnatal growth and in fracture repair. Mech Dev 1998;71:65-76.

[22] Fanayan S, Firth SM, Butt AJ, Baxter RC. Growth inhibition by insulin-like growth factor-binding protein-3 in T47D breast cancer cells requires transforming growth factor-beta (TGF-beta) and the type II TGF-beta receptor. J Biol Chem 2000;275(50):39146-51.

[23] De LRP, Hill DJ. Expression and release of insulin-like growth factor binding proteins in isolated epiphyseal growth plate chondrocytes from the ovine fetus. J Cell Physiol 2000;183(2):172-81.

[24] Erlebacher A, Derynck R. Increased expression of TGF-beta 2 in osteoblasts results in an osteoporosis-like phenotype. J Cell Biol 1996;132(1-2):195-210.

[25] Erlebacher A, Filvaroff EH, Ye JQ, Derynck R. Osteoblastic responses to TGF-beta during bone remodeling. Mol Biol Cell 1998;9(7):1903-18.

[26] Vailhe B, Tranqui L. The role of transforming growth factor-beta 1 (TGF-beta 1) and of vascular endothelial growth factor (VEGF) on the in vitro angiogenesis process. C R Acad Sci [III] 1996;319(11):1003-10.

[27] Piek E, Heldin C-H, Ten Diek P. Specificity, diversity, and regulation in TGF-b superfamily signaling. FASEB J 1999;13:2105-24.

[28] Barbara NP, Wrana JL, Letarte M. Endoglin is an accessory protein that interacts with the signaling receptor complex of multiple members of the transforming growth factor-b superfamily. J Biol Chem 1999;274:584-94.

[29] Azuma H. Genetic and molecular pathogenesis of hereditary hemorrhagic telangiectasia. Journal of Medical Investigation 2000;47:81-90.

[30] Li DY, Sorensen LK, Brooke BS, et al. Defective angiogenesis in mice lacking endoglin. Science 1999;284:1534-7.

[31] Roberts AB. Molecular and Cell Biology of TGF-beta. Miner Electrolyte Metab 1998;24:111-9.

[32] Saharinen J, Keski-Oja J. Specific sequence motif of 8-cys repeats of TGF-beta binding proteins, LTBPs, creates a hydrophobic interaction surface for binding of small latent TGF-beta. Mol Biol Cell 2000;11:2691-704.

[33] Sanford LP, Ormsby I, Gittenberger-de GAC, et al. TGFbeta2 knockout mice have multiple developmental defects that are non-overlapping with other TGFbeta knockout phenotypes. Development 1997;124(13):2659-70.

[34] Dunker N, Krieglstein K. Targeted mutations of transforming growth factor-beta genes reveal important roles in mouse development and adult homeostasis. Eur J Biochem 2000;267(24):6982-8.

[35] Gibson G. Active role of chondrocyte apoptosis in endochondral ossification. Microsc Res Technique 1998;43(2):191-204.

[36] D'Angelo M, Pacifici M. Articular chondrocytes produce factors that inhibit maturation of sternal chondrocytes in serum-free agarose cultures: A TGF-beta independent process. J Bone Miner Res 1997;12(9):1368-77.

[37] Gelb DE, Rosier RN, Puzas JE. The production of transforming growth factor-beta by chick growth plate chondrocytes in short term monolayer culture. Endocrinology 1990;127(4):1941-7.

[38] Gibson GJ, Lin D-L, Wang X, Xhang L. The release and activation of TGF-beta 2 associated with the apoptosis of chick hypertrophic chondrocytes. J Bone Min Res 2001;in press.

[39] Taipale J, Saharinen J, Keskioja J. Extracellular matrix-associated transforming growth factor-beta: Role in cancer cell growth and invasion. Advances in Cancer Research 1998;75:87-134.

[40] Oklu R, Hesketh R. The latent transfroming growth factor beta binding protein (LTBP) family. Biochem J 2000;352:601-10.

[41] Munger JS, Harpel JG, Gleizes PE, Mazzieri R, Nunes I, Rifkin DB. Latent transforming growth factor-beta: Structural features and mechanisms of activation. Kidney Int 1997;51(5):1376-82.

[42] Crawford SE, Stellmach V, Murphyullrich JE, et al. Thrombospondin-1 is a major activator of TGF-beta 1 in vivo. Cell 1998;93(7):1159-70.

[43] Saito T, Kinoshita A, Yoshiura K, et al. Domain-specific mutations of a transforming growth factor (TGF)-beta 1 latency-associated peptide cause Camurati-Engelmann disease because of the formation of a constitutively active form of TGF-beta 1. J Biol Chem 2001;276(15):11469-72.

[44] Grey AC, Wallace R, Crone M. Engelmann's disease: a 45-year follow-up. J Bone Joint Surg [Br] 1996;78(3):488-91.

[45] Gibson GJ, Flint MH. Type x collagen synthesis by chick sternal cartilage and its relationship to endochondral development. J Cell Biol 1985;101:277-84.

[46] Gibson G, Lin DL, Roque M. Apoptosis of terminally differentiated chondrocytes in culture. Exp Cell Res 1997;233(2):372-82.

[47] Mosmann T. Rapid colocrimetric assay for cellular growth and survival: application to proliferation and cytotoxicity assays. J Immunol Methods 1983;65:5-63.

[48] Enari M, Talanian RV, Wong WW, Nagata S. Sequential activation of ICE-like and CPP32-like proteases during Fas-mediated apoptosis. Nature 1996;380(6576):723-6.

[49] Earnshaw WC, Martins LM, Kaufmann SH. Mammalian caspases: structure, activation, substrates, and functions during apoptosis. Annu Rev Biochem 1999;68:383-424.

[50] Reddien PW, Cameron S, Horvitz HR. Phagocytosis promotes programmed cell death in C. elegans. Nature 2001;412(6843):198-202.

[51] Chandler JM, Cohen GM, Macfarlane M. Different subcellular distribution of caspase-3 and caspase-7 following Fas-induced apoptosis in mouse liver. J Biol Chem 1998;273(18):10815-8.

[52] Hogquist KA, Nett MA, Unanue ER, Chaplin DD. Interleukin-1 Is Processed and Released During Apoptosis. Proc Natl Acad Sci USA 1991;88(19):8485-9.

[53] Behrensdorf HA, van de CM, Knies UE, Vandenabeele P, Clauss M. The endothelial monocyte-activating polypeptide II (EMAP II) is a substrate for caspase-7. FEBS Lett 2000;466(1):143-7.

The Growth Plate
I.M. Shapiro et al. (Eds.)
IOS Press, 2002

Cell Death and Transdifferentiation in the Growth Plate

H.I. Roach
University Orthopaedics, University of Southampton, CF86, General Hospital, Southampton SO16 6YD, U.K.
Corresponding author and address for reprints: Dr. H.I. Roach MSc PhD, University Orthopaedics, CF86, MP817, General Hospital, Southampton, SO16 6YD, U.K.

Abstract. It is generally assumed that growth plate chondrocytes are irrevocably pre-programmed to die by apoptosis even though a chondrocyte with the ultrastructure of classical apoptosis has not yet been identified *in vivo* and in spite of evidence that chondrocytes have the capacity to transdifferentiate to bone-forming cells. To clarify the fate of chondrocytes *in vivo*, avian and mammalian growth plates at different stages of growth were studied with the electron, confocal and light microscope. Chondrocytes with the morphology of classical apoptotic cells were identified *in vitro*, but the cells undergoing programmed cell death *in vivo,* (termed 'dark chondrocytes') had a non-apoptotic morphology and apoptotic bodies were not present. This death involved early loss of membrane integrity, extensive vacuole formation, partial expulsion of cellular content into the extracellular space and autophagocytosis. The incidence of 'dark chondrocytes' was low and the majority of terminal chondrocytes were not condensed, but were large, hydrated cells (termed 'hydropic'). These cells contained a pale nucleus within pale, hydrated cells whose organelles appeared to disintegrate by an unknown mechanism. In the growth plates of rapidly growing mammals, there was no evidence of transdifferentiation to bone-forming cells. This process only took place when hypertrophic chondrocytes remained inside intact lacunae while adjacent cartilage matrix was resorbed. This situation does not arise in the mammalian growth plates during stages of rapid growth, since the non-resorbed struts of cartilage onto which bone is deposited in the primary spongiosa do not contain cells. However, in avian growth plates or in growth plates of old rats or during the endochondral ossification of the facture callus, the above conditions are fulfilled and evidence of transdifferentiation was found. The results suggest that death of terminal chondrocytes *in vivo* includes alternative, non-apoptotic forms of programmed cell death. It is possible that confinement within the lacunae, which would prevent phagocytosis of apoptotic bodies, necessitates different mechanisms of elimination. Transdifferentiation to bone-forming cells only takes place under special circumstances.

The fate of hypertrophic chondrocytes has been debated for nearly half a century. Much confusion has arisen because of the expectation that the fate of hypertrophic chondrocytes should be the same *in vitro* and *in vivo,* and without species- or age-dependent variations. The following will discuss the evidence for different fates of hypertrophic chondrocytes as well as various modes of programmed cell death (PCD). Studies on the growth plates of rapidly growing mammals invariably concluded that hypertrophic chondrocytes die, whereas studies of chick growth plates found evidence suggesting that some chondrocytes became bone-forming cells.

Is Programmed Cell Death Always Pre-Programmed?

When the concept of programmed cell death emerged, it seemed to make intuitive sense that the hypertrophic chondrocytes of the growth plate, which need to be eliminated to enable longitudinal growth, should undergo PCD. The term 'programmed' refers to the fact that the process of death occurs according to a sequential 'program' of molecular and cellular events. However, PCD is frequently interpreted as 'pre-programmed' in the sense that a particular cell is irreversibly committed to cell death at a particular time during development. In some cases, this interpretation would be correct, for example pre-programmed cell death always occurs in specific cells at a specific time in C. elegans, the nematode in which the fate of each individual cell has been mapped. Does the same apply to hypertrophic chondrocytes or is their fate dependent on local signals and circumstances? When chick sternal chondrocytes from the caudal region were isolated and put in suspension culture, they remained viable, whereas chondrocytes from the cephalic region become hypertrophic and apoptotic in culture.[1] Since caudal chondrocytes were not committed to terminal differentiation whereas cephalic chondrocytes were, the authors concluded that hypertrophy irreversibly committed the cells to die. This might be true for sternal chondrocytes *in vitro*, but does not necessarily apply to growth plate chondrocytes *in vivo*. First, hypertrophy is not always a pre-requisite for cell death, because some epiphyseal chondrocytes and proliferating chondrocytes also undergo PCD.[2,3] Secondly, cell death of terminal cells may be accelerated or delayed. For example, when the metaphyseal vasculature was interrupted, a dramatic increase in the incidence of dying hypertrophic cells was found,[4] whereas absence of MMP-9 delayed cell death so that the hypertrophic region increased 8-fold in length.[5] This suggests that cell death is not pre-determined, but dependent on local stimuli.

Cell Death of Terminal Chondrocytes

At first it appeared an easy task to verify that terminal hypertrophic chondrocytes underwent PCD. Studies in other tissues had established that the principal mode of PCD was apoptosis, and the morphological and biochemical changes associated with the death process had been well defined.[6,7,8,9] However, apoptosis is a short-lived event (a few hours at most), hence the number of cells identifiable in any one section is likely to be small. It has been calculated that the expected incidence apoptotic cells in any one histological section would be 0.4%, even if all cells of a tissue died via apoptosis within 20 days.[10]

Ultrastructure of Chondrocytes Undergoing PCD

Classical Apoptosis. Classical apoptotic cells have a well-fined and unmistakable structure: the chromatin is compacted into solid circular or half-moon shaped masses, while the organelles remain intact, albeit within condensed cytoplasm.[9] The plasma membrane initially remains intact, even when the cell breaks up into apoptotic bodies. When viewed with the light microscope, cellular and nuclear condensations may be suggestive of apoptosis, but only the presence of apoptotic bodies provides convincing evidence. In culture, condensed chondrocytes with the typical morphology of classical apoptosis were identified (Fig. 1A). The chromatin had condensed into spheres, the nucleus had fragmented into apoptotic bodies, while the organelles were intact within shrunk cytoplasm. With the light microscope, apoptotic bodies could be detected in a cultured rabbit epiphysis

(Fig. 1B). So far, such chondrocytes with the ultrastructural morphology of classical apoptotic cells have only been identified *in vitro*.

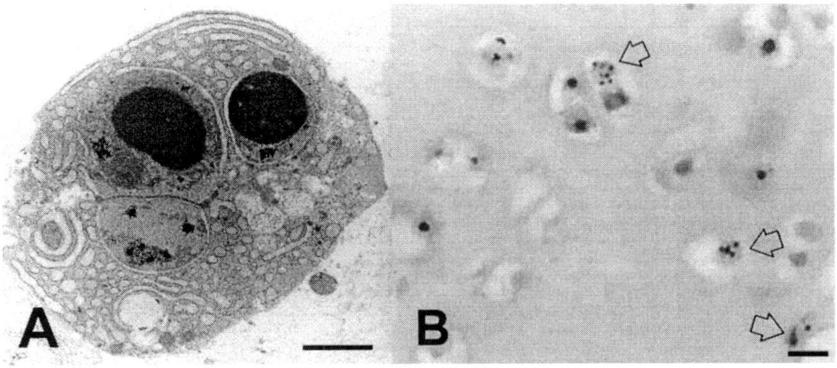

Figure 1. Morphology of chondrocytes undergoing classical apoptosis in culture.
(A) The nucleus has fragmented into three membrane-bound bodies, still contained within the cell, while organelles are intact. Tibial rat growth plate, cultured for 5h with etoposide, scale bar = 1μm, micrograph by courtesy of Dr. H. Clarke Anderson, KUMC, Kansas City.
(B) Apoptotic bodies (arrows) can also be recognised with the light microscope. Femoral head from 4-day old rabbit, cultured on the chorio-allantoic membrane for 7 days, scale bar = 10μm.

'Dark Chondrocytes'. At the vascular front of growth plates from 4-week old Yucatan swines, Farnum and Wilsman [11]identified condensed cells, but with a morphology different from classical apoptosis. Since the cytoplasm and nucleus were very electron-dense, these cells were referred to as "dark chondrocytes".[12] We also studied the femoral growth plates of rabbits, aged 5-20 weeks, with the electron microscope and found that ~ 2-5% of terminal chondrocytes were 'dark chondrocytes' (Fig. 2A,B). These cells had condensed into the centre of the lacunae, but the vacated space was filled with proteoglycan-containing material, whereas condensation during classical apoptosis left empty spaces.[3] This suggested that the process of condensation was gradual, since matrix material was produced as the cell shrunk. There were other crucial differences from apoptotic cells (Table 1). Dark cells contained unusually large amounts of endoplasmic reticulum and many vacuoles, some of which were autophagic, others contained secreted material. The plasma membrane was no longer intact and partial expulsion of cellular contents into the lacunar space had taken place.[3] The nucleus had a characteristic convoluted shape, and the chromatin condensations were patchy, as opposed to the solid masses at the perimeter of apoptotic cells.[3] Similar 'dark chondrocytes' had been identified in other EM studies.[13,12,14]

Table 1. The main features of various modes of cell death

	Physiological cell death			**Pathological cell death**
	Apoptosis	**Dark chondrocytes**	**Hydropic chondrocytes**	**Necrosis**
Pattern of chromatin condensation	Large rounded or crescent-shaped masses, usually at nuclear perimeter	Small patchy condensations, nuclear membrane convoluted	No nuclear condensations, pale disintegrating nucleus	Pyknotic nucleus, usually rounded, patchy chromatin condensations
Membrane integrity	Remains intact, at least initially	Early loss of membrane integrity	Not known	Immediate loss of membrane integrity
Cytoplasmic digestion	No digestion, except by phagocytosis in other cells	Partial digestion within cell, partial extrusion of cellular debris	Gradual disintegration and disappearance of all cellular material	Initial swelling and bursting, vacuolation
Blebbing vs budding*	Budding into apoptotic bodies	Extensive blebbing	No budding or blebbing	Extensive blebbing
Proteolytic enzymes released?	No, lysosomes remain intact within apoptotic bodies	Probably not outside of the cell	Not known	Yes, uncontrolled release of lysosomal enzymes
Elimination of cell	Hetero-phagocytosis	? autophagocytosis	Not known ? disintegration	Inflammatory reaction

* These two terms have been used interchangeably. The author prefers to follow Majno and Joris [7] and distinguish between "budding" of apoptotic bodies from the much smaller, blister-like, vacuolated extrusions, which are more aptly described as "blebbing".

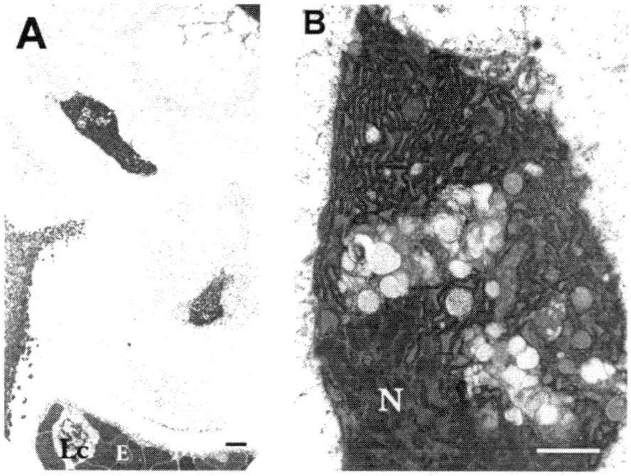

Figure 2. 'Dark chondrocytes' at the vascular front. The vascular front, with erythrocytes (E) and one leucocyte (Lc) is at the bottom of the figure. The cells in the ultimate and pen-ultimate lacuna (A) have a morphology different from classical apoptosis. At higher magification (B), the dark cytoplasm containing excessive endoplasmic reticulum and the dark nucleus (N) with patchy chromatin condensations can be seen. From the growth plate of a femoral head of a 6-week old rabbit, scale bar = 5μm in (A) and 1μm in (B).

Secretory Dark Chondrocytes. In a subset of dark chondrocytes, Golgi apparatus was extremely prominent (arrows in Fig. 3A) together with condensing vacuoles (cv) and secretory vesicles (sv). When dark chondrocytes were attached to the last transverse septum (Fig. 3B), transport vesicles (Tv) were extruding from the cell towards the last septum (Fig. 3B, inset). Strong tartrate-resistant acid phosphatase activity (TRAP) could be demonstrated in and beneath those chondrocytes that were attached to the last septum (not shown), suggesting that chondrocytes themselves might contribute to the resorption of this final barrier.

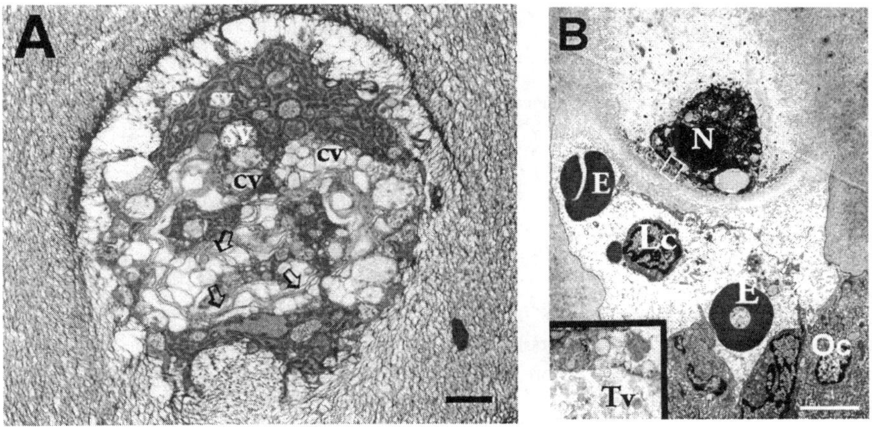

Figure 3. 'Secretory' chondrocytes within the growth plate. (A) Very occasionally, a chondrocyte contained abnormally high Golgi apparatus (arrows) as well as secretory (sv) and condensing (cv) vacuoles was found. Located within the upper hypertrophic zone of the femoral head growth plate from an 8-day old rabbit. (B) A 'dark chondrocyte' with a dark nucleus (N) attached to the last transverse septum. Erythrocytes (E), a leucocyte (Lc) and an osteoclasts (Oc) are present within the vascular space. Transport vesicles (Tv, see inset) are emerging from the basal membrane towards the septum. Femoral head growth plate from 20-week old rabbit, scale bars = 1μm.

'Hydropic' Chondrocytes. In the growth plates of rapidly growing mammals, the vast majority of terminal chondrocytes are not condensed, but are large, hydrated cells. These cells were termed "hydropic" by H. Clarke Anderson (personal communication), which is an appropriate descriptive term that reflects their high water content. These cells were viable and metabolically active as indicated by intracellular fluorescence with the viability marker fluorescein diacetate [15] or CellTracker green (not shown). When viewed with the light or electron microscope, these cells appear fragmented (Fig. 4A,B), partly due to fixation artefacts associated with the dehydration during processing. Yet fixation artefacts cannot explain the fact that hydropic chondrocytes contain hardly any organelles. Cell remnants, which could have originated from a hydropic chondrocytes were found within opened lacunae (Fig. 4C), suggesting the possibility of cellular disintegration, although furthers investigations are required.

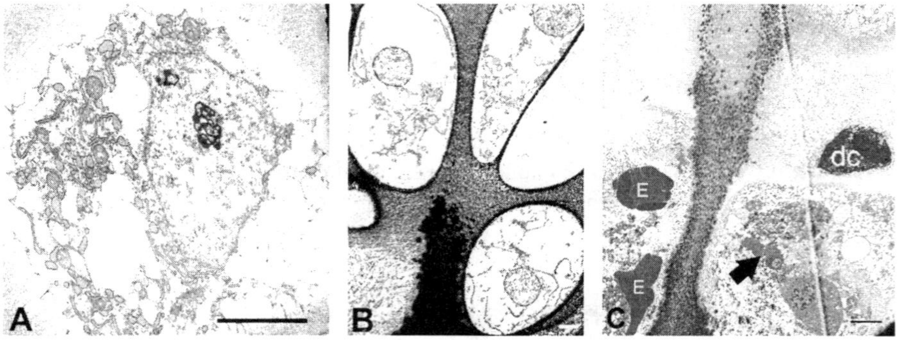

Figure 4. Pale 'hydropic' hypertrophic chondrocytes. The majority of hypertrophic chondrocytes were non-condensed, pale cells which appeared fragmented with few organelles. From the femoral head growth plate of a 6-week old rabbit (A) or a 4-day old rabbit (B). Fragmented cell remnants (arrow in C) can be seen within the vascular space (C), while a dark chondrocyte (dc) is seen attached to the last septum. Scale bars = 51μm, E= erythrocytes.

TUNEL Staining of Growth Plate Chondrocytes

Since electron microscopy does not permit survey of a large number of sections, methods were developed to study apoptotic cells *in situ* with the light microscope. Since apoptosis involves double stranded DNA breaks, Gavrieli et al. [16] developed the **T**erminal-transferase d**U**TP **N**ick **E**nd **L**abelling (TUNEL) method. When applied to sections of growth plates, the results were very variable. Some studies found a high percentage (>30 %) of TUNEL-positive hypertrophic cells with more labelled cells in the maturing and upper hypertrophic region than at the vascular front.[17,18,19] In the author's experience, the usual pre-treatment with proteinase K often resulted in TUNEL-staining of viable cells, [3] which raised doubts about the reliability of TUNEL staining and/or whether the TUNEL-positive cells in the upper hypertrophic zone were actually dying. Other studies found far fewer TUNEL-positive cells and these were located only at the vascular front: In 3-9 day old mice, Bronckers [20] identified TUNEL-positive cells in some *open* lacunae. In the tibial growth plates from 9-week rats, only 1.3 % of terminal chondrocytes were TUNEL-labelled [21]. Hence the TUNEL method has produced inconsistent results.

Annexin-V Detection of Phosphotidylserine Flipover

An alternative *in situ* method exploits the fact that an early event of the apoptotic program is the 'flip-flop' of phosphatidylserine from the inside of the plasma membrane to the outside.[22] This takes place while the membrane is still intact and before DNA fragmentation. Annexin-V (anx-V) has a high affinity to phosphatidylserine so that apoptotic cells can be detected with labelled anx-V. Bronckers [23] studied the growth plates from 8-week old mice after injection of anx-V-biotin and found a positive anx-V label in 2% of hypertrophic chondrocytes in closed lacunae and up to 17% in cells of opened lacunae. If one assumes that the anx-V labelled cells in opened lacunae were released chondrocytes rather than vascular cells, then the results suggest that the majority of hypertrophic chondrocytes started the apoptotic program either in the terminal lacunae or after release into the vascular space. Interestingly, Bronckers [23] also found that most of the labelled cells in opened lacunae were in contact with osteoclasts, consistent with phagocytic removal of the apoptotic remnants.

Labelling Growth Plate Chondrocytes with Fluorescent Live/Dead Markers

The fluorescent markers ethidium homodimer-1 (EtH-1) and 5-chloromethylfluorescein diacetate (CMFDA, CellTracker green) are useful to distinguish viable cells from dead or dying cells. EtH-1 identifies cells with a damaged cell membrane, CellTracker green labels the cytoplasm of viable, metabolically active cells. Although apoptotic cells initially maintain an intact plasma membrane, its integrity is lost during the later stages of apoptosis. Since 'dark chondrocytes' also lose membrane integrity, EtH-1 labels late apoptotic and dark chondrocytes as well as necrotic cells. When the femoral growth plates from 2-week old rats were pre-loaded with these markers, EtH-1 labelled cells were easily identified by the intense red fluorescence. In Fig. 5, two cells were labelled at the interface between the epiphyseal chondrocytes and the proliferating zone of the growth plate (open arrows) and three labelled cells are visible at the vascular front (solid arrows). Only one EtH-1 labelled cell was within a closed lacuna, the other two were located within opened lacunae. Overall, the incidence of EtH-1 labelled cells at the vascular front was 1-2% of terminal chondrocytes and *no* EtH-1 labelled cells were detected in the proliferating/upper hypertrophic zones. These results are in agreement with the findings of Silvestrini et al. [21] and Bronckers et al. [23]

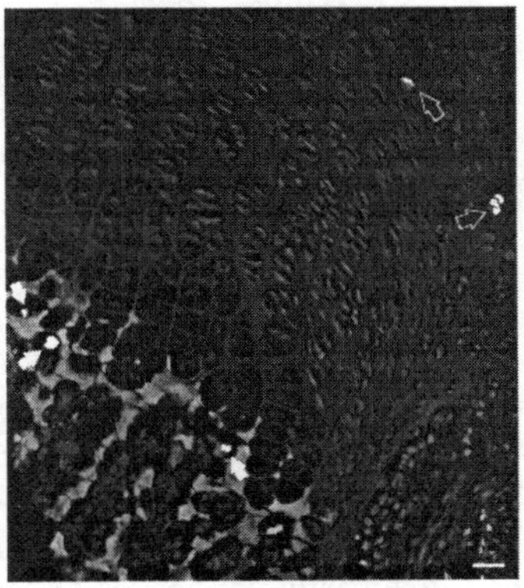

Figure 5. A confocal image of the femoral head from a 2-week old rat, pre-treated with CellTracker green and Ethidium homodimer-1 (EtH-1). All the growth plate zones are visible, and the vascular front can be distinguished from the bright auto-fluorescence of calcified matrix. In colour images, the bright red fluorescence of EtH-1 labelled cells (white arrows) clearly distinguishes these dying cells from the viable chondrocytes.

Fas Protein and c-Myc in the Growth Plate

Two proteins with a possible function in the cell death process are Fas/CD95 and the transcription factor c-Myc. The presence of Fas protein (CD95) in the membrane confers responsiveness to death-inducing Fas ligand.[24] Death of articular chondrocytes could be induced by anti-Fas.[25] To investigate whether the death of growth plate chondrocytes was Fas-mediated, CD95 was immuno-localized in the femoral growth plates of 5-20 week old rabbits (closure takes place at ~23 weeks).[26] At 5-weeks, only a few chondrocytes were immuno-positive for Fas, but at 15-20 weeks, almost 50% of hypertrophic chondrocytes showed positive immunostaining for CD95.[26] This suggested an increased susceptibility to Fas ligand with age, although whether Fas ligand is actually present in the growth plates is not yet known.

Expression of c-Myc has been associated with apoptosis as well as proliferation.[27] In the growth plates of rapidly growing rabbits, c-Myc was predominantly present in proliferating/maturing chondrocytes consistent with its role in proliferation, whereas the immuno-localization changed to the lower hypertrophic zone in the growth plates of older rabbits.[28] The changes in Fas and c-Myc protein suggest that the chondrocytes in older growth plates may have a greater susceptibility to death-inducing factors than the hypertrophic chondrocytes in younger growth plates, but do not provide any information regarding the incidence of cell death.

Non-Apoptotic Modes of PCD in Chondrocytes

More than 20 years ago, the various types of physiological cell death were recognized and succinctly classified [29]. With the explosion of research into apoptosis, these variations were largely forgotten and the term apoptosis became a synonym of the term programmed cell death. Only recently has our group [2,3] drawn attention to the significance of non-apoptotic modes of PCD in chondrocytes. Sperandio et al.[30] also defined a non-apoptotic mode of PCD, termed 'paraptosis', in neurodegeneration. Indeed, non-apoptotic modes of PCD have been beautifully summarized in a monograph by Andrew Wyllie, who recognises that there is '*More than one way to go*'[31].

Most of the *in situ* methods referred to above do not distinguish apoptosis from other modes of PCD. The most unambiguous way to identify classical apoptotic cells is ultrastructure – and this has clearly differentiated 'dark' and 'hydropic' chondrocytes from classical apoptosis. Are 'dark chondrocytes' simply a variant of apoptotic cells or is their different morphology due to an alternative mechanism of PCD? In classical apoptosis, the plasma membranes remain initially intact, whereas in dark cells there is early loss of membrane integrity. In apoptotic cells, caspases initiate the cytoplasmic and nuclear condensations, but the cellular remnants are ultimately destroyed by phagocytosis by other cells. In dark cells, the abnormally large quantities of endoplasmic reticulum suggest an initial increase in protein synthesis, although it is not known what this synthesis consists of. The presence of autophagic vacuoles and extrusion of cellular material into the extracellular space suggests some mechanism of self-elimination. Autophagocytosis is known to play a role in the cell death during insect metamorphosis [32], ovarian atrophy [33] and other situations.[34] Consistent with the notion of self-destruction is the observation that empty, intact lacunae can be found at the vascular front. In single sections, it is impossible to establish whether such lacunae are truly empty, but by carefully examining serial sections Farnum & Wilsman [35,11] established beyond doubt that at least some terminal chondrocytes disappear, leaving an empty lacuna, prior to invasion of the lacuna by an endothelial cell. When chondrocyte death and vascular invasion were uncoupled by the absence of galectin-3, many empty lacunae were present.[36] This provides further evidence for a mode of PCD that involves self-destruction rather than phagocytic removal of apoptotic bodies.

On the other hand, some cells in opened lacunae, especially those close to osteoclasts, stained with Annexin-V [23] and EtH-1 labelled cells were mainly present in opened lacunae. Both these findings are consistent with classical apoptosis, although it is impossible to prove that the labelled cells were released chondrocytes.

The least understood mode of death is that of 'hydropic' cells, yet the majority of hypertrophic chondrocytes are hydropic cells. Do these cells ultimately condense and undergo one of the cell death processes discussed above or does hydropic degeneration represents a yet another, completely different mode of cell death?

In summary, the data suggest that non-apoptotic modes of PCD might be more important than classical apoptosis *in vivo*, although the possibility that some chondrocytes undergo classical apoptosis not only *in vitro*, but also *in vivo* remains.

Transdifferentiation to Bone-Forming Cells

Even more controversial than non-apoptotic modes of PCD has been the suggestion that not all hypertrophic chondrocytes die, but that some survive and become bone-forming cells. The notion that a fully differentiated cell, such as the hypertrophic chondrocytes, could change phenotype was counter-intuitive to what is known about cellular differentiation processes, hence there was an understandable resistance to this idea. Yet the evidence that chondrocytes have the potential to become bone-forming cells was presented repeatedly by different groups. Holtrop [37] labelled rat rib cartilage cells with ^{3}[H]-thymidine and found that the label was present in osteoblasts and –cytes after intramuscular transplantation of the rib cartilage. Thesingh [38] co-cultured cartilaginous mouse long bones with brain tissue and found bone matrix present within chondrocytic lacunae. To exclude the possibility that stem cells from the perichondrium had given rise to the bone-forming cells, Groot [38] demonstrated that osteoblasts could develop from *isolated* fetal mouse chondrocytes when co-cultured with brain tissue. Roach [39] found that a mineralised bone matrix was present within intact chondrocytic lacunae after cutting through the hypertrophic region of embryonic chick femurs and culturing the explants for up to 12 days. Further work identified that the crucial cellular event in the switch from chondrocytes to bone-forming cells was an asymmetric cell division:[40] One daughter cell underwent PCD, while the other cell re-differentiated along the osteogenic pathway and, following further mitotic divisions, became a cell with all the characteristics of an osteoblast.

All the above studies were carried out in organ or cell cultures, so was trans-differentiation merely a culture phenomenon or could it be demonstrated *in vivo*? In the chick embryo growth plate, resorption does not occur synchronously across the plate as it does in mammals, but specialized regions develop and the fate of the chondrocytes depended on its location within the growth plate:[41] Where resorption predominated, chondrocytes underwent PCD. In regions of first endochondral bone formation, some chondrocytes underwent an asymmetric cell division resulting in one apoptotic and one viable cell, the latter re-entering the cell cycle, just as it had been observed in the bone culture system. These cells were released into the vascular space and could theoretically contribute to the pool of endochondral osteoblasts. Where a layer of endochondral bone covered the cartilage and thus 'protected' the subjacent cartilage from further resorption, chondrocytes differentiated into bone-forming cells and deposited bone matrix within their lacunae.[41] Hence the conditions for transdifferentiation were that chondrocytes remained within intact lacunae while adjacent cartilage was resorbed.

In mammalian growth plates from rapidly growing animals, bone-formation within chondrocytic lacunae has not been demonstrated. This is not surprising, since *no* chondrocytes remain within the struts of calcified cartilage onto which endochondral bone is deposited. However, mammalian hypertrophic chondrocytes express bone-typical genes, such as osteocalcin, osteopontin and bone sialoprotein [42] as well as cbfa-1, the transcription factor essential for osteoblast differentiation [43]. This suggests that these mammalian hypertrophic chondrocytes have the potential to transdifferentiate but that the cells, under normal circumstances, undergo PCD before the process is complete.

A different situation arises in the stationary growth plates of aged rats, in which longitudinal growth has ceased. In these growth plates several features were found that were not present in the growth plates of rapidly growing animals. Some of aged rats must have received a strong osteogenic stimulus, because large amounts of new trabecular bone had formed within the medullary canal and the spongiosa. In these rats, bone matrix was present in apparently intact lacunae. However, the best evidence for bone formation by former chondrocytes can be found during the endochondral ossification of the fracture callus, in which chondrocytes remain within the struts onto which endochondral bone is deposited [44]. These chondrocytes become bone-forming cells that directly replace the cartilaginous central part of the spicules of hard callus with bone matrix, thus contributing to the strength of the fracture callus.

In summary, the fate of hypertrophic chondrocytes is not pre-programmed, but depends on the local environment. Chondrocytes have the potential to transdifferentiate, but this potential is not realised in rapidly growing growth plates. The programmed cell death of hypertrophic chondrocytes involves alternative modes to classical apoptosis, which exist in parallel with apoptosis, but may be subject to different controls.

References

[1] Gibson G, Lin DL, Roque M 1997 Apoptosis of terminally differentiated chondrocytes in culture. Exp Cell Res **233** :372-382.

[2] Roach HI, Clarke NMP 1999 "Cell paralysis" as an intermediate stage in the programmed cell death of epiphyseal chondrocytes. J Bone Miner Res **14**:1367-1378.

[3] Roach HI, Clarke NMP 2000 Physiological cell death of chondrocytes in vivo is not confined to apoptosis: New observations on the mammalian growth plate. J Bone Joint Surg [Br] **82-B**:601-613.

[4] Trueta J, Amato VP 1960 The vascular contribution to osteogenesis. III. Changes in the growth cartilage caused by experimentally induced ischemia. J Bone Joint Surg [Br] **42**:571-581.

[5] Vu TH, Shipley JM, Bergers G, Berger JE, Helms JA, Hanahan D, Shapiro SD, Senior RM, Werb Z 1998 MMP-9/Gelatinase B is a key regulator of growth plate angiogenesis and apoptosis of hypertrophic chondrocytes. Cell **93**:411-422.

[6] Lockshin RA, Zakeri Z, Tilly JL 1998 When Cells Die: A comprehensive evaluation of apoptosis and programmed cell death.

[7] Majno G, Joris I 1995 Apoptosis, oncosis, and necrosis. Am J Pathol **146**:3-15.

[8] Wyllie AH 1992 Apoptosis and the regulation of cell numbers in normal and neoplastic tissues: an overview. Canc Metastasis Rev **11**:95-103.

[9] Kerr JFR, Wyllie AH, Currie AR 1972 Apoptosis: a basic biological phenomenon with wide-ranging implications in tissue kinetics. Br J Cancer **26**:239-257.

[10] Tidball JG, Albrecht DE 1998 Regulation of apoptosis by cellular interactions with the extracellular matrix.411-427.

[11] Farnum CE, Wilsman NJ 1989 Condensation of hypertrophic chondrocytes at the chondro-osseous junction of growth plate cartilage in Yacatan swine:relationship to long bone growth. Am J Anat **186**:346-358.

[12] Wilsman NJ, Farnum CE, Hilley HD 1981 Ultrastructural evidence of a functional heterogeneity among physeal chondrocytes in growing swine. Am J Vet Res **42**:1547-1553.

[13] Silbermann M, Frommer J 1973 Heterogeneity among chondrocytes of the mandibular condyle in foetal and postnatal mice. Arch Oral Biol **18**:1549-1554.

[14] Erenpreisa J, Roach HI 1998 Aberrant death in dark chondrocytes of the avian growth plate. Cell Death Diff **5**:60-66.

[15] Farnum CE, Turgai J, Wilsman NJ 1990 Visualization of living terminal hypertrophic chondrocytes of growth plate cartilage in situ by differential interference contrast microscopy and time-lapse cinematography. J Orthop Res **8**:750-763.

[16] Gavrieli Y, Sherman Y, Ben-Sasson SA 1992 Identification of programmed cell death in situ via specific labeling of nuclear DNA fragmentation. J Biochem **119**:493-501.

[17] Fujita I, Hirata S, Ishikawa H, Mizuno K, Itoh H 1997 Apoptosis of hypertrophic chondrocytes in rat cartilaginous growth plate. Journal of Orthopaedic Science **2**:328-333.

[18] Aizawa T, Kokubun S, Tanaka Y 1997 Apoptosis and proliferation of growth plate chondrocytes in rabbits. J Bone Joint Surg [Br] **79-B**:483-486.

[19] Hatori M, Klatte KJ, Teixeira CC, Shapiro IM 1995 End labeling studies of fragmented DNA in the avian growth plate:evidence of apoptosis in terminally differentiated chondrocytes. J Bone Joint Surg **10**:1960-1968.

[20] Bronckers ALJJ, Goei W, Luo G, Karsenty G, D'Souza RN, Lyaruu DM, Burger EH 1996 DNA fragmentation during bone formation in neonatal rodents assessed by transferase-mediated end labeling DNA fragmentation during bone formation in neonatal rodents assessed by transferase-mediated end labeling. J Bone Miner Res **11**:1281-1291.

[21] Silvestrini G, Mocetti P, Ballanti P, Di Grezia R, Bonucci E 1998 In vivo incidence of apoptosis evaluated with the TdT FragEL™ DNA fragmentation detection kit in cartilage and bone cells of the rat tibia. Tissue & Cell **30**:627-633.

[22] Bratton DL, Fadok VA, Richter DA, Kailey JM, Guthrie LA, Henson PM 1997 Appearance of phosphatidylserine on apoptotic cells requires calcium-mediated non-specific flip-flop and is enhanced by loss of the aminophospholipid translocase. J Biol Chem **272**:26159-26165.

[23] Bronckers AL, Goei W, van Heerde WL, Dumont EA, Reutelingsperger CP, van den Eijnde SM 2000 Phagocytosis of dying chondrocytes by osteoclasts in the mouse growth plate as demonstrated by annexin-V labelling. Cell Tissue Res **301**:267-272.

[24] Pinkoski MJ, Green DR 1999 Fas ligand, death gene. Cell Death Differ **6**:1174-1181.

[25] Hashimoto S, Setareh M, Ochs RL, Lotz M 1997 The Fas/Fas ligand expression and induction of apoptosis in chondrocytes. Arthritis Rheum **40**:1749-1755.

[26] Aizawa T, Roach HI, Kokubun S, Tanaka Y 1998 Changes in the expression of fas, osteonectin and osteocalcin with age in the rabbit growth plate. J Bone Joint Surg [Br] **80B**:880-887.

[27] Evan GI, Wyllie AH, Gilbert CS, Littlewood TD, Land H, Brooks M, Waters CM, Penn LZ, Hancock DC 1992 Induction of apoptosis in fibroblasts by c-myc protein. Cell **69**:119-128.

[28] Aizawa T, Roach HI, Kokubun S, Kawamata T, Tanaka Y 1999 c-Myc protein in the rabbit growth plate: Changes in immunolocalization with age and possible roles in proliferation and apoptosis. J Bone Joint Surg **81-B**:921-925.

[29] Schweichel J-U, Merker H-J 1973 The morphology of various types of cell death in prenatal tissues. Teratology **7**:253-266.

[30] Sperandio S, de B, I, Bredesen DE 2000 An alternative, nonapoptotic form of programmed cell death. Proc Natl Acad Sci U S A **97**:14376-14381.

[31] Wyllie AH, Golstein P 2001 More than one way to go. Proc Natl Acad Sci U S A **98**:11-13.

[32] Lockshin RA, Zakeri Z 1994 Programmed cell death:early changes in metamorphosing cells. Biochem Cell Biol **72**:589-596.

[33] D'Herde K, De Prest B, Roels F 1996 Subtypes of active cell death in the granulosa of ovarian atretic follicles in the quail. Reprod Nutr Dev **36**:175-189.

[34] Zakeri ZF, Bursch W, Tenniswood M, Lockshin RA 1995 Cell death: programmed, apoptosis, necrosis, or other? Cell Death Diff **2**:87-96.

[35] Farnum CE, Wilsman NJ 1989 Cellular turnover at the chondro-osseous junction of growth plate cartilage:analysis by serial sections at the light microscopical level Cellular turnover at the chondro-osseous junction of growth plate cartilage:analysis by serial sections at the light microscopical level. J Orthop Res **7**:654-666.

[36] Colnot C, Sidhu SS, Balmain N, Poirier F 2001 Uncoupling of chondrocyte death and vascular invasion in mouse galectin 3 null mutant bones. Dev Biol **229**:203-214.

[37] Holtrop ME 1966 The origin of bone cells in endochondral ossification.32-36.

[38] Thesingh CW, Groot CG, Wassenaar AM 1991 Transdifferentiation of hypertrophic chondrocytes into osteoblasts in murine fetal metatarsal bones, induced by co-cultured cerebrum. Bone Min **12**:25-40.

[39] Roach HI 1992 Transdifferentiation of hypertrophic chondrocytes into cells capable of producing a mineralized bone matrix. Bone Min **19**:1-20.

[40] Roach HI, Erenpreisa J, Aigner T 1995 Osteogenic differentation of hypertrophic chondrocytes involves asymmetic cell divisions and apoptosis. J Cell Biol **131**:483-494.

[41] Roach HI 1997 New aspects of endochondral ossification in the chick: chondrocyte apoptosis, bone formation by former chondrocytes, and acid phosphatase activity in the endochondral bone matrix. J Bone Miner Res **12**:795-805.

[42] Gerstenfeld LC, Shapiro FD 1996 Expression of bone-specific genes by hypertrophic chondrocytes:Implications of the complex functions of the hypertrophic chondrocyte during endochondral bone development. J Cell Biochem **62**:1-9.

[43] Inada M, Yasui T, Nomura S, Miyake S, Deguchi K, Himeno M, Sato M, Yamagiwa H, Kimura T, Yasui N, Ochi T, Endo N, Kitamura Y, Kishimoto T, Komori T 1999 Maturational disturbance of chondrocytes in cbfa1-deficient mice. Dev Dynam **214**:279-290.

[44] Scammell BE, Roach HI 1996 A new role for the chondrocyte in fracture repair: Endochondral ossification includes direct bone formation by former chondrocytes. J Bone Miner Res **11**:737-745.

Matrix Vesicles Contain Metalloproteinases that Are Released into the Matrix by Treatment with $1\alpha,25(OH)_2D_3$ and Are Capable of Activating Latent Transforming Growth Factor-β1

David D. Dean[1], Shingo Maeda[1,2], Zvi Schwartz[1,3], and Barbara D. Boyan[1]
[1]The University of Texas Health Science Center at San Antonio, San Antonio, TX 78229-3900; [2]Tohoku University School of Medicine, Sendai, Japan; and [3]Hebrew University Hadassah School of Dental Medicine, Jerusalem, Israel 91010

Abstract. Matrix vesicles isolated from cultures of costochondral growth zone chondrocytes and treated with $1\alpha,25(OH)_2D_3$ can activate recombinant human latent transforming growth factor beta-1 (rhTGF-β1). This effect was not found with $24R,25(OH)_2D_3$ or with matrix vesicles from resting zone cell cultures. Matrix vesicles contain a number of matrix processing enzymes, including plasminogen activator, 72 kDa gelatinase (MMP-2), and stromelysin-1 (MMP-3), suggesting that one or more of these enzymes may play a role in the activation of latent TGF-β1. This paper summarizes experiments testing the hypothesis that enzymes present in matrix vesicles can activate latent TGF-β1 and that $1\alpha,25(OH)_2D_3$ plays a role in breakdown of the matrix vesicle membrane and enzyme release. Resting zone and growth zone chondrocytes were isolated from the costochondral junction of male Sprague Dawley rats and cultured to fourth passage. Matrix vesicles were isolated from these cultures and the enzymes extracted with buffered guanidine using a Polytron device. Extracts were mixed with small latent rhTGF-β1 for varying periods of time and the production of active TGF-β1 measured by an ELISA specific for active TGF-β1. In addition, enzymes previously determined to be present in matrix vesicles were screened for their ability to activate small latent rhTGF-β1. The results demonstrated that extracts of matrix vesicles produced by both growth zone and resting zone chondrocytes dose-dependently activated small latent rhTGF-β1. Of the enzymes known to be present in matrix vesicles, only rhMMP-3 was able to activate small latent rhTGF-β1, and the effect of rhMMP-3 was time- and dose-dependent. Anti-MMP-3 antibody blocked the ability of matrix vesicle extracts to activate latent TGF-β1. Neither $1\alpha,25(OH)_2D_3$ nor $24R,25(OH)_2D_3$ had a direct effect on activation of small latent rhTGF-β1 by the extracts. However, when intact matrix vesicles were treated with $1\alpha,25(OH)_2D_3$, their ability to activate small latent rhTGF-β1 was increased. Inhibition of phospholipase A_2 with quinacrine blocked the $1\alpha,25(OH)_2D_3$-dependent effect. These results suggest that $1\alpha,25(OH)_2D_3$ had an effect on the matrix vesicle membrane and not on the enzymes in the matrix vesicles. Since matrix vesicles isolated from growth zone chondrocytes display increased phospholipase A_2 activity after treatment with $1\alpha,25(OH)_2D_3$, it is likely that this seco-steroid promotes loss of membrane integrity, resulting in the release of MMP-3 into the matrix, where latent TGF-β1 is stored. Taken together, the results show that matrix vesicles produced by growth plate chondrocytes contain MMP-3. Further, this enzyme appears to be at least partially responsible for activation of small latent TGF-β1 in the matrix, and suggest that $1\alpha,25(OH)_2D_3$ regulates MMP release from the matrix vesicles.

Introduction

During endochondral development, chondrocytes produce a complex extracellular matrix that undergoes extensive maturation and remodeling as the cells progress through their lineage cascade, culminating in calcification. While it has been accepted for some time that matrix vesicles are important in determining the rate and extent of calcium phosphate deposition by providing nucleation sites for initial crystal formation, their role as modulators of matrix maturation is less well understood. Many in vitro studies have shown that proteoglycan aggregates (aggrecan) are inhibitors of mineralization, [1] suggesting that these macromolecules must be removed for calcification to occur. [2-4] This can occur by breaking down the aggregates, proteolysis of the core protein, or glycosaminoglycan chain degradation. Hirschman et al. [5] found neutral proteases in matrix vesicles and suggested that they may participate in matrix maturation. Katsura and Yamada [6] have made similar observations using matrix vesicles from chick growth plates, and Einhorn et al. [7] have postulated that matrix vesicle-derived proteases may also be involved in fracture callus remodeling.

We have found that matrix vesicles are enriched in proteoglycan-degrading metalloproteinases and that the amount of proteinase activity incorporated into the matrix vesicles varies with the stage of cell maturation in the endochondral cascade. [8] Matrix vesicles produced by growth zone cells (prehypertrophic/upper hypertrophic zones) contain higher levels of both acid and neutral metalloproteinase activity than matrix vesicles produced by resting zone cells. Most of the metalloproteinase in the organelles is in active form, even though the tissue inhibitor of metalloproteinases (TIMP) is present. Plasminogen activator and β-glucuronidase are also present; however, matrix vesicles do not contain collagenase, lysozyme, plasmin, or hyaluronidase. [8] Subsequent studies showed that the two main metalloproteinases in matrix vesicles were 72 kDa gelatinase (MMP-2) and stromelysin-1 (MMP-3). [9]

The amount of metalloproteinase activity found in extracts of matrix vesicles is regulated by vitamin D metabolites in a cell maturation- and vitamin D metabolite-specific manner. $24R,25(OH)_2D_3$ has been shown to dose-dependently inhibit the amount of active and total proteoglycan-degrading metalloproteinase activity found in matrix vesicles, but not plasma membranes, from resting zone chondrocyte cultures. [10] In contrast, matrix vesicles from growth zone chondrocyte cultures treated with $1\alpha,25(OH)_2D_3$ were found to contain increased amounts of active and total proteoglycan-degrading metalloproteinase activity. In addition, plasminogen activator activity in these extracts was also affected by vitamin D metabolite treatment. Matrix vesicles from resting zone cell cultures treated with $24R,25(OH)_2D_3$ contained increased amounts of plasminogen activator, while those from growth zone cell cultures treated with $1\alpha,25(OH)_2D_3$ contained decreased amounts of this protease activity.

These observations suggest that the two metabolites play very different roles. Given that the effects of $1\alpha,25(OH)_2D_3$ and $24R,25(OH)_2D_3$ are cell-specific, this differential regulation of matrix vesicle enzyme activity is of interest. One possibility is that the composition of the matrix vesicles is regulated at the genomic level during organelle biogenesis. This appears to be the case for MMP-3, [11] which we have shown can be phosphorylated by PKC and display changes in activity. Another possibility is that the vitamin D metabolites interact directly with the matrix vesicle membrane once the organelles are resident in the extracellular matrix, leading to release of the enzymes at specific sites.

If this latter hypothesis is correct, then there must be a mechanism by which the cell can modulate matrix vesicle activity in the matrix. A number of recent studies have shown that growth plate chondrocytes produce and secrete both $1,25(OH)_2D_3$ and $24,25(OH)_2D_3$ in a regulated manner. [12-14] These hormones can act directly on the matrix vesicle membrane via specific membrane receptors to modulate activity of the organelle through nongenomic mechanisms (see Boyan et al. [15-17] for reviews).

The value of this kind of control to the cell is considerable. The cell is able to regulate whether the matrix vesicles contain more or less of certain types of enzymes and then regulate when they are released into the matrix. That this is important physiologically has been shown in a number of studies. When matrix vesicles are incubated in gelatin gels into which Ca^{++} and inorganic phosphate (Pi) ions are perfused, they exhibit only weak hydroxyapatite nucleation ability, but in gelatin gels containing aggrecan, matrix vesicles significantly enhance the rate of mineral formation by degrading the inhibitory proteoglycans. [18] Matrix vesicles also participate in the release and activation of growth factors that are produced by growth plate chondrocytes and stored in their extracellular matrix. This is the case for transforming growth factor beta-1 (TGF-β1).

Costochondral chondrocytes store TGF-β1 in the extracellular matrix as a three-component complex (large latent TGF-β1) composed of TGF-β1 homodimer (25 kDa), its latency associated peptide (LAP) (75 kDa) and latent TGF-β1 binding protein-1 (LTBP1) (190 kDa). [19,20] The TGF-β1 homodimer together with LAP are often referred to as small latent TGF-β1. In vivo, activation of latent TGF-β1 probably occurs predominately via enzymatic mechanisms. Earlier work has implicated plasmin, [21,22] cathepsin D, [21] and calpain [23] in this process. In addition to this list of candidate enzymes, thrombospondin, which is found in platelet alpha-granules and in the extracellular matrix, has also been shown to activate latent TGF-β1. [24-26]

The sensitivity of costochondral growth plate chondrocytes to TGF-β1 [27,28] suggests that activation of latent TGF-β1 in the matrix must be tightly regulated in time and space. Matrix vesicles are likely candidates because they contain proteinases, including matrix metalloproteinases (MMPs) such as stromelysin-1 (MMP-3) and 72 kDa gelatinase (MMP-2), and plasminogen activator. [8,9] When matrix vesicles isolated from growth zone chondrocyte cultures are treated directly with $1\alpha,25(OH)_2D_3$, [29] they are able to activate small latent TGF-β1, suggesting that they contain an enzyme or enzymes that can proteolytically cleave the LAP, releasing the active homodimer. This effect of $1\alpha,25(OH)_2D_3$ is cell maturation-dependent. Activation of small latent TGF-β1 by matrix vesicles treated with $1\alpha,25(OH)_2D_3$ is specific for this metabolite and for matrix vesicles from growth zone cell cultures; $24R,25(OH)_2D_3$ has no effect on matrix vesicles from either cell type. [29] Although exogenous plasmin has been shown to release and activate small latent TGF-β1, [30] when it is incubated with extracellular matrix produced by chondrocyte cultures, plasmin is not present in matrix vesicles, indicating that other enzymes are responsible for the ability of these organelles to activate the growth factor.

In the present paper, we review recent studies in our lab that examined the following hypotheses: (1) enzymes present in matrix vesicles can activate small latent TGF-β1; and (2) $1\alpha,25(OH)_2D_3$ regulates the activity of these enzymes in matrix vesicles from growth zone cell cultures, but not those from resting zone cell cultures. To test these hypotheses, the ability of matrix vesicle extracts to activate small latent TGF-β1 was determined as a function of dose and time. We also examined the ability of proteolytic enzymes known to be present in matrix vesicles to activate small latent TGF-β1. We assessed the in vivo relevance of our observations by determining whether specific antibodies to the proteinases could block the activation seen with the extracts. To assess the role of cell maturation, we used the rat costochondral growth plate chondrocyte culture model referred to above. Chondrocytes are derived from two distinct zones of rat costochondral cartilage, the resting zone and the growth zone, enabling us to determine the contribution of cell maturation state in the endochondral lineage to the role of matrix vesicles in management of the extracellular matrix.

Materials and Methods

Materials

All tissue culture reagents were purchased from Gibco (Grand Island, NY). Recombinant human latent TGF-β1 (rhTGF-β1) was purchased from R&D Systems, Inc. (Minneapolis, MN) and the ELISA kit for measuring active TGF-β1 was purchased from Promega (Madison, WI). Active recombinant human matrix metalloproteinase-2 (MMP-2) and sheep polyclonal anti-human MMP-3 antibody were purchased from Oncogene Research Products (Cambridge, MA) and the Binding Site (Birmingham, England), respectively. Recombinant human active MMP-3 was obtained from Dr. Scott Wilhelm of Bayer AG (West Haven, CT). Plasminogen from human serum was purchased from Boehringer Mannheim (Indianapolis, IN), and urokinase from human urine was purchased from Sigma Chemical Company (St. Louis, MO). All other reagents were of the highest quality available and purchased from Sigma.

Methods

Preparation of Matrix Vesicle Enzymes

Chondrocytes were isolated by enzymic digestion from the resting zone (RC; reserve zone) or growth zone (GC; prehypertrophic and upper hypertrophic zones) of the costochondral cartilage of 125-g male Sprague Dawley rats and cultured in Dulbecco's modified Eagle medium (DMEM) containing 10% fetal bovine serum (FBS), antibiotics, and ascorbic acid as described previously. [31] Matrix vesicles were isolated by differential centrifugation of the trypsin-digested matrix. Enzymes present in the matrix vesicles were extracted by a modification of a previously published method. [8,10] Matrix vesicles (20-30 T-75 flasks, yielding 0.5 to 1.0 mg matrix vesicle protein) were extracted by homogenization in 50 mM Tris buffer, pH 7.5, containing 2 M guanidine-HCl and 0.01 M $CaCl_2$. To maximize recovery of the matrix vesicle enzymes, the homogenate was centrifuged at 100,000 x g for 40 minutes at 4°C. The pellet was re-extracted using the same procedure, and the resulting supernatants were pooled, and then dialyzed against the assay buffer containing 0.05 M Tris, 0.01 M $CaCl_2$, 0.2 M NaCl, 0.0001% Brij 35, and 0.02% sodium azide. Protein content of the extract was determined by micro BCA. To account for any traces of extraction buffer that might not have dialyzed out, as well as for any Brij 35 that might not have dialyzed into the samples, control extracts were prepared by starting with a 1.0 mL aliquot of 0.9% NaCl (diluent for the matrix vesicle preparations) that was carried through the complete extraction and dialysis procedures.

Activation of Small Latent rhTGF-β1 by Matrix Vesicle Extracts

To measure the ability of the matrix vesicle enzymes to activate TGF-β1, matrix vesicle extracts were diluted with control extract and incubated for 3 hours at 37°C with DMEM containing a known quantity of small latent rhTGF-β1. Active TGF-β1 was measured by ELISA immediately after the incubation. Activation of small latent TGF-β1 was calculated by subtracting the amount of active TGF-β1 in control extracts from that in the matrix vesicle extracts and then dividing by the protein content in the matrix vesicle extracts. Previous studies showed that matrix vesicles produced by the chondrocytes contained no latent or active TGF-β1. [29] Therefore, any active TGF-β1 could only come from the substrate provided. Information provided by R&D Systems indicated that the substrate was 97% small latent rhTGF-β1 based on SDS-PAGE.

Effect of MMP Inhibition on Activation of Small Latent rhTGF-β1

Specific matrix vesicle enzymes were inhibited to determine their contribution to TGF-β1 activation. Initially, matrix vesicle extracts were incubated with the MMP inhibitor, a zinc ion chelator, 1,10-phenanthroline. MMP-2, MMP-3, and plasminogen activator are present in matrix vesicles, so we also studied the ability of these enzymes to activate latent TGF-β1. Latent TGF-β1 was incubated with active rhMMP-2, active rhMMP-3, urokinase, or plasminogen. Based on the results, we used anti-MMP-3 antibodies to determine if stromelysin-1 was the proteinase in the extracts responsible for activation of small latent TGF-β1. Matrix vesicle extracts were first pre-incubated with anti-human MMP-3 antibody and then the ability of the extracts to activate the latent growth factor was determined.

To test the hypothesis that $1\alpha,25(OH)_2D_3$ stimulates matrix vesicles to activate latent TGF-β1 by direct action on the organelles, we performed two different sets of experiments. The first set of experiments examined whether treatment of matrix vesicles with $1\alpha,25(OH)_2D_3$ altered the activity of enzymes in the matrix vesicle extracts. To do this, matrix vesicles isolated from growth zone chondrocytes were incubated directly with 10^{-8} M $1\alpha,25(OH)_2D_3$. Matrix vesicles from resting zone chondrocytes were incubated directly with vehicle alone or 10^{-7} M $24R,25(OH)_2D_3$ as a negative control. Enzyme extracts were prepared, pretreated with anti-MMP-3 antibodies, and examined for their ability to activate latent growth factor. The second set of experiments determined if the effect of $1\alpha,25(OH)_2D_3$ on the ability of matrix vesicles to activate latent TGF-β1 was on the membrane and mediated by phospholipase A_2. For these studies, matrix vesicles isolated from growth zone chondrocyte cultures were treated with $1\alpha,25(OH)_2D_3$ plus quinacrine. Quinacrine is a general inhibitor of phospholipase A_2 [32] and has been shown by us to block the membrane-mediated effects of $1\alpha,25(OH)_2D_3$ on growth zone chondrocytes. [33,34] In addition, as a control, small latent TGF-β1 was incubated directly with 10^{-8} M $1\alpha,25(OH)_2D_3$. Active TGF-β1 in the samples was measured by ELISA.

Results

The results of these experiments showed that extracts of matrix vesicles isolated from both growth zone and resting zone chondrocyte cultures dose-dependently activated small latent rhTGF-β1. When extracts were pre-incubated for 1.5 hours with 1,10-phenanthroline, an MMP inhibitor, activation of latent rhTGF-β1 was reduced by 23% (Table 1). Analysis of several enzymes known to be present in matrix vesicles showed that only rhMMP-3 caused an increase in activation of small latent rhTGF-β1 (Table 2). This effect was dose- and time-dependent. The percentage of the activated rhTGF-β1 against the total latent rhTGF-β1 was about 1% following a three-hour incubation. Anti-MMP-3 antibody reduced the ability of matrix vesicle extracts to activate small latent rhTGF-β1, whether the matrix vesicles were isolated from cultures of growth zone cells or resting zone cells (Table 3).

Table 1. Effect of the MMP inhibitor, 1,10-phenanthroline, on activation of small latent rhTGF-β1 by matrix vesicle extracts

Treatment	Active TGF-β1 (pg/μg extract protein)
Extract alone	19.1 ± 5.9
TGF-β1 alone	45.0 ± 9.8
Extract + TGF-β1	1,534.0 ± 139.0*
Extract + TGF-β1 + phe	1,190.0 ± 116.5*#

Resting zone matrix vesicle extracts were tested for their ability to activate small latent rhTGF-β1. Each value is the mean ± SEM of four extracts. *$P < 0.05$, vs. extract alone or TGF-β1 alone; #$p < 0.05$, vs. extract + TGF-β1. TGF-β1 = small latent rhTGF-β1; phe = 1,10-phenanthroline.

Treatment of matrix vesicles with either $1\alpha,25(OH)_2D_3$ or $24R,25(OH)_2D_3$ prior to extraction had no effect on the ability of the extracted enzymes to activate small latent rhTGF-β1 (data not shown). However, treatment of intact matrix vesicles with $1\alpha,25(OH)_2D_3$ enhanced the ability of the organelles to activate small latent rhTGF-β1 and this effect was dependent on phospholipase A_2 (Table 4).

Table 2: Activation of small latent rhTGF-β1 by MMPs

	Active TGF-β1 (pg/sample)
Control/vehicle	15.0 ± 1.1
15 ng rhMMP-2	18.9 ± 4.5
25 ng rhMMP-2	20.4 ± 2.4
35 ng rhMMP-2	14.4 ± 2.0
15 ng rhMMP-3	31.3 ± 3.4*
25 ng rhMMP-3	51.8 ± 3.1*#
35 ng rhMMP-3	61.5 ± 2.8*#

Small latent TGF-β1 was incubated for 3 hrs. at 37°C with 15, 25, or 35 ng active rhMMP-2 or rhMMP3. Active TGF-β1 was measured by ELISA. Each value is the mean ± SEM of six samples. *$P < 0.05$, vs. control/vehicle; #$P < 0.05$, vs. 15 ng rhMMP-3.

Table 3: Activation of small latent rhTGF-β1 by matrix vesicle extracts and inhibition by anti-MMP-3 antibody

	Active TGF-β1 (pg/μg extract protein)	**% Inhibition**
MV extract alone (GC cells)	77.6 ± 7.7	—
+ 0.5 μg anti-MMP-3	62.7 ± 4.3*	19%
+ 5.0 μg anti-MMP-3	44.9 ± 6.9*#	42%
+ 12.5 μg anti-MMP-3	26.5 ± 7.8*#	66%
+ 12.5 μg non-immune IgG	78.0 ± 10.6	—
MV extract alone (RC cells)	43.4 ± 6.8	—
+ 0.5 μg anti-MMP-3	33.4 ± 2.5*	23%
+ 5.0 μg anti-MMP-3	25.8 ± 3.2*#	41%
+ 12.5 μg anti-MMP-3	19.7 ± 3.2*#	55%
+ 12.5 μg non-immune IgG	40.0 ± 4.2	—

Matrix vesicle extracts were preincubated with varying concentrations of anti-MMP-3 antibody or non-immune sheep IgG for 1.5 hrs. at 37°C, followed by incubation with small latent rhTGF-β1 for 3 hrs. at 37°C. Active TGF-β1 was measured by ELISA. Each value is the mean ± SEM of six extracts. * $P < 0.05$, vs. extract not incubated with anti-MMP-3 antibody, # $P < 0.05$, vs. 0.5 μg of anti-MMP-3 antibody.

Discussion

These observations show that matrix vesicles produced by growth zone cells in culture can activate latent rhTGF-β1. [29] The results indicate that the matrix vesicle components responsible for the activation are present in guanidine-HCl extracts of the organelles and that metalloproteinases are involved, based on the dose-dependent inhibition of the activation with a specific antibody to MMP-3 and a general inhibition with 1,10-phenanthroline, an inhibitor of MMPs. The effect of extracts from matrix vesicles produced by growth zone cells was greater than the effect of extracts from matrix vesicles produced by resting zone cells. This is consistent with the fact that matrix vesicles from growth zone chondrocyte cultures are enriched with more neutral MMP activity than those from resting zone cells. [8]

Both MMP-2 and MMP-3 are present in matrix vesicles, [9] but when we examined the ability of these candidate enzymes to activate small latent TGF-β1, only MMP-3 was able to do so. The ratio of conversion from latent to active TGF-β1 by MMP-3 was approximately 1%, which is consistent with the ratio of conversion by plasmin in co-cultures of endothelial cells and smooth muscle cells, [35] and by calpain in cultures of endothelial cells. [23]

Our observations [10,30] indicate that the plasminogen activator-plasmin system is not responsible for the activation of small latent rhTGF-β1 by the matrix vesicle extracts, even though plasminogen activator is present in the organelles. Indeed, other reports have shown that latent TGF-β1 can be physiologically activated by mechanisms other than the plasminogen activator-plasmin system, such as cathepsin D, [21] thrombospondin, [24] calpain, [23] and furin. [36] Related studies in our lab indicate that neither thrombospondin nor the active thrombospondin peptide [25,26] activate growth plate chondrocyte extracellular matrix-associated latent TGF-β1 (unpublished data).

Table 4: Inhibition of phospholipase A_2 blocks the $1,25(OH)_2D_3$-dependent activation of small latent rhTGF-β1 by matrix vesicles

	Active TGF-β1 (pg/sample)
Control vehicle	0.07 ± 0.02
10^{-8} M $1,25(OH)_2D_3$ alone	0.04 ± 0.01
MVs alone	0.33 ± 0.02*
MVs + 10^{-8} M $1,25(OH)_2D_3$	0.47 ± 0.02*#
MVs + 10^{-8} M $1,25(OH)_2D_3$ + 10 μM quinacrine	0.21 ± 0.05*•

Matrix vesicles from growth zone cells were incubated with small latent rhTGF-β1 for 3 hrs. at 37°C in the presence or absence of 10^{-8} M $1\alpha,25(OH)_2D_3$ ± 10 μM quinacrine. Active TGF-β1 was measured by ELISA. Each value is the mean ± SEM of six matrix vesicle preparations. *$P < 0.05$, vs. latent TGF-β1 or $1,25(OH)_2D_3$ + latent TGF-β1; #$P < 0.05$, vs. matrix vesicle alone; •$P < 0.05$, vs. matrix vesicle + $1,25(OH)_2D_3$.

These results also support the hypothesis that the role of $1\alpha,25(OH)_2D_3$ in the activation mechanism is to promote release of matrix vesicle enzymes, and not to specifically alter the activity of the latent TGF-β1-activating enzyme(s). Pretreatment of matrix vesicles with $1\alpha,25(OH)_2D_3$ had no effect on the ability of the extracts to activate small latent rhTGF-β1, whereas $1\alpha,25(OH)_2D_3$ directly causes matrix vesicles to activate latent TGF-β1. [29] The hypothesis that $1\alpha,25(OH)_2D_3$ stimulates the release of active MMP-3 from matrix vesicles relies upon the assumption that $1\alpha,25(OH)_2D_3$ causes a loss of membrane integrity. The effect of $1\alpha,25(OH)_2D_3$ on the ability of matrix vesicles to activate small latent TGF-β1 was blocked by the inhibition of phospholipase A_2, suggesting that formation of lysophospholipids via the action of phospholipase A_2 might result in destabilization of the phospholipid bilayer. [37]

Pretreatment of matrix vesicles produced by resting zone cells with $24R,25(OH)_2D_3$ had no effect on the ability of the extracted enzymes to activate latent rhTGF-β. This metabolite was shown previously not to stimulate matrix vesicles to activate latent rhTGF-β1. [29] The results described here suggest that this may be due to the inhibition of matrix vesicle phospholipase A_2 by $24R,25(OH)_2D_3$, [38] thereby preventing release of the enzyme(s) capable of activating the latent growth factor.

It should be noted that the integrity of the matrix vesicle membrane may be a critical determinant in the regulation of TGF-β1 activation in the cartilage extracellular matrix. Matrix vesicles produced by resting zone cells contain proteolytic enzymes that are fully functional, [8] and when extracted from the organelle, are capable of activating latent TGF-β1. The intact organelle lacks this ability. [8] Since $1\alpha,25(OH)_2D_3$ has no effect on phospholipase A_2 activity in matrix vesicles produced by resting zone cells [39] and $24R,25(OH)_2D_3$ inhibits the activity of this enzyme, [38,39] the integrity of the membrane is retained in vivo. In the growth zone, however, $1\alpha,25(OH)_2D_3$-dependent increases in matrix vesicle phospholipase A_2 may lead to release of MMP-3 and increased activation of TGF-β1. Active TGF-β1 can then act on the cells to stimulate alkaline phosphatase activity [27] and further production of vitamin D metabolites, [12,13] thereby modulating the rate of growth plate maturation. Interestingly, TGF-β1 downregulates phospholipase A2 activity [27] and subsequent calcification, [40] providing a potential feedback mechanism.

In summary, our observations show that matrix vesicle MMP-3 is responsible for some, if not all, of the ability of matrix vesicles to activate small latent TGF-β1. This is consistent with a multi-step activation of matrix-associated latent TGF-β1 in cartilage. The work described here did not address the mechanisms involved in presenting latent TGF-β1 to the activating enzymes in the context of the complex, three-dimensional extracellular matrix in vivo. Other enzymes, like plasmin, may participate in the release of the 100 kD homodimer from the large latent TGF-β1 complex. [30] $1\alpha,25(OH)_2D_3$ contributes to the mechanism by altering the integrity of the matrix vesicle membrane, releasing active MMP-3 into the matrix where it can act on the small latent TGF-β1.

References

[1] Chen CC, Boskey AL 1985 Mechanisms of proteoglycan inhibition of hydroxyapatite growth. Calcif Tissue Int **37**:395-400.

[2] Cuervo LA, Pita JC, Howell DS 1973 Inhibition of calcium phosphate mineral growth by proteoglycan aggregate fractions in a synthetic lymph. Calcif Tissue Res **13**:1-10.

[3] Buckwalter JA, Rosenberg LC, Ungar R 1987 Changes in proteoglycan aggregates during cartilage mineralization. Calcif Tissue Int **41**:228-236.

[4] Kawabe N, Ehrlich MG, Mankin HJ 1986 In vivo degradation systems of the epiphyseal cartilage. Clin Orthop Rel Res **211**:244-251.

[5] Hirschman A, Deutsch D, Hirschman M, Bab IA, Sela J, Muhlrad A 1983 Neutral peptidase activities in matrix vesicles from bovine fetal alveolar bone and dog osteoscarcoma. Calcif Tissue Int **35**:791-797.

[6] Katsura N, Yamada K 1986 Isolation and characterization of a metalloprotease associated with chick epiphyseal cartilage matrix vesicles. Bone **7**:137-143.

[7] Einhorn TA, Hirschman A, Kaplan C, Nashed R, Devlin VJ, Warman J 1989 Neutral protein-degrading enzymes in experimental fracture callus: a preliminary report. J Orthop Res **7**:792-805.

[8] Dean DD, Schwartz Z, Muniz OE, Gomez R, Swain LD, Howell DS, Boyan BD 1992 Matrix vesicles are enriched in metalloproteinases that degrade proteoglycans. Calcif Tissue Int **50**:342-349.

[9] Schmitz JP, Dean DD, Schwartz Z, Cochran DL, Grant GM, Klebe RJ, Nakaya H, Boyan BD 1996 Chondrocyte cultures express matrix metalloproteinase mRNA and immunoreactive protein: Stromelysin-1 and 72kDa gelatinase are localized in extracellular matrix vesicles. J Cell Biochem **61**:375-391.

[10] Dean DD, Boyan BD, Muniz OE, Howell DS, Schwartz Z 1996 Vitamin D metabolites regulate matrix vesicle metalloproteinase content in a cell maturation-dependent manner. Calcif Tissue Int **59**:109-116.

[11] Schmitz JP, Schwartz Z, Sylvia VL, Dean DD, Calderon F, Boyan BD 1996 Vitamin D_3 regulation of stromelysin-1 (MMP-3) in chondrocyte cultures is mediated by protein kinase C. J Cell Physiol **168**:570-579.

[12] Schwartz Z, Brooks BP, Swain LD, Del Toro F, Norman AW, Boyan BD 1992 Production of 1,25-dihydroxyvitamin D_3 and 24,25-dihydroxyvitamin D_3 by growth zone and resting zone chondrocytes is dependent on cell maturation and is regulated by hormones and growth factors. Endocrinology **130**:2495-2504.

[13] Pedrozo HA, Boyan BD, Mazock J, Dean DD, Gomez R, Schwartz Z 1999 TGF-β1 regulates 25-hydroxyvitamin D_3 1α- and 24R-hydroxylase activity in cultured growth plate chondrocytes in a maturation-dependent manner. Calcif Tissue Int **64:**50-56.
[14] Schwartz Z, Pedrozo HA, Sylvia VL, Gomez R, Dean DD, Boyan BD 2001 1α,25-$(OH)_2D_3$ regulates 25-hydroxyvitamin D_3 24R-hydroxylase activity in growth zone costochondral growth plate chondrocytes via protein kinase C. Calcif Tissue Int, in press.
[15] Boyan BD, Dean DD, Sylvia VL, Schwartz Z 1997 Cartilage and vitamin D: Genomic and nongenomic regulation by 1,25-$(OH)_2D_3$ and 24,25-$(OH)_2D_3$. In: Feldman D, Glorieux FH, Pike JW (eds) Vitamin D, Academic Press, San Diego, CA, pp. 395-421.
[16] Boyan BD, Sylvia VL, Dean DD, Del Toro F, Schwartz Z 2001 Differential regulation of growth plate chondrocytes by 1α,25-$(OH)_2D_3$ and 24R,25-$(OH)_2D_3$ involves cell maturation specific membrane receptor activated phospholipid metabolism. Crit Rev Oral Biol Med, in press.
[17] Boyan BD, Bonewald LF, Sylvia VL, Nemere I, Larsson D, Norman AW, Rosser J, Dean DD, Schwartz Z 2001 Evidence for distinct membrane receptors for 1α,25-$(OH)_2D_3$ and 24R,25-$(OH)_2D_3$ in osteoblasts. Steroids, in press.
[18] Boskey AL, Boyan BD, Schwartz Z 1997 Matrix vesicles promote mineralization in a gelatin gel. Calcif Tissue Int **60:** 309-315.
[19] Pedrozo HA, Schwartz Z, Gomez R, Ornoy A, Xin-Sheng W, Dallas SL, Bonewald LF, Dean DD, Boyan BD 1998 Growth plate chondrocytes store latent TGF-β1 in their matrix through latent TGFβ binding protein-1. J Cell Physiol **177:**343-354.
[20] Bonewald LF, Wakefield LM, Oreffo RO, Escobedo A, Twardzik DR, Mundy GR 1991 Latent forms of transforming growth factor-β (TGFβ) derived from bone cultures: Identification of a naturally occurring 100 kDa complex with similarity to recombinant latent TGFβ. Molec Endocrinol **5:**741-751.
[21] Lyons RM, Keski-Oja J, Moses HL 1988 Proteolytic activation of latent transforming growth factor-β from fibroblast-conditioned media. J Cell Biol **106:**1659-1665.
[22] Sato Y, Rifkin DB 1989 Inhibition of endothelial cell movement by pericytes and smooth muscle cells: Activation of a latent transforming growth factor-β1-like molecule by plasmin during co-culture. J Cell Biol **109:**309-315.
[23] Flaumenhaft R, Kojima S, Abe M, Rifkin DB 1993 Activation of latent transforming growth factor β. Advances in Pharmacology, 24. Academic Press, Inc., pp. 51-77.
[24] Schultz-Cherry S, Ribeiro S, Gentry L, Murphy-Ullrich JE 1994 Thrombospondin binds and activates the small and large forms of latent transforming growth factor-β in a chemically defined system. J Biol Chem **269:**26775-26782.
[25] Ribeiro SM, Poczatek M, Schultz-Cherry S, Villain M, Murphy-Ullrich JE 1999 The activation sequence of thrombospondin-1 interacts with the latency-associated peptide to regulate activation of latent transforming growth factor-beta. J Biol Chem **274:**13586-13593.
[26] Murphy-Ullrich JE, Poczatek M 2000 Activation of latent TGF-β by thrombospondin-1: mechanisms and physiology. Cytokine Growth Fac Rev **11:**59-69.
[27] Schwartz Z, Bonewald LF, Caulfield K, Brooks BP, Boyan BD 1993 Direct effects of transforming growth factor-β on chondrocytes are modulated by vitamin D metabolites in a cell maturation-specific manner. Endocrinology **132:**1544-1552.
[28] Schwartz Z, Sylvia VL, Liu Y, Dean DD, Boyan BD 1998 Treatment of resting zone chondrocytes with transforming growth factor-β1 induces differentiation into a phenotype characteristic of growth zone chondrocytes by downregulating responsiveness to 24,25-$(OH)_2D_3$ and upregulating responsiveness to 1,25-$(OH)_2D_3$. Bone **23:**465-470.
[29] Boyan BD, Schwartz Z, Park-Snyder S, Dean DD, Yang F, Twardzik D, Bonewald LF 1994 Latent transforming growth factor-β is produced by chondrocytes and activated by extracellular matrix vesicles upon exposure to 1,25-$(OH)_2D_3$. J Biol Chem **269:**28374-28381.
[30] Pedrozo HA, Schwartz Z, Robinson M, Gomez R, Dean DD, Bonewald LF, Boyan BD 1999 Potential mechanisms for the plasmin mediated release and activation of latent TGF-β1 from the extracellular matrix of growth plate chondrocytes. Endocrinology **140:**5806-5816.
[31] Boyan BD, Schwartz Z, Swain LD, Carnes DL, Jr., Zislis T 1988 Differential expression of phenotype by resting zone and growth region costochondral chondrocytes *in vitro*. Bone **9:**185-194.
[32] Church D, Braconi S, Vallotton M, Lang U 1993 Protein kinase C-mediated phospholipase A_2 activation, platelet-activating factor generation and prostacyclin release in spontaneously beating rat cardiomyocytes. Biochem J **290:**477-482.
[33] Boyan BD, Sylvia VL, Curry D, Chang Z, Dean DD, Schwartz Z 1998 Arachidonic acid is an autocoid mediator of the differential action of 1,25-$(OH)_2D_3$ and 24,25-$(OH)_2D_3$ on growth plate chondrocytes. J Cell Physiol **176:**516-524.

[34] Sylvia VL, Schwartz Z, Curry DB, Chang Z, Dean DD, Boyan BD 1998 1,25-$(OH)_2D_3$ regulates protein kinase C activity through two phospholipid-independent pathways involving phospholipase A_2 and phospholipase C in growth zone chondrocytes. J Bone Miner Res **13:**559-569.

[35] Sato Y, Tsuboi R, Lyons R, Moses H, Rifkin DB 1990 Characterization of the activation of latent TGF-β by co-cultures of endothelial cells and pericytes or smooth muscle cells: A self-regulating system. J Cell Biol **111:**757-763.

[36] Dubois CM, Laprise MH, Blanchette F, Gentry LE, Leduc R 1995 Processing of transforming growth factor beta 1 precursor by human furin convertase. J Biol Chem **270:**10618-10624.

[37] Ginsburg I, Misgav R, Pinson A, Varani J, Ward PA, Kohen R 1992 Synergism among oxidants, proteinases, phospholipases, microbial hemolysins, cationic proteins, and cytokines. Inflammation **16:**519-538.

[38] Schwartz Z, Schlader DL, Swain LD, Boyan BD 1988 Direct effects of 1,25-dihydroxyvitamin D_3 and 24,25- dihydroxyvitamin D_3 on growth zone and resting zone chondrocyte membrane alkaline phosphatase and phospholipase-A_2 specific activities. Endocrinology **123:**2878-2884.

[39] Schwartz Z, Boyan BD 1988 The effects of vitamin D metabolites on phospholipase A_2 activity of growth zone and resting zone cartilage cells *in vitro*. Endocrinology **122:**2191-2198.

[40] Lian JB, Stein GS 1993 The developmental stages of osteoblast growth and differentiation exhibit selective responses of genes to growth factors (TGFβ1) and hormones (vitamin D and glucocorticoids). J Oral Implantol **19:**95-105.

The Growth Plate
I.M. Shapiro et al. (Eds.)
IOS Press, 2002

Mechanisms that Regulate Normal Bone Mineral Deposition: A Hypothesis on the Role of Antagonistic Pathways in Preventing Hypo- and Hyper-Mineralization

Lovisa Hessle[1], Sonoko Narisawa[1], Arata Iwasaki[1], Kristen Johnson[2], Robert Terkeltaub[2], and José Luis Millán[1]

[1]*The Burnham Institute, 10901 North Torrey Pines Road, La Jolla, CA 92037, USA and* [2]*VA Medical Center, UCSD, La Jolla, CA 92037*

Abstract. Three molecules present in osteoblasts that affect PPi and Pi metabolism have been identified, by means of gene knock-out models, as affecting the controlled deposition of bone mineral, i.e., alkaline phosphatase (TNAP); PC-1 (or *Npps*, a nucleoside triphosphate pyrophosphate hydrolase isozyme, NTPPPH) and the *ank* gene product (a PPi transporter). Mice deficient in TNAP mimic the most severe forms of hypophosphatasia, i.e., perinatal and infantile hypophosphatasia. The TNAP-/- mice are growth impaired, develop epileptic seizures, apnea, and die before weaning. Skeletal preparations clearly show poor mineralization in the parietal bones, scapulae, vertebral bones, and ribs at approximately 8-10 days of age. PC-1 knockout mice develop progressive ossification of spinal and peripheral joint ligaments and also articular and meniscal cartilage calcification. A remarkably similar hypermineralizing phenotype has been found in *ank/ank* mice that lack a normal functioning ANK molecule. Matrix vesicles derived from primary osteoblasts from TNAP-/- hypophosphatasia mice contain increased levels of inorganic pyrophosphate (PPi), a known inhibitor of mineralization. PPi is produced by NTPPPH activity (due preferentially to the action of PC-1 in MVs) and is exported outside the cell by the action of the ANK protein. Our work centers on testing the hypotheses that TNAP's key function in bone is degradation of PPi to remove this mineralization inhibitor and provide free phosphate for apatite deposition. We further hypothesize that PC-1 and ANK are direct antagonist of TNAP-dependent matrix calcification. We are currently testing whether loss of function of PC-1 or ANK will ameliorate TNAP deficiency-associated osteomalacia *in vivo*.

Introduction

Osteoblasts mineralize the pericellular matrix by promoting initial formation of crystalline hydroxyapatite. The sheltered interior of membrane-limited matrix vesicles (MVs) shed from osteoblasts and sequestered zones between collagen fibrils are believed to be the principal sites where osteoblast-mediated matrix calcification is initiated. Osteoblasts further promote propagation of apatite outside of the calcification initiation sites by modulating matrix composition. Among the molecules present in osteoblasts that have been identified as affecting the controlled deposition of bone mineral are species that affect PPi and Pi metabolism, i.e., alkaline phosphatase (TNAP); PC-1 (or *Npps*, a nucleoside triphosphate pyrophosphate hydrolase isozyme, NTPPPH) and the *ank* gene product (a PPi transporter). A deficiency in the TNAP isozyme causes hypophosphatasia and the study of this disease has provided the best evidence of the importance of TNAP for bone mineralization. TNAP is the only tissue-nonrestricted isozyme of a family of four homologous human AP genes (EC. 3.1.3.1) [1]. Expressed as an ecto-enzyme transported to the osteoblast plasma membrane and anchored via a phosphatidylinositol glycan moiety, TNAP has been demonstrated to play an essential physiological role during osteoblastic bone matrix mineralization [2,3,4]. Specifically, defective bone mineralization (osteomalacia) occurs in TNAP deficiency (hypophosphatasia) [2]. The severity of hypophosphatasia is variable and modulated by the nature of the TNAP mutation [5,6,7,3,4]. The different syndromes are: perinatal hypophosphatasia, infantile hypophosphatasia, childhood hypophosphatasia, adult hypophosphatasia, odontohypophosphatasia and pseudohypophosphatasia [8]. These phenotypes range from complete absence of bone mineralization and stillbirth to spontaneous fractures and loss of decidual teeth in adult life. Unlike most types of rickets or osteomalacia neither calcium nor inorganic phosphate levels in serum are subnormal in hypophosphatasia and and hypercalciuria is common in infantile hypophosphatasia [8].

Physiologic bone matrix mineralization is hypothesized to be dependent on the availability of Pi released from a variety of substrates by certain MV ecto-enzymes [9, 10]. For example, ATP is hypothesized to drive the initiation of calcification by MVs *in vivo*, and a specific bone and cartilage ATPase appears to be responsible for the ATP-dependent calcium and Pi-depositing activity of bone and cartilage-derived MVs *in vitro* [10, 11]. Skeletal TNAP can catalyze Pi release from ATP [10, 11], TNAP catalyzes several transphosphorylation reactions [2, 12] and TNAP can also function as a pyrophosphatase [13, 14]. Though TNAP does not appear to dephosphorylate membrane proteins [15], TNAP has been hypothesized to modulate bridging of MVs to matrix collagen [2, 16]. TNAP has been demonstrated to bind calcium [17]. Moreover, TNAP degrades at least 3 phosphocompounds, i.e., phosphoethanolamine, pyridoxal 5' phosphate, and PPi, that accumulate endogenously in hypophosphatasia [18]. The central function or functions of TNAP in conditioning mineralization have not been completely defined [2]. Importantly, aberrant localization of TNAP can occur, including defective transport of TNAP to the plasma membrane associated with hypophosphatasia [12, 6]. This has suggested that TNAP might act at the level of plasma membrane-derived structures such as MVs. In subjects with perinatal hypophosphatasia MVs were present in approximately normal numbers and distribution, and these MVs contained internal hydroxyapatite crystals [19]. However, propagation of hydroxyapatite crystals outside of isolated MVs was impaired but by an undefined mechanism [19].

The ability of TNAP to hydrolyze PPi to Pi [18] has been hypothesized to be central to the role of TNAP in promoting osteoblastic mineralization [19, 2, 11]. A major action of PPi is to suppress both the deposition and propagation of hydroxyapatite crystals *in vitro*

[20, 21]. Thus, critically timed removal or exclusion of PPi at sites of mineralization appears to be necessary for active crystal deposition to proceed [20, 21]. Since TNAP functions as a PPi-ase *in vitro* [13, 14], the finding that NTPPPH activity is normal in fibroblasts from hypophosphatasia patients further supported the hypothesis that accumulation of PPi in this disease is the result of defective degradation [22]. *In vitro* studies have shown that PPi promotes formation of amorphous calcium phosphate, while the subsequent transformation into hydroxyapatite and growth of hydroxyapatite crystals are inhibited [23]. The physiological importance of the former finding is however not certain, since it is unclear if amorphous calcium phosphate represents a hydroxyapatite precursor *in vivo*. In any case, PPi has been hypothesized to act as a physiological regulator of mineralization, where TNAP's specific role might be to hydrolyze this presumed calcification inhibitor.

Plasma Cell Membrane Glycoprotein-1 (PC-1) is a nucleotide triphosphate pyrophosphate hydrolase (NTPPPH) isozyme expressed by cultured osteoblastic cells [24, 20, 25]. NTPPPH activity is a property of several members of a phosphodiesterase nucleotide pyrophosphatase (PDNP) family of ecto-enzymes that also includes B10 and autotaxin [20]. Similar to the case for TNAP and other ecto-enzymes, PC-1 expression, and the extent of PC-1 distribution to MVs, are regulated by certain growth factors and calciotropic hormones, including TGFβ, bFGF, and 1,25 dihydroxyvitamin D3 [26, 20, 27, 25]. Osteoblast-derived MV PC-1 appears to function directly to increase MV fraction PPi content and to restrain mineralization by isolated MVs *in vitro* [20]. In this regard, a 2-4 fold increase in osteoblast PC-1 expression decreases, by > 80%, the amount of hydroxyapatite deposited in the pericellular matrix of osteoblasts *in vitro* [20]. On the other hand a dysregulated increase in chondrocyte PPi production is a central feature of idiopathic chondrocalcinosis (or primary calcium pyrophosphate dihydrate, CPPD, crystal deposition disease) whose prevalence appears to be greater than 15% at age 65 and rises progressively with age. Mean cartilage PPi-generating NTPPPH activity doubles, promoting PPi supersaturation that stimulates CPPD crystal deposition in the pericellular matrix of chondrocytes in articular cartilage and fibrocartilaginous menisci [28]. Focal up-regulation of PC-1 expression appears to be intimately linked to cartilage calcification in this disease [28].

Interestingly, it appears that both up-regulation as well as inactivation of PC-1 leads to osteoarthitic disease albeit by different molecular mechanisms. The role of PC-1 on mineralization have been confirmed to be physiologically significant in ttw/ttw (formerly known as "tiptoe walking Yoshimura") mice [29], which are homozygous for a naturally occurring PC-1 truncation mutation. In early life, ttw/ttw mice develop not only progressive ossification of spinal and peripheral joint ligaments but also articular and meniscal cartilage calcification. A remarkably similar hypermineralizing phenotype has been characterized in *ank/ank* mice that lack expression of ANK, a trans-membrane protein that appears to serve as plasma membrane PPi channel and is needed to maintain physiologic extracellular PPi concentrations [30].

Results and Discussion

Phenotypic Abnormalities in Alkaline Phosphatase-Deficient Mice In Vivo

We have generated and characterized [3, 31] mice deficient in bone alkaline phosphatase (TNAP). These mice mimic the most severe forms of hypophosphatasia, i.e., perinatal and infantile hypophosphatasia. The TNAP-/- mice are growth impaired, develop epileptic seizures secondary to vitamin B6 deficiency [32, 33], apnea, and die before weaning with evidence of cranial and pulmonary hemorrhages. Examination of the tissues indicate abnormal bone mineralization, morphological changes in the osteoblasts, aberrant development of the lumbar nerve roots, disturbances in intestinal physiology, increased apoptosis in the thymus and abnormal spleen. As in human patients, there is a striking elevation of pyrophosphate (PPi) in the urine of the TNAP knock-out mice, elevated levels of urinary phosphoethanolamine and a striking accumulation of plasma pyridoxal-5'-phosphate. Skeletal preparations of embryos and newborns revealed no differences between the TNAP+/+, TNAP+/- and TNAP-/- mice. However, the staining of 8-day old TNAP-/- bones clearly showed poor mineralization in the parietal bones, scapulae, vertebral bones, and ribs. Evidence of spontaneous fractures was evident in the fibulae. Fractures in the rib bones and broken incisors were also observed. The bone abnormalities worsen progressively with age.

Bone Nodule Formation In Vitro

To evaluate the ability of primary osteoblasts to form and mineralize bone nodules *in vitro*, wild-type (wt), heterozygous and knock-out (ko) osteoblasts were cultured in media supplemented with ascorbic acid, using ß-glycerophosphate as phosphate source [33]. At different time points (day 4, 6 and 8), cultures were fixed and stained with the von Kossa procedure to visualize mineralized nodules. Staining of post-confluent cultures of TNAP ko osteoblasts showed that these cells were able to form cellular nodules, typical of long-term calvarial osteoblast cultures. However, in contrast to cultures of TNAP positive osteoblasts, mineralization by TNAP-/- osteoblasts was never initiated. Similar results were found using ATP as phosphate source. Calcium measurements further confirmed the lack of mineral deposition in these TNAP ko cultures. The amounts and sizes of the nodular structures did not vary significantly between the different genotypes, only mineralization of the nodules appeared to be affected by the lack of TNAP. Of particular interest is the finding that initiation of mineralization was delayed in the TNAP heterozygous osteoblast cultures compared to wt osteoblasts. This was correlated with a delayed increase in the levels of TNAP activity in TNAP+/- in comparison with TNAP+/+ osteoblasts. The extent of bone mineral deposition in these cultures was confirmed by quantifications of deposited calcium. These results were compatible with the von Kossa stainings, showing clear phenotypic differences between wt, heterozygous and ko osteoblasts concerning bone nodule mineralization. Mineralization of ko osteoblast cultures could be restored by exposure to conditioned media from wt osteoblast cultures. This mineralization was induced both by untreated conditioned media and by media that had been ultracentrifuged to exclude matrix vesicles. Mineralization was also induced by adding purified soluble recombinant human TNAP enzyme to the ko osteoblast culture medium. As in control cultures, the deposition of mineral in these cultures was restricted to bone nodules. In contrast, neither heat-inactivated recombinant TNAP, nor enzymatically inactive mutants of TNAP, such as [R54C]TNAP or [V365I]TNAP, were able to induce mineralization [33]. These data suggest that a certain level of TNAP activity has to be reached for calcium deposition to be initiated. Our data suggest that even a moderate reduction in the levels of expression of AP protein and enzyme activity can be sufficient to impair the mineralization process and cause dominant phenotypic abnormalities in some cases of hypophosphatasia.

Pathologic Calcifications in PC-1-Deficient Mice and Humans and in Mice Expressing Truncated ANK

In long bones of wild-type mice, PC-1 is expressed in osteoblasts, osteocytes, chondrocytes in articular hyaline and meniscal cartilages, and in periarticular and intra-articular ligaments. In growth plates, PC-1 is best detected in epiphyseal regions in late hypertrophic chondrocytes in the calcifying zone, a region in which "trans-differentiation" to osteoblasts may occur. PC-1 also is strongly expressed at entheses (e.g. sites of insertion of intra-articular ligaments, and the junction of synovial membrane with periosteum). In addition, perispinal ligaments are markedly calcified with amorphous calcium phosphate, the mineral phase seen during active bone formation. Calcification is particularly intense around intervertebral disks, where there is an unrestrained regenerative osteoblastic hyperplasia of the periosteum. The PC-1 knockout mice demonstrated abnormal development of cartilage and bone at sites where PC-1 is normally distributed [29]. There is extension of endochondral growth plates, and progressive ossific fusion of synovium and the lateral edges of growth plates. There is calcification of fibrocartilages, knee cruciate ligaments, and the Achilles tendon. Thus, PC-1 expression by osteoblasts, chondrocytes and ligament fibroblasts modulates skeletal cell differentiation and mineralization. Mice deficient in PC-1 develop both periarticular and arterial apatite calcification in early life. The PC-1 null mice have proven to be a useful model for a subset of human Idiopathic Infantile Arterial Calcification (IIAC), in which there is hydroxyapatite deposition with concomitant stenosing smooth muscle cell proliferation in large arteries by early infancy, and dense periarticular calcifications of wrists and ankles associated with deficiency of PC-1 and extracellular PPi [35]. Paradoxically, even though the PC-1 null mice hyperossify they are osteopenic in association with high turnover osteoporosis. Here, the disorganized trabecular architecture may result from widespread hypercalcification of the matrix of the marrow and the disorganized bone architecture contributes to the osteopenia in trabecular bone. But altered bone mineral resorption also may play a role.

The *ank/ank* mutant mice develop phenotypic abnormalities remarkably similar in timing, localization and extent to those manifested by the PC-1 null mice [29]. Using primers designed from the recently cloned murine ank gene we were able to detect ank mRNA in primary osteoblasts and in osteoblastic cell lines.

Activities of TNAP and NTPPPHs in Matrix Vesicles

In previous studies the majority of both NTPPPH activity [36] and TNAP activity [37] have been found to be active on the external face of MVs. Thus, we assessed the relationship between TNAP and PC functions in MV fractions [38]. Cell-associated NTPPPH decreased over time in culture in osteoblasts from TNAP-/- mice, relative to cells from TNAP+/- and TNAP+/+ mice, and was significantly less in TNAP-/- cells than in TNAP+/+ cells. Despite the presence of the lowest MV fraction NTPPPH specific activity, it was in the TNAP-/- state that the highest MV fraction-associated concentration of the mineralization inhibitor PPi was observed. We then transfected MC3T3-E1 cells with cDNAs encoding wild-type TNAP and a catalytically inactive mutant of TNAP, i.e. [R54C]TNAP. Transfection of wild-type TNAP significantly elevated cell-associated and MV fraction-associated TNAP activity in MC3T3-E1 cells and decreased the PPi associated with MV fractions derived from MC3T3-E1 cells transfected with PC-1. Paradoxically, transfection of wild type TNAP was associated with a significant increase in cell-associated NTPPPH, and an even greater increase in MV fraction-associated NTPPPH activity in MC3T3 cells. Because TNAP appeared to regulate the NTPPPH activity of osteoblastic MC3T3-E1 cells, we assessed if this effect was dependent on TNAP enzymatic activity. Wild type TNAP, but not the enzyme-inactive mutant of TNAP induced an increase in both MV fraction NTPPPH and AP activity. In contrast to wild type TNAP, the mutant TNAP failed to decrease MV fraction PPi [38].

A Hypothetical Model of the Antagonistic Action of TNAP, PC-1 and ANK in Controlling Matrix Calcification

The results summarized above indicate that inactivation of TNAP results in an increased accumulation of PPi inside the matrix vesicles and in the extracellular matrix surrounding the matrix vesicles. Furthermore we demonstrated that transfection of TNAP into osteoblastic cell lines decreases the basal levels of PPi in the matrix vesicles. This observation is consistent with the hypothesis that the specific and dominant function of TNAP in bone is to degrade PPi (a potent inhibitor of mineralization) while concomitantly producing free inorganic phosphate (Pi) to promote hydroxyapatite deposition. While this central hypothesis is not novel, the availability of the TNAP knockout mice, as well as mice deficient in molecules that appear to converge on this pathway of regulation of PPi levels, makes it possible for us to test this central hypothesis and clarify its basic mechanism.

Our initial analysis of the PC-1 knockout mice indicates that the absence of PC-1 leads to a decreased concentration of intracellular and extracellular PPi levels as well as a decrease in PPi inside the matrix vesicles. PPi inhibits crystallization of calcium phosphate from solution, slows the transformation of amorphous calcium phosphate to it's crystalline form and slows the aggregation of seed crystals into larger clusters [21]. Thus we hypothesize that a central function of PC-1 may be to maintain a high enough level of PPi inside the matrix vesicles to help regulate the rate of intramembranous formation of apatite crystals and to, thereby, control the first phase of crystal formation in the matrix vesicles.

The phenotypic abnormalities of the *ank/ank* mutant mice are surprisingly similar to those of the PC-1 knockout mice. They show a generalized progressive form of arthritis, progressive ankylosis, accompanied by increased mineral deposition, bony outgrowths and joint destruction. However while the phenotypic abnormalities appear to overlap, the mechanism may be different. We hypothesize that the ANK molecule causes progressive ankylosis by interfering with the extravesicular step, or phase II, of bone mineral deposition. The ANK protein has been shown to be a transmembrane protein that most likely functions as a component of a PPi transporter, shuttling PPi from inside the cell to the outside extracellular fluid. Notably in the ANK-deficient mice intracellular PPi levels are increased to twice the normal levels while extracellular PPi levels are reduced three to five fold. Thus a normal function of the ANK protein may be to transport PPi to the outside of the cell to be able to regulate the rate of bone mineral deposition in the extracellular fluid affecting the second phase of mineralization.

It appears clear from the description of the phenotypic abnormalities of the TNAP null, PC-1 null and ANK mutant (*ank/ank*) mice that the function of these three molecules converge on a pathway regulating intracellular and extracellular PPi levels. Our experiments have showen that introduction of TNAP cDNA into an osteoblastic cell line induces NTPPPH activity due to specific induction of PC-1 and also that the reduction of TNAP activity in the TNAP knockout osteoblasts is followed by a reduction in PC-1 levels. Furthermore manipulations that elevate extracellular PPi induce TNAP activity in normal fibroblasts. These data suggest the hypothesis that both PC-1 and ANK either directly or indirectly regulate TNAP expression and that TNAP is also able to regulate PC-1 expression and ANK expression. How is this achieved? Our data have not suggested a physical interaction between TNAP and PC-1 [38]. Thus, we hypothesize that TNAP

Figure 1. A model of the concerted action of TNAP, PC-1 and ANK to regulate extracellular steady-state levels of PPi which in turn control the amount of hydroxyapatite being deposited in the matrix. The production of extracellular PPi is proposed to be mainly contributed by the action of the NTPPPH activity of PC-1 using ATP as substrate. Additionally the ANK molecule transports intracellular PPi to the outside of the cell and thus contributes to this PPi pool. TNAP is needed to break down PPi and thus help establish a proper steady-state concentration of this mineralization inhibitor. In so doing, TNAP also contributes Pi to the extracellular Pi pool needed for hydroxyapatite deposition.

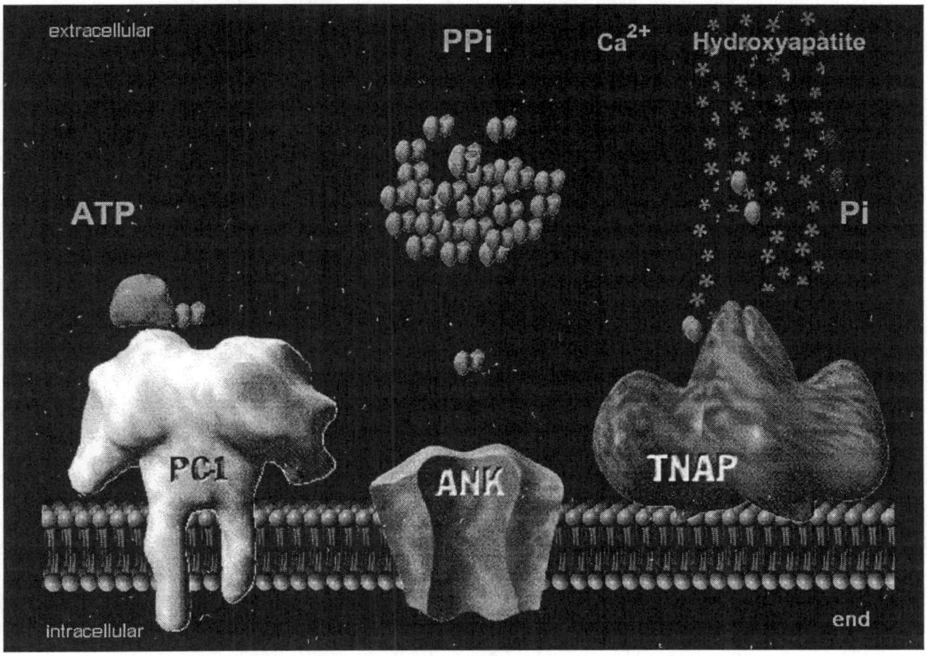

regulation by PC-1 and ANK is mediated by extracellular PPi, which in turn is known to be regulated in the same manner by physiologic PC-1 and ANK function. We also hypothesize that ANK and PC-1 expression levels are regulated by PPi. A significant precedent for the contention that a small metabolite such as PPi can up-regulate gene expression is the recent demonstration that the TNAP-derived byproduct of PPi hydrolysis, inorganic phosphate (Pi), induces the osteoblast protein osteopontin [39]. If this hypothesis proves correct it would mean that PPi levels can modulate the expression of the genes that control its production, degradation and secretion.

Acknowledgement

This study was supported by grants CA42595. DE 12889, AR47347 and PO1 AGO-7996 from the National Institutes of Health, USA and by the Research Service of the Department of Veterans Affairs.

References

[1] Millán JL, Fishman WH 1995 Biology of human alkaline phosphatases with special reference to cancer. Crit Rev Clin Lab Sci **32:** 1-39.

[2] Whyte MP 1994 Hypophosphatasia and the role of alkaline phosphatase in skeletal

[3] mineralization. Endocrine Rev **15:** 439-461.

[4] Narisawa S, Fröhlander N, Millán JL 1997 Inactivation of two mouse alkalinephosphatase genes and establishment of a model of infantilehypophosphatasia. Dev Dyn **208:** 432-446.

[5] Zurutuza L, Muller F, Gibrat JF, Taillandier A, Simon-Bouy B, Serre JL, Mornet E 1999 Correlations of genotype and phenotype in hypophosphatasia. Hum Mol Genet **8:** 1039-1046.

[6] Henthorn PS. Whyte MP 1992 Missense mutations of the tissue-nonspecific alkaline phosphatase gene in hypophosphatasia. Clin Chem **38:** 2501-5.

[7] Fukushi M, Amizuka N, Hoshi K, Ozawa H, Kumagai H, Omura S, Misumi Y, Ikehara Y, Oda K 1998 Intracellular retention and degradation of tissue-nonspecific alkaline phosphatase with a Gly317-->Asp substitution associated with lethal hypophosphatasia. Biochem Biophys Res Comm **246:** 613-618.

[8] Shibata H, Fukushi M, Misumi Y, Ikehara Y, Ohashi Y, Oda K 1998. Defective intracellular transport of tissue-nonspecific alkaline phosphatase with an Ala162Thr mutation associated with lethal hypophosphatasia. J Biochem **123:** 968-977.

[9] Whyte MP 1995 Hypophosphatasia. In: *The metabolic and molecular bases of inherited disease.* Ed. Scriver CR, Beaudet AL, Sly SW, Valle D. McGraw-Hill Inc., New York, pp. 4095-4112.

[10] Anderson HC 1995 Molecular biology of matrix vesicles. Clin Orthopaed Rel Res **314**: 266-280.

[11] Hsu HHT, Anderson HC 1996 Evidence of the presence of a specific ATPase responsible for ATP-initiated calcification by matrix vesicles isolated from cartilage and bone. J. Biol. Chem. **271:** 26383-26388.

[12] Pizauro JM, Demenis MA, Ciancaglini P, Leone FA. 1998 Kinetic characterization of a membrane-specific ATPase from rat osseous plate and its possible significance on endochondral ossification. Biochim. Biophys. Acta **1368:** 108-114.

[13] Fedde KN, Lane CC, Whyte MP 1990 Alkaline phosphatase: (tissue nonspecific isoenzyme) is a phosphoethanolamine and pyridoxal 5'-phosphate ectophosphatase: normal and hypophosphatasia fibroblast study. Am Hum J Genet **47**: 767-775.

[14] Moss DW, Eaton RH, Smith JK, Whitby LG 1967 Association of inorganic pyrophosphatase activity with human alkaline phosphatase preparations. Biochem J **102:** 53-57.

[15] Rezende AA, Pizauro JM, Ciancaglini P, Leone FA. 1994 Phosphodiesterase activity is a novel property of alkaline phosphatase from osseous plate. Biochem. J. **301:** 517-522.

[16] Fedde KN, Michel MP, Whyte MP 1993 Evidence against a role for alkaline phosphatase in the dephosphorylation of plasma membrane proteins: hypophosphatasia fibroblast study. J. Cell. Biochem. **53:** 43-50.

[17] Henthorn P, Millán JL, Leboy P 1999 Acid and alkaline phosphatases. In Principles of Bone Biology (eds. Seibel, Robins and Bilezikian). Academic Press. Chapter 9, pgs 127-137.

[18] de Bernard B, Bianco P, Bonucci E, Constantini M, Lunazzi GC, Martinuzzi P, Mocricky C, Moro L, Panfili E, Pollesello P 1986 Biochemical and immunohistochemical evidence that in cartilage an alkaline phosphatase is a Ca2+-binding glycoprotein. J. Cell Biol. **103:** 1615-1623.

[19] Whyte MP, Landt M, Ryan LM, Mulivor RA, Henthorn PS, Fedde KN, Mahuren JD, Coburn SP 1995 Alkaline phosphatase: placental and tissue-nonspecific isoenzymes hydrolyze phosphoethanolamine, inorganic pyrophosphate, and pyridoxal 5'-phosphate. Substrate accumulation in carriers of hypophosphatasia corrects during pregnancy. J Clin Invest **95:** 1440-1445.

[20] Anderson HC, Hsu HH, Morris DC, Fedde KN, Whyte MP 1997 Matrix vesicles in osteomalacic hypophosphatasia bone contain apatite-like mineral crystals. Am J Pathol **151:** 1555-1561.

[21] Johnson K, Moffa A, Pritzker K, Chen Y, Goding J, Terkeltaub R 1999 Matrix vesicle Plasma Cell Membrane Glycoprotein-1 (PC-1) regulates mineralization by murine osteoblastic MC3T3 cells. J Bone Miner Res **14:** 883-892.

[22] Meyer JL 1984 Can biological calcification occur in the presence of pyrophosphate? Arch. Biochem. Biophys. **231:** 1-8.

[23] Caswell AM, Whyte MP, Russell RG 1986 Normal activity of nucleoside triphosphate pyrophosphatase in alkaline phosphatase-deficient fibroblasts from patients with infantile hypophosphatasia. J Clin Endocrinol Metab **63:** 1237-1241.

[24] Caswell AM, Whyte MP, Russell RGG 1991 Hypophosphatasia and the extracellular metabolism of inorganic pyrophosphate: clinical and laboratory aspects. Crit Rev Clin Lab Sci **28:** 175-232.

[25] Goding J, Terkeltaub R, Maurice M, Deterre P, Sali A, Belli S 1998 Ecto-phosphodiesterase/pyrophosphatase of lymphocytes and nonlymphoid cells: structure and function of the PC-1 family. Immunol Reviews **161:** 11-26.

[26] Solan J, Deftos LJ, Goding JW, Terkeltaub RA 1996. Expression of the Nucleoside Triphosphate Pyrophosphohydrolase PC-1 is Induced by Basic Fibroblast Growth Factor (bFGF) and Modulated by Activation of the Protein Kinase A and C Pathways in Osteoblast-like Osteosarcoma Cells. J. Bone Miner. Res. **11:** 183-192.

[27] Bonewald LF., Schwartz Z, Swain LD, Boyan BD 1992 Stimulation of matrix vesicle enzyme activity in osteoblast-like cells by 1,25 $(OH)_2D3$ and transforming growth factor β (TGFβ). Bone and Mineral **17:** 139-144.

[28] Oyajobi BO, Caswell AM, Russell RG 1994 Transforming growth factor beta increases ecto-nucleoside triphosphate pyrophosphatase activity of human bone-derived cells. J. Bone Miner. Res. **9:** 99-109.

[29] Johnson K, Hashimoto S, Lotz M, Pritzker K, Goding J, Terkeltaub R. Up-Regulated Expression of the Phosphodiesterase Nucleotide Pyrophosphatase Family Member Plasma Cell Membrane Glycoprotein-1 (PC-1) is Both a Marker and Pathogenic Factor for Knee Meniscal Cartilage Matrix Calcification. Arthritis Rheum. **44**:1071-81, 2001

[30] Okawa A, Nakamura I, Goto S, Moriya H, Nakamura Y, Ikegawa S 1998 Mutation in *Npps* in a mouse model of ossification of the posterior longitudinal ligament of the spine. Nature Gen **19:** 271-273.

[31] Ho AM; Johnson MD; Kingsley DM 2000 Role of the mouse ank gene in control of tissue calcification and arthritis. Science, **289:** 265-70.

[32] Fedde KN, Blair L, Silverstein J, Weinstein RS, Waymire K, MacGregor GR, Narisawa S, Millán JL, Whyte MP 1999 Alkaline phosphatase knock-out mice recapitulate the metabolic and skeletal defects of infantile hypophosphatasia. J. Bone Min. Res. **14:** 2015-2026.

[33] Waymire KG, Mahuren JD, Jaje JM, Guilarte TR, Coburn SP, MacGregor GR 1995 Mice lacking tissue non-specific alkaline phosphatase die from seizures due to defective metabolism of vitamin B-6. Nat Genet **11:** 45-51.

[34] Narisawa S. Wennberg C. Millan JL 2001 Abnormal vitamin B6 metabolism in alkaline phosphatase knock-out mice causes multiple abnormalities, but not the impaired bone mineralization. Journal of Pathology. **193:** 125-33.

[35] Wennberg C, Hessle L, Lundberg P, Mauro S, Narisawa S, Lerner UH, Millán JL 2000 Functional characterization of osteoblasts and osteoclasts from alkaline phosphatase knock-out mice. J. Bone Min. Res. **15:** 1879-1888.

[36] Rutsch F, Vaingankar S, Johnson K, Schauerte P, Kalhoff H, Goldfine I, Maddux B, Superti-Furga A, Terkeltaub R 2001 Deficiency of the PPi-Generating Nucleoside Triphosphate Pyrophosphohydrolase (NTPPPH) Isozyme PC-1 in Idiopathic Infantile Arterial Calcification (IIAC) Associated with Periarticular Calcification. Am J Pathol **158**: 543-554.

[37] Clair T, Lee HY, Liotta LA, Stracke ML 1997 Autotaxin is an exoenzyme possessing 5'-nucleotide phosphodiesterase/ATP pyrophosphatase and ATPase activities. J Biol Chem **272:** 996-1001.

[38] Rachow J, Ryan L 1988 Inorganic pyrophosphate metabolism in arthritis. Rheum Dis Clin N Am **14:** 289-302.

[39] Johnson KA, Wennberg C, Hessle L, Mauro S, Narisawa S, Goding J, Millán JL, Terkeltaub R 2000 Tissue-nonspecific alkaline phosphatase antagonizes and regulates PC-1. Am J Physiol. Regulatory Integrative & Comparative Physiology **279:** R1365-1377.

[40] Beck GR, Zerler B, Moran E 2000 Phosphate is a specific signal for induction of osteopontin gene expression. Proc Natl Acad Sci USA **97**: 8352-8357.

In Vitro Differentiation and Matrix Vesicle Biogenesis in Primary Cultures of Rat Growth Plate Chondrocytes

Rama Dhanyamraju, Joseph B. Sipe, H. Clarke Anderson

Department of Pathology and Laboratory Medicine, University of Kansas Medical Center, Kansas City, KS 66160, USA

Abstract. During endochondral ossification cartilage is replaced by bone. Chondrocytes of growth plate undergo proliferation, maturation, hypertrophy, matrix vesicle (MV) biogenesis and programmed cell death (apoptosis). It has been suggested that not all chondrocytes are destined to die. Some may become bone-forming cells, and secrete bone matrix within existing chondrocytic lacunae. The *in vitro* system presented here provides a potential experimental model for studying *in vitro* differentiation and MV biogenesis in chondrocyte cultures. Chondrocytes were obtained from collagenase digested tibial and femoral growth plate cartilage of 4 week old rachitic rats. The isolated chondrocytes were plated as monolayers at a density of 0.5 x 10^6 cells/35mm plate and grown for 17 days in BGJ_b medium supplemented with 10 % fetal bovine serum, 50 μg/ml ascorbic acid. Light microscopy revealed sirius red-positive, apparent bone matrix in layers at the surfaces of cartilaginous nodules that developed in the cultures. The central matrix was largely alcian blue staining thus resembling cartilage matrix. Electron microscopy revealed superficial areas of bone like matrix with large banded collagen fibrils, consistent with type I collagen. Most of the central matrix was cartilaginous, with small fibrils, randomly arranged consistent with type II collagen. The presence of peripheral Type-I and central Type-II collagen was confirmed by Immunohistochemical staining. Immunohistochemistry with anti-Bone morphogenetic proteins 2, 4 and 6 showed that BMP expression is associated with maturing hypertrophic central chondrocytes, many of which were TUNEL positive and undergoing cell death with plasma membrane breaks, hydropic swelling and cell fragmentation. During early mineralization small radial clusters of hydroxyapatite-like mineral were associated with matrix vesicles. Collagenase digestion-released MVs from the cultures showed a high specific activity for alkaline phosphatase and demonstrated a pattern of AMP-stimulated non-radioactive $CaPO_4$ deposition similar to that observed with native MVs. These studies confirm that primary cultures of rat growth plate chondrocytes are a reasonable *in vitro* model of growth plate histotype, MV biogenesis and programmed cell death.

Introduction

The epiphyseal growth plate is located at the ends of long bones and is mainly involved in the regulation of longitudinal bone growth by the process of endochondral ossification. During endochondral ossification, cartilage is replaced with bone. Chondrocytes of growth plate undergo proliferation, maturation, matrix vesicle biogenesis, hypertrophy and programmed cell death (apoptosis). Chondrocytes undergo proliferation by secreting an extracellular matrix, that is composed of Type-II collagen [40] and proteoglycan [6,30]. Maturation of chondrocytes is associated with an increase in the cellular size i.e. hypertrophy, and the expression of phenotypic markers of hypertrophy such as alkaline phosphatase activity and Type-X collagen [13]. The onset of maturation is believed to be promoted by bone morphogenetic proteins [11,21,29,54], c-myc [22,34] and Indian hedgehog [48]. Matrix vesicles (MVs), which later will initiate calcification, are released into the newly forming longitudinal septal matrix by budding from the lateral edges of maturing and early hypertrophic chondrocytes [10]. MVs are extracellular membrane invested particles about 100 nm in size [4]. MVs initiate calcification through the action of MV-associated phosphatases [3,5,8,33,53] and calcium binding phospholipids and proteins [10,24,41,51]. Finally, chondrocytes die by programmed cell death [26,56]. Some chondrocytes do survive and transdifferentiate into phenotypic bone cells [15,44,45]. Although, it has been shown that cultured rat and chick chondrocytes [9,27,28,37,53] and Saos-2 osteoblastic cells [23,43] are capable of releasing calcifiable matrix vesicles, the mechanism of MV biogenesis needs to be characterized and the in vivo factors regulating MV biogenesis need to be identified. Hence a reliable cell culture system is needed to test the effect of cellular differentiation and/or apoptosis on biogenesis and calcifiability of matrix vesicles. We are also interested in knowing whether MV biogenesis results from vesiculation due to apoptosis of hypertrophic chondrocytes [36] or is a consequence of normal, non-apoptotic skeletal cell differentiation.

In this study, we were able to show that primary cultures of rat growth plate chondrocytes are a reasonable *in vitro* model of growth plate histotype showing proliferation, maturation, hypertrophy, MV biogenesis, programmed cell death and calcification.

Materials and Methods

Cell Culture

Chondrocytes were obtained from collagenase digested tibial and femoral growth plate cartilage of 4 week-old rachitic rats as described previously [9]. The isolated chondrocytes were plated as monolayers at a density of 0.5 x 10^6 cells/35 mm plate and grown for 17 days in BGJ_b culture medium supplemented with 10% fetal bovine serum, 100 units/ml Penicillin, 100 μg/ml streptomycin, 50 μg/ml ascorbic acid and in some instances 10^{-8} M dexamethasone. At confluence on day 7, 5mM β-glycerophosphate was added for 24 h to some cultures to enhance mineralization. Cultures were harvested on day 17.

MV Isolation

MVs were isolated from chondrocyte cultures grown in the absence of dexamethasone and β-glycerophosphate by collagenase digestion as described previously [9]. Briefly, adherent chondrocytes were washed with HBSS and incubated in 2.5 mg/ml collagenase solution for 90 min. at 37°C. MVs were harvested by two-step differential ultra-centrifugation as described previously . The yield of MVs was estimated by measuring the protein content and alkaline phosphatase activity of the micro-vesicle fractions released from cultures upon collagenase digestion [33].

In Vitro Calcifiability of MVs
Calcifiability of MVs isolated from cultures was assessed by non-radioactive $CaPO_4$ deposition assay. Briefly, this assay involves the incubation of 30 μg samples of MV protein in a calcifying solution [33] containing 2.2 mM Ca^{2+} and 1.6 mM PO_4^{2-} in the presence of 0 to 3 mM phosphoester substrate e.g. ATP, AMP or β-GP for 5.5 h at 37°C. The incubation was terminated after 5.5 h by centrifugation at 8800g for 30 min. to precipitate MVs and $CaPO_4$ mineral formed during incubation. The pellet containing matrix vesicles and $CaPO_4$ mineral was then solubilized with 0.6N HCl for 24 h. The calcium content of the HCl supernatant was then determined colorimetrically by the O-cresolpthalein complexone method (Calcium Kit, Sigma).

Light Microscopy
Cultures were fixed immediately in 4% phosphate buffered paraformaldehyde (pH 7.4) for 2 h at room temperature. After fixation, the cultures were embedded in paraffin wax and sectioned. Paraffin embedded 5-micron thick sections of chondrocyte cultures were dewaxed and rehydrated by exposure to xylene followed by incubation in ethanol solution series (100%, 90%, 80%). Deparaffinized and rehydrated sections were stained with hematoxylin & eosin (H & E) and Weigert's hematoxylin/alcian blue/sirius red [35,39].

Immunohistochemistry
The following polyclonal antibodies were used: rabbit Anti-Rat collagen Type I and rabbit Anti-Human Collagen Type II (Chemicon), anti BMP-2, 6 polyclonal antibodies (Santa Cruz Biotechnology, Inc.) and anti BMP-4 monoclonal antibody (Novacastra Laboratories Ltd.). These primary antibodies were visualized using either Rabbit ABC staining system, goat ABC staining system or mouse ABC staining system (Santa Cruz Biotechnology, Inc.) according to the manufacturer's instructions.

In Situ *Detection of Apoptosis*
Detection of DNA fragmentation by terminal deoxynucleotidyl transferase (Tdt) – mediated dUTP nick end labeling was performed at the light microscopic level using an Apoptag peroxidase *in situ* apoptosis detection kit (S7100, Intergen, Purchase, NY) according to the manufacturer's instructions.

Electron Microscopy
Cultures were fixed in 2.5% glutaraldehyde, post-fixed in 1% osmium tetraoxide, dehydrated, embedded in Spurr's low viscosity epoxy resin, cut with diamond knives and stained with uranyl acetate and lead citrate [4]. Thin sections were examined and photographed using a Zeiss EM IOA electron microscope.

Results

Light Microscopy, Immunohistochemistry and TUNEL Staining
Light microscopy revealed that confluence was attained by Day 7. The cells were mature with well-defined matrix by day 11 and by day 17, chondrogenic nodules were observed. At this stage, mineralization was also detected around central large hypertrophic chondrocytes (Fig. 1). Staining with Weigert's hematoxylin/alcian blue/sirius red dyes revealed sirius red-

positive, apparent bone matrix in layers at the surfaces of multi-layered areas of the cultures [19]. The central matrix was largely alcian blue staining thus resembling cartilage matrix.

Immunohistochemical staining was positive for both Type-I at the outer surfaces of chondrogenic nodules, while most of the central matrix stained positively for Type-II collagen [19]. This suggests that the rat growth plate chondrocytes in primary cultures not only undergo proliferation, maturation, hypertrophy and express Type II collagen but also some chondrocytes may enter an osseous pathway of differentiation as indicated by the expression of Type-I collagen. Immunohistochemical staining also localized the expression of BMP-2, 4 and 6 in the cytoplasm of central hypertrophic chondrocytes (Fig. 2).

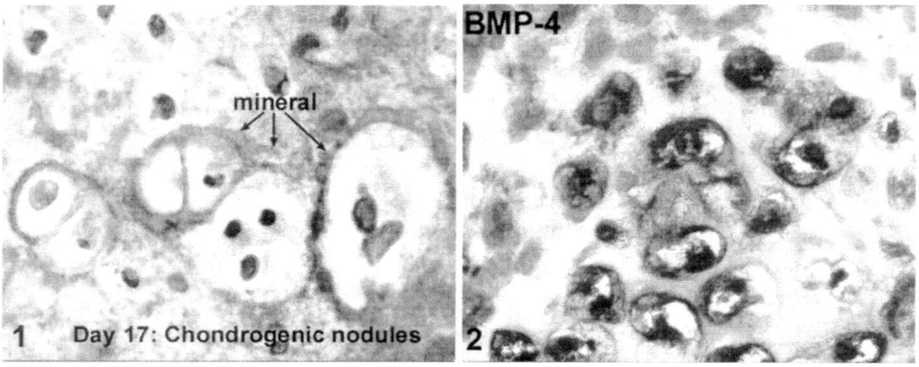

Figure 1. Photomicrograph of primary cultures of 17 day rat growth plate chondrocytes stained with Hematoxylin & Eosin, showing mature and hypertrophic central chondrocytes surrounded by cartilaginous matrix and encircled by coalescing mineral deposits. Less advanced unfused mineral clusters are seen as dot-like deposits in the inter-territorial matrix. Central chondrocytes show nuclear chromatin condensation, indicative of programmed cell death. (X 2150)

Figure 2. Photomicrograph of primary culture of rat growth plate chondrocytes, showing dark immunostaining for BMP-4 in the cytoplasm of central hypertrophic chondrocytes. (X 2150)

Positive TUNEL staining was observed in the nuclei of central hypertrophic chondrocytes (Fig. 3A). Central chondrocytes were hypertrophic and many were undergoing cell death with chromatin margination and condensation typical of early stages of apoptosis (Fig. 3C). These cells also exhibited plasma membrane breaks, hydropic swelling and cell fragmentation. No TUNEL staining could be detected in the negative control, lacking terminal deoxynucleotidyl transferase (Fig. 3B).

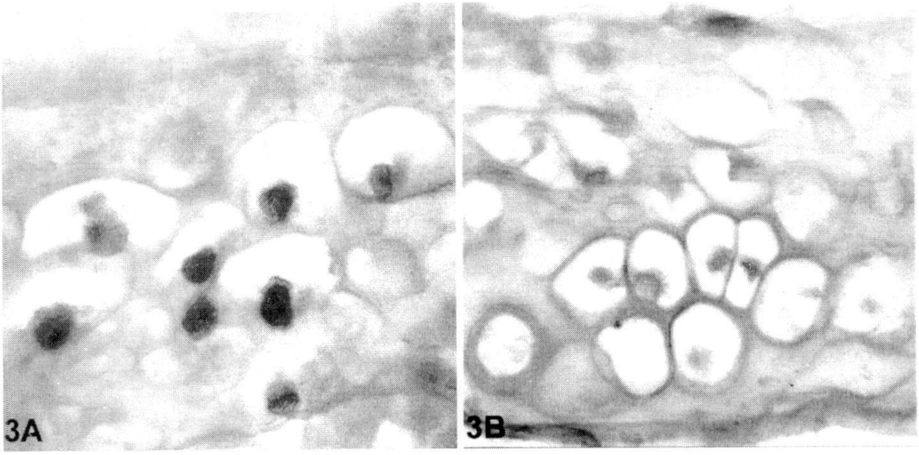

Figure 3A. Photomicrograph of primary culture of rat growth plate chondrocytes, showing dark TUNEL positive staining in nuclei of central chondrocytes indicating that programmed cell death is occurring in these chondrocytes. (X 2,500)

Figure 3B. Photomicrograph of primary culture of rat growth plate chondrocytes, showing negative control for TUNEL staining in which terminal deoxynucleotidyl transferase enzyme was omitted. (X 2,500)

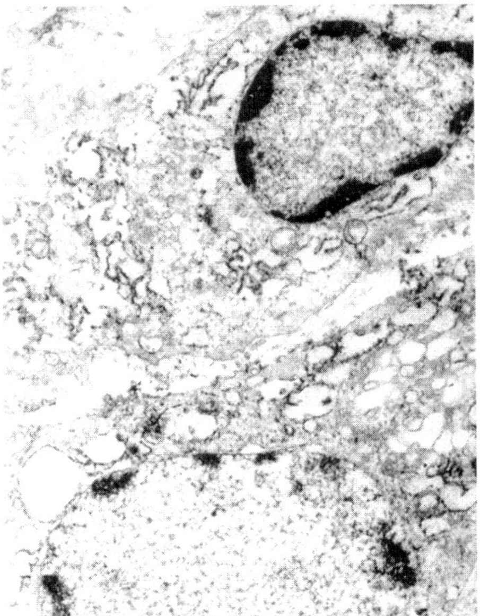

Figure 3C. Electronmicrograph of primary culture of rat growth plate chondrocytes, showing chromatin margination characteristic of early stages of programmed cell death (apoptosis). (X 13,000)

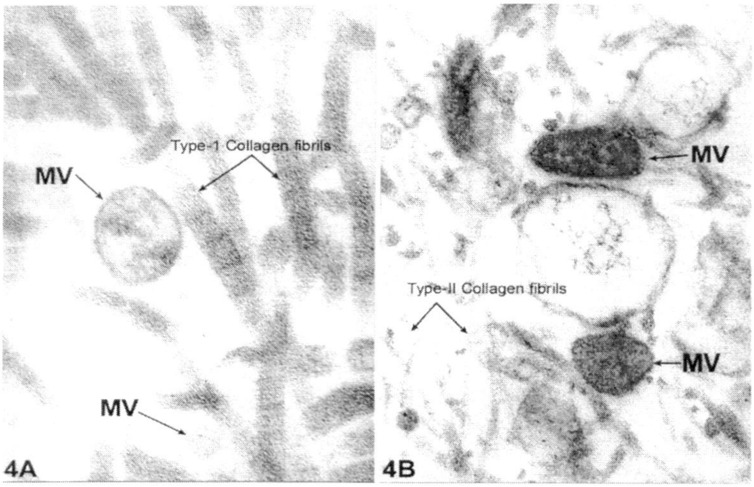

Figure 4A. Electronmicrograph of primary culture of periphery of rat growth plate chondrocytes, showing bone-like matrix with large banded, Type-I collagen fibrils. (X 112,300)

Figure 4B. Electronmicrograph of primary culture of rat growth plate chondrocytes, showing cartilaginous matrix with small, randomly arranged Type-II collagen fibrils and matrix vesicles. (X 96,000)

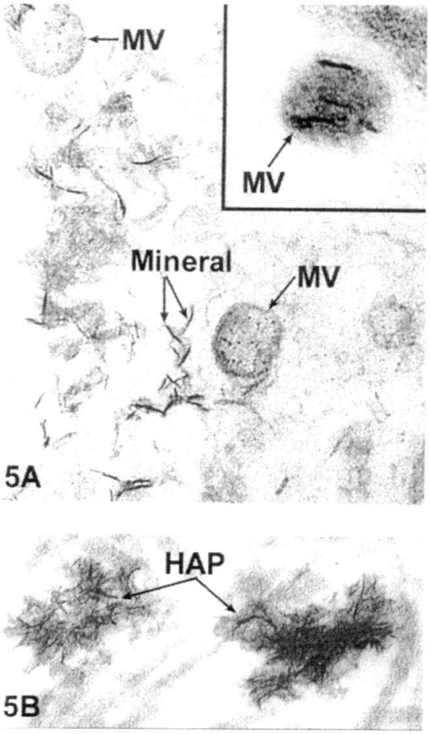

Figure 5A. Electronmicrograph of primary culture of rat growth plate chondrocytes, showing at low and high magnification needle like profiles of hydroxyapatite-like mineral associated with matrix vesicles (MV) (early stages of mineralization). (X 96,000; insert: X 190,000)

Figure 5B. Electronmicrograph of primary culture of rat growth plate chondrocytes, showing hydroxyapatite (HAP) deposition in fused perivesicular clumps (late stages of mineralization). (X 115,000).

Transmission Electron Microscopy
Electron microscopy revealed superficial areas of bone-like matrix with large banded collagen fibrils, consistent with Type-I collagen (Fig. 4A). Most of the central matrix was cartilaginous, with small fibrils, randomly arranged, onsistent with Type-II collagen (Fig. 4B). Hydroxyapatite-like mineral deposition was associated initially with matrix vesicles (Fig. 5A) and later formed fused perivesicular lumps between collagen fibrils (Fig. 5B).

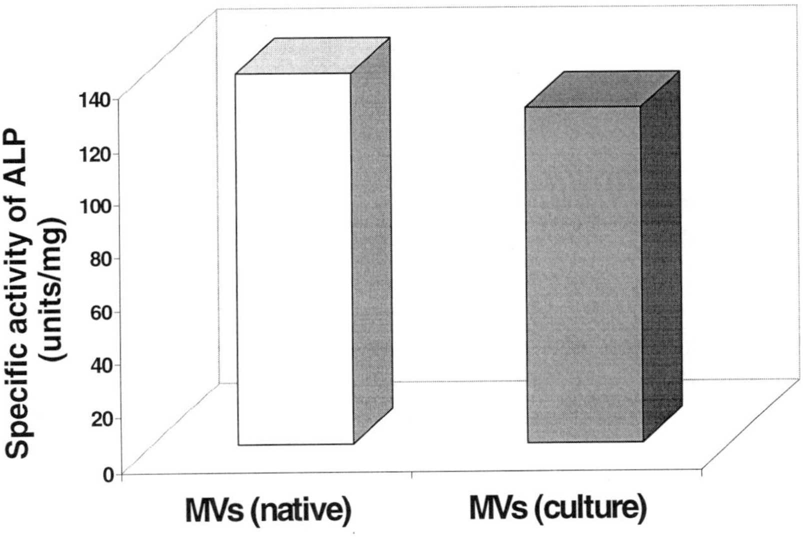

Figure 6. Bar diagram comparing alkaline phosphatase activity of matrix vesicles isolated from cultures vs. native matrix vesicles

In Vitro Calcification of Matrix Vesicles:
MVs isolated from cultures by collagenase digestion not only had a very high alkaline phosphatase (ALP) specific activity (Fig. 6) but also significant calcifying activity comparable to that of native MVs (Figs. 7A & 7B). Like the native MVs, MVs from cultures were able to hydrolyze phosphoester substrates including ATP, AMP and β-GP thus leading to calcium deposition (Figs. 7A & 7B). It was also observed that AMP and β-GP are better substrates to support MV initiated mineralization than is ATP (Fig. 7A & 7B).

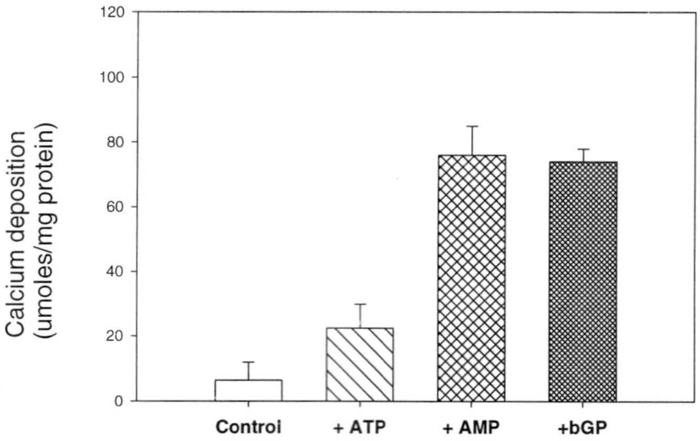

Figure 7A. Bar diagram comparing the effect of 1 mM ATP, 3 mM AMP or 3 mM β-GP on the *in vitro* calcification of matrix vesicles isolated from cultures. Values are mean ± SEM from four cultures. (control = no substrate added.)

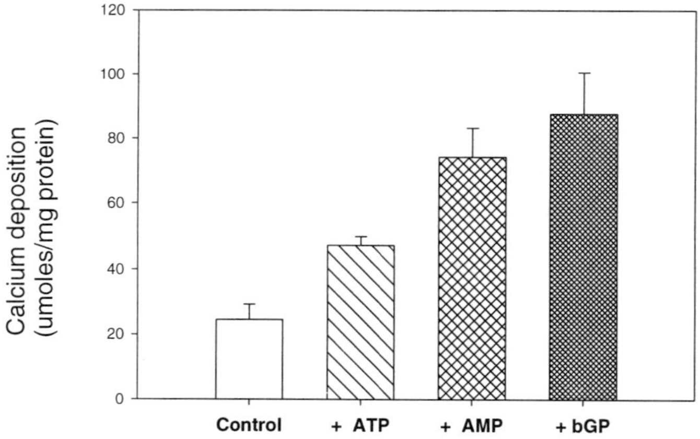

Figure 7B. Bar diagram comparing the effect of 1 mM ATP, 3 mM AMP or 3 mM β-GP on the *in vitro* calcification of native matrix vesicles. Values are mean ± SEM from four samples. (control = no substrate added.)

Discussion

Our findings indicate that rat growth plate chondrocytes in primary cultures undergo proliferation, maturation, hypertrophy and programmed cell death (PCD). Chondrogenic differentiation was associated with hypertrophy and the expression of Type-II collagen. That some hypertrophic chondrocytes underwent osteogenic differentiation was suggested by the expression of Type-I collagen and the presence of bone-like matrix with mineralizing matrix vesicles (Fig. 4A). This finding is consistent with the notion that during endochondral ossification, not all hypertrophic chondrocytes die. Some may survive, and become bone forming cells, and secrete bone-like matrix within existing chondrocytic lacunae [15,44,45].

We also report here the expression of BMP-2, 4, and 6 in this culture system. BMP-2, 4 and 6 were concentrated in hypertrophic chondrocytes. This finding is consistent with several reports of localization of BMP-2, 4 and 6 in hypertrophic chondrocytes of growth plate [11,21,29,54]. BMPs have been implicated in promoting not only skeletal cell differentiation [11,21,29,54], but also apoptosis during skeletal cell development [13,20,31,55]. We also observed evidence of apoptosis in primary cultures of rat growth plate chondrocytes, as indicated by the presence of TUNEL positive nuclei and nuclear chromatin margination in hypertrophic chondrocytes, thereby suggesting that chondrocyte programmed cell death begins by DNA fragmentation, a hallmark of apoptosis. However, the subsequent morphologic changes seen in cultured chondrocytes undergoing programmed cell death, i.e. plasma membrane breaks, hydropic swelling and cell fragmentation, are unlike those detected in thymocytes undergoing apoptosis, and are more consistent with the sequence of events occuring in the rat growth plate, *in vivo* [12]. MV biogenesis, which characteristically occurs by budding from the lateral sides of hypertrophic chondrocytes [4,16], may also be an early reflection of a process that begins with apoptotic DNA fragmentation in maturing hypertrophic chondrocytes but culminates in hydropic cell death during late stages of hypertrophy.

We also demonstrate here that primary cultures of rat growth plate chondrocytes produce matrix vesicles, which readily undergo mineralization. The alkaline phosphatase activity of culture-derived matrix vesicles was comparable to that of native matrix vesicles isolated from growth plate. The mineralization capacity of culture derived matrix vesicles was also comparable to that of native MVs. ALP and other MV phosphatases mediate the hydrolysis of phosphoester substrates such as ATP, AMP & PPi, thereby increasing the local concentration of orthophosphate (Pi) and thus facilitating precipitation of calcium phosphate [2,10]. These enzymes may also be involved in the removal of inhibitors of mineralization, including PPi [3,7,42,53]. Our electron micrographs revealed the presence of biological apatite-like mineral to be associated with the vesicles generated in cultures. Currently, the nature of the mineral produced by the matrix vesicles isolated from cultures upon exposure to phosphoester substrates viz., AMP, ATP and β-GP is being investigated by FTIR.

In conclusion, these studies confirmed that primary cultures of rat growth plate chondrocytes are a reasonable *in vitro* model of growth plate histotype, with chondrocyte differentiation, MV biogenesis, calcification and programmed cell death (apoptosis). Some of the cultured chondrocytes appeared to enter an osseous pathway of differentiation as indicated by Type-I collagen synthesis and secretion at the surfaces of the chondrogenic nodules *in vitro*.

References

[1] Ali SY, Sajdera SW, and Anderson HC 1970 Isolation and characterization of calcifying matrix vesicles from epiphyseal cartilage. Proc Nat. Sci.(USA) **67:** 1513-1520.

[2] Ali SY, Evans L 1973 The uptake of (^{45}Ca) Calcium ion by matrix vesicles isolated from calcifying cartilage. Biochem J. **134:** 647-650.

[3] Ali SY. 1976 Analysis of matrix vesicles and their role in the calcification of epiphyseal cartilage. Fed. Proc **35:** 135-142.

[4] Anderson HC 1969 Vesicles associated with calcification in the matrix of epiphyseal cartilage. J. Cell Biol. **41:** 59-72.

[5] Anderson HC, Matasuzawa T, Sajdera SW, Ali SY 1970 Membranous particles in calcifying cartilage matrix Trans. N.Y. Acad Sci **32:** 619-630.

[6] Anderson HC, Sajdera SW 1971. The fine structure of bovine nasal cartilage. Extractin as a technique to study proteoglycans and collagen in cartilage matrix. J Cell Biol 1971 **49:** 650-663.

[7] Anderson HC, Reynolds JJ 1973 Pyrophosphate stimulation of calcium uptake into cultured embryonic bones. Fine structure of matrix vesicles and their role in calcification. Dev Biol **34:** 211-227.

[8] Anderson HC, Hsu HHT 1978 A new method to measure ^{45}Ca accumulation by matrix vesicles in slices of rachitic growth plate cartilage. Metab. Bone Dis Rel Res. **1:** 193-198.

[9] Anderson HC, Stechschulte Jr. DJ, Collins DE, Jacobs DH, Morris DC, Hsu HHT, Redford PA, Zeiger S 1990 Matrix vesicle biogenesis in vitro by rachitic and normal rat chondrocytes. American Journal of Pathology 31: 391-398.

[10] Anderson HC 1995 Molecular biology of matrix vesicles. Clin. Orthop Relat Res **314:** 266-280.

[11] Anderson HC, Hodges PT, Aguilera XM, Missana L & Moylan P 2000. Bone Morphogenetic protein (BMP) localization in developing human and rat growth plate, metaphysis, epiphysis and articular cartilage. J Histochem Cytochem **48:** 1493-1502.

[12] Anderson HC, Sipe JB, Moylan P 2000. Ultrastructure of programmed cell death (apoptosis) in chondrocytes of actively growing rat growth plate. Trans European Symp Calcif Tiss.

[13] Buckland RA, Collinson JM, Graham E, Davidson DR 1998 Antagonistic effects of FGF4 on BMP induction of apoptosis and chondrogenesis in chick limb bud. Mech Dev 71: 143-150.

[14] Buckwalter JA, Mower D, Ungar R, Schaeffer J, Ginsberg B 1986 Morphogenetic analysis of chondrocyte hypertrophy. J. Bone Joint Surg.**68 A:** 243-255.

[15] Cancedda FD, Gentili C, Manduca P, Cancedda R 1992 Hypertrophic chondrocytes undergo further differentiation in culture. J Cell Biol. **117:** 427-435.

[16] Cecil RNA, Anderson HC 1978 Freeze-fracture studies of matrix vesicle calcification in epiphyseal growth plate. Metab Bone Dis & Rel Res. **1:** 89-95.

[17] Closs CI, Murray AB, Strauss JPG 1990 C-fos expression precedes osteogenic differentiation of the cartilage cells *in vitro*. J Cell Biol **111:** 1313-1323.

[18] Descalzi Cancedda F, Gentili C, Manduca P, Cancedda R 1992 Hypertrophic chondrocytes undergo further differentiation in culture. J Cell Biol **117:**427-435.

[19] Dhanyamraju R, Sipe JB, Anderson HC 2000 Chondrogenic and osteogenic differentiation of primary cultures of rat growth plate chondrocytes. J Bone Min Res. **15** (Suppl.1): S 381.

[20] Ekanayake S, Hall BK 1997 The *in vivo* and *in vitro* effects of bone morphogenetic protein-2 on the development of the chick mandible. Int J Dev Biol **41:** 67-81.

[21] Enomoto-Iwamoto M, Iwamoto M, Mukudai Y, Kawakami Y, Nohno T, Higuchi Y, Takemoto S, Ohuchi H, Noji S, Kurisu K 1998 Bone morphogenetic protein signaling is required for maintenance of differentiation phenotype, control of proliferation, and hypertrophy in chondrocytes. J Cell Biol **140:** 409-418.

[22] Farquharson C, Hesketh JE, Loveridge N 1992 The protooncogene c-myc is involved in cell differentiation as well as cell proliferation: studies on growth plate chondrocytes *in situ*. J. Cell Physiol. **152**: 135-144.

[23] Fedde KN 1992 Human osteosarcoma cells spontaneously release matrix-vesicle-like structures with the capacity to mineralize. Bone & Min. **17:** 145-151.

[24] Genge BR, Cao X, Wu LN, Buzzi WR, Showman RW, Arsenault AL, Ishikawa Y, Wuthier RE 1992 Establishment of the primary structure of the major lipid-dependent Ca2+ binding proteins of chicken growth plate cartilage matrix vesicles: identity with anchorin CII (annexin V) and annexin II. J Bone Miner Res. **7:** 807-819.

[25] Gentili C, Bianco P, Neri M, Malpeli M, Campanile G, Castagnola P, Cancedda R, Cancedda FD 1993 Cell proliferation, extracellular matrix mineralization and ovotransferrin transient expression during *in vitro* differentiation of chick hypertrophic chondrocytes into osteoblast-like cells. J Cell Biol **122:**703-712.

[26] Gibson G 1998 Active role of chondrocyte apoptosis in endochondral ossification. Microsc. Res. **43:** 191-204.

[27] Glaser JH, Conrad HE 1981 Formation of matrix vesicles by cultured chick embryo chondrocytes. J. Biol Chem. **256:** 2607-2611.

[28] Golub EE, Schattschneider SC, Berthold P, Burke A, Shapiro IM 1983 Induction of chondrocyte vesiculation *in vitro*. J. Biol Chem. **258:** 616-621.

[29] Grimsrud CD, Romano PR, D'Souza M, Puzas JE, Reynolds PR, Rosier RN, O'Keefe RJ 1999 BMP-6 is an autocrine stimulator of chondrocyte differentiation. J. Bone Miner Res. **14:** 475-482.

[30] Heinegard D 1977 Polydispersity of cartilage proteoglycans. J. Biol. Chem. **252:**1980-1989.

[31] Hogan BLM 1996 Bone morphogenetic proteins: multifunctional regulators of vertebrate development. Genes Dev **10:** 1580-1594.

[32] Horton WE, Feng LX, Adams C 1998 Chondrocyte apoptosis in development, aging and disease. Matrix Biol. **17:** 107-115.

[33] Hsu HH, Anderson HC 1977 A simple and defined method to study calcification by isolated matrix vesicles. Effect of ATP and vesicle phosphatase. Biochim Biophys Acta. **500**:162-72.

[34] Iwamoto M, Yagami K, Valle PL, Olsen BR, Petropoutous CJ, Ewert DL, Pacifici M 1993 Expression and role of c-myc in chondrocytes undergoing endochondral ossification. J. Biol. Chem. **268:** 9645-9652.

[35] Junquiera LCU, Bignolas G, Brentant RR 1979 Picosirius staining plus polarization microscopy, a specific method for collagen detection in tissue sections. Histochem J. **11**: 447-455.

[36] Kardos TB, Hubbard MJ 1982 Are matrix vesicles apoptotic bodies? In: Factors and mechanisms influencing bone growth. AD Dixon and BG Sarnat eds. Alan R. Liss, Inc. NY. pp 45-60.

[37] Kirsch T, Nah HD, Shapiro IM, Pacifici M 1997 Regulated production of mineralization competent matrix vesicles in hypertrophic chondrocytes. J. Cell Biol. **137:** 1149-1160.

[38] Lee K, Lanske B, Karaplis AC, Deeds JD, Kohno H, Nissenson RA, Kronenberg HM, Segre GV 1996 Parathyroid hormone-related peptide delays terminal differentiation of chondrocytes during endochondral bone development. Endocrinology **173:** 5109-5118.

[39] Lison L 1954 Alcian blue 8G with chorantine fast red 5B: A technique for selective staining of mucopolysaccharides. Stain Tech **28**: 131-138.

[40] Miller EJ 1969 Chick cartilage collagen: a new type of α1 chain not present in bone or skin of the species. Proc Nat Acad Sci (USA) **64:** 1264-1268.

[41] Peress, NS, Anderson HC, Sajdera, SW 1974 The lipids of matrix vesicles from bovine fetal epiphyseal cartilage. Calcif Tiss Res **14:** 275-281.

[42] Register TC and Wuthier RE 1985 Effect of pyrophosphate and two diphosphonates on ^{45}Ca and ^{32}Pi uptake and mineralization by matrix vesicle-enriched fractions and by hydroxyapatite. Bone (NY) **6:** 307-312.

[43] Ringbom-Anderson T, Jantti J, Akerman KEO 1994, Production and release of matrix vesicle in the cell processes of TPA-treated human osteoblast-like cells. J. Bone Min. Res. **9:** 661-670.

[44] Roach HI 1992 Trans-differentiation of hypertrophic chondrocytes into cells capable of producing a mineralized bone matrix. Bone & Min. **19:** 1-20.

[45] Roach HI, Erenpreisa J, Aigner T 1995 Osteogenic differentiation of hypertrophic chondrocytes involves asymmetric cell divisions and apoptosis. J. Cell Biol **131:** 438-494.

[46] Sampath TK, Maliakal JC, Hauschka PV, Jones WK, Sasak H, Tucker RF, White KH, Coughlin JE, Tucker MM, Pang RH, et al. 1992 Recombinant human osteogenic protein-1 (hOP-1) induces new bone formation in vivo with a specific activity comparable with natural bovine osteogenic protein and stimulates osteoblast proliferation and differentiation in vitro. J Biol Chem. **267**:20352-62.

[47] Strauss JPG, Closs EI, Schmidt J, Erfle V 1990 Gene expression during osteogenic differentiation in mandibular condyles *in vitro*. J Cell Biol **110:**1369-1377.

[48] Vortkamp A, Kaechoong L, Lanske B, Segre GV, Kronenberg HM, Tabin CJ 1996 Regulation of rate of cartilage differentiation by Indian hedgehog and PTH-related protein. Science **173:** 613-623.

[49] Wang Y, Toury R, Hauchecorne, Balmain N 1977 Expression of Bcl-2 protein in the epiphyseal plate cartilage and trabecular bone of growing rats. Histochem Cell Biol. **108:** 45-55.

[50] Wezeman FH, Bollnow MR 1997 Immunohistochemical localization of fibroblast growth factor II in normal and brachymorphic mouse tibial growth plate and articular cartilage. Histochem J. **29:** 505-514.

[51] Wuthier RE 1975 Lipid composition of isolated cartilage cell membranes and matrix vesicles. Biochem Biophys Acta **409:** 128-143.

[52] Wuthier RE, Gore ST 1977 Partition of inorganic ions and phospholipids in isolated cell, membrane and matrix vesicle fraction: Evidence for Ca-Pi acidic phospholipid complexes. Calcif. Tiss Res. **24:** 163-171.

[53] Wuthier RE, Chin JE, Hale JE, Register TC, Hale LV, Ishikawa Y 1985 Isolation and characterization of calcium accumulating matrix vesicles from chondrocytes of chicken epiphyseal growth plate cartilage in primary culture. J. Biol Chem. **260:** 15972-15979.

[54] Yazaki Y, Matsunaga S, Onisi T, Nagamine T, Origuchi N, Yamamoto T, Ishidou Y, Imamura T, Sakou T 1998 Immunohistochemical localization of bone morphogenetic proteins and the receptors in epiphyseal growth plate. Anti. Cancer Res. **18:** 2339-2344.

[55] Zou H, Niswander L 1996 Requirements for BMP signaling in interdigital apoptosis and scale formation. Science **272:** 738-741.

[56] Zenmyo M, Komiya S, Kawabata R, Sasaguri Y, Inoue A, Moriatsu M 1996 Morphological and biochemical evidence for apoptosis in the terminal hypertrophic chondrocytes of the growth plate. J. Pathol **180:** 430-433.

The Growth Plate
I.M. Shapiro et al. (Eds.)
IOS Press, 2002

Growth Plate Proteins and Biomineralization

Adele L. Boskey, Lyudmilla Spevak, Stephen B. Doty, and Itzhak Binderman

Hospital for Special Surgery, New York, NY 10021.

Abstract. Extracellular matrix proteins are multifunctional; they regulate matrix properties, cell-matrix interactions, and biomineralization. In this review, the current understanding of how extracellular matrix proteins in the growth plate regulate calcification is presented. Lessons are derived from cell-free solution and cell culture studies, and from mechanical and infrared microscopic studies of knockout animals. The proteins to be discussed are the phosphorylated glycoproteins (osteopontin (OPN), bone sialoprotein (BSP), and osteonectin (ON)), the gamma carboxylated proteins (matrix gla-protein (MGP) and osteocalcin), the large (aggregating) and small non-aggregating (biglycan, decorin, and related) proteoglycans, as well as the collagens and the matrix proteins which define matrix structure. In vitro studies show that BSP is the most effective apatite nucleator, and that biglycan and the more phosphorylated variants of OPN can also cause apatite formation. The proteoglycan aggregates and MGP are the most effective inhibitors of apatite formation in the growth plate. FTIR microscopic and FTIR imaging studies of murine knockouts of MGP, biglycan, decorin, osteocalcin, ON, and OPN provide confirmation of the effects predicted from *in vitro* studies. Specifically, in the absence of MGP cartilage calcification is increased relative to age matched controls. Animals with targeted disruption of OPN have more bone mineral and the mineral crystals are larger than those in the wildtype. ON knockouts have larger bone mineral crystals, but compared to age-matched controls, the most significant alteration is in the knockout animals' mature collagen cross-links. Osteocalcin knockouts have an increased mineral density, but the mineral contains immature crystals relative to controls. Biglycan knockouts contain less mineral, but compared to the wildtype, the crystal size of this mineral is increased.

Introduction

Mineral deposition in the lower half of the epiphyseal growth plate is governed by processes similar to those in other physiologically controlled extracellular biomineralization sites (bones, teeth, shells and other exoskeletons) [1-7]. These processes include both physicochemical events and cell and matrix mediated events. From the physical chemical point of view, elevations in the solution concentration of precipitating ions reduce the energy required for both homogeneous nucleation (without a preformed substrate) and heterogeneous and epitaxial nucleation (on surfaces resembling the structure of the mineral crystal). The cells regulate the initial mineralization process by producing extracellular matrix vesicles (ECMVs) which provide sites for mineral deposition [8-13]. Cells also regulate the flux of ions into the extracellular matrix [5]. Both ECMVs and cells provide enzymes that modify the extracellular matrix, preparing it for calcification [14-17].

In bones, teeth, and calcifying cartilage, the mineral is a micro-crystalline analogue of geologic hydroxyapatite [4-6]. This apatite is hydroxide deficient [18] and contains acid phosphate and carbonate substituents [6]. The majority of this apatitic mineral is formed at discrete sites by heterogeneous nucleation processes [3,4]. There are numerous proteins within the growth plate which influence the calcification process by orchestrating the differentiation, proliferation, and maturation of cells, and the organization of the matrix [19]. This manuscript reviews the extracellular matrix proteins of the growth plate that affect both the physicochemical process and the matrix mediated nucleation and growth of apatitic crystals, along with a brief discussion of systems used to demonstrate these functions. Growth factors and intracellular proteins, except as they affect the extracellular matrix proteins, will not be discussed.

Fig. 1 outlines the proteins found in each of the morphologically defined zones of the growth plate, indicating where their synthesis is maximum. Because the distributions of these proteins vary with species and age, they are presented in a qualitative format. Cytokines and cellular proteins that are phenotypic markers for these cells are also noted. The extracellular matrix proteins probably have more than one function, including, but not limited to: matrix organization and turnover; recruitment and binding of cells and growth factors; maintaining appropriate mechanical properties and tissue hydration, and controlling the mineralization process. The actions of these proteins have been studied in cell and organ culture in which emphasis has predominately been on their expression. In vitro, in the absence of cells, their effects on apatite nucleation and growth have been assessed in a variety of systems, described in detail elsewhere [1,2,20-33]. Animal models, as illustrated below, generated by genetic manipulation (transgenics/ knockouts) and naturally occurring mutants have been evaluated in terms of morphology of the growth plate, mineral content, and in some cases, ability to repair. Since fracture healing mimics endochondral ossification [34], models of fracture healing in older animals provide great insights into the roles of these proteins in earlier stages of development.

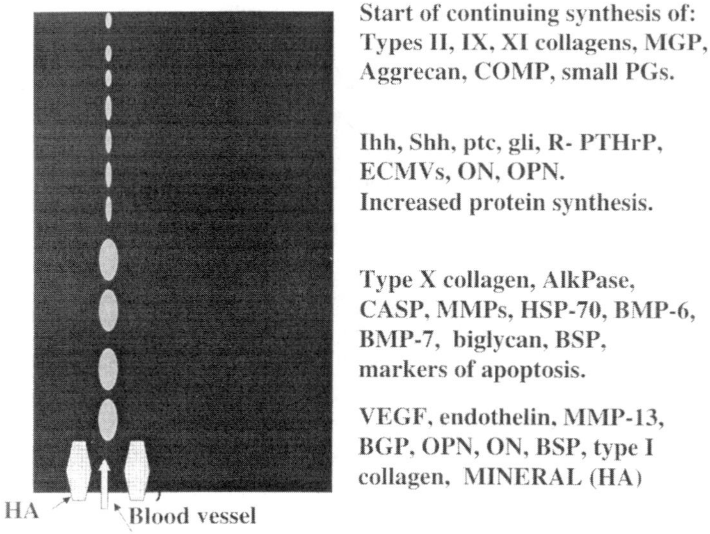

Figure 1. Cartoon illustrating the relative distribution of extracellular matrix proteins within the different zones of the growth plate. Proteins shown are phenotypic markers and include growth factors and their regulators such as the BMPs, alkaline phosphatase (APase), and the PTHrP receptor (R-PTHrP). The dark octagons represent mineral, the light ovals, chondrocytes. Sites of maximal enzyme activities of alkaline phosphatase (Apase), and metalloproteinases (MMPs) are indicated. Cytokines and receptors that are phenotypic markers of the stage of chondrocyte development are also shown. These include indian hedgohog (Ihh), sonic hedgehog (Shh), the receptors (ptc, gli), the bone morphogenetic proteins (BMPs), as well as vascular endothelial growth factor (VEGF), and heat shock protein (HSP-70).

Matrix Proteins and Cartilage Calcification

Collagens

The collagens are the most abundant of the growth plate matrix proteins, types II, IX, and XI being widely distributed throughout the entire growth plate [35-37]. There are no examples of viable animals in which there is no type II collagen expression, but overexpression of type II collagen results in impaired fracture healing, and by analogy, altered endochondral ossification [38]. Additionally, abnormalities in type II collagen have been associated with a variety of chondrodysplasias in humans and mice. In these conditions, growth plate development is impaired and shortness of stature is the predominant phenotype [39-44]. The significance of type II collagen for matrix mineralization is less clear, as there are several studies suggesting that in the growth plate and in cultures of hypertrophic chondrocytes expressing a mineralizing matrix, it is the type I collagen that provides the template for matrix mineralization [45-47]. It is not known whether, type II collagen by itself can support apatite formation in solution.

Type I collagen is expressed in the growth plate [35] and in cultures mimicking the growth plate [45] and this protein, as reviewed elsewhere, has been shown to provide a template for apatite deposition [5,8], although as discussed below, the matrix proteins associated with this collagen are believed to act as the apatite nucleators. Co-expressed with type II collagen are types XI and type IX collagen. Animals with targeted disruption (knockouts) of type IX collagen have forms of osteoarthritis [48-49]. Although in its end stage osteoarthritis is associated with pathologic calcification, direct effects on biomineralization have not been documented. A type XI knockout has not been reported.
Type X collagen is a relatively unique product of hypertrophic chondrocytes [50], however there is a report that hypertrophic chondrocytes in tracheal cartilage do not express this "phenotypic marker" protein [51]. This small chain collagen forms hexamer-like structures which appear to provide a template for the deposition of type I collagen during endochondial ossification. In solution type X collagen has no direct effect on mineralization [1], and type X collagen knockout animals do not appear to have any mineral abnormality [52]. In contrast, transgenic animals which produce an abnormal type X collagen with a "kinked" structure, show major disruption of the mineral organization, although the mineral properties themselves are not effected [53].

Proteoglycans

The most abundant of the noncollagenous proteins in the growth plate are the proteoglycans [54-55]. These can be subdivided into the large aggregating proteoglycans, aggrecan and epiphican [55-56], the smaller proteoglycans such as decorin, biglycan, lumican, fibromodulin, and osteoadherin [56], the cell surface proteoglycans such as syndecan and glypican, and the heparin sulfate proteoglycans such as perlecan [57]. In cell culture, proteoglycans have been reported both to inhibit mineralization [58] and to promote it [59], however the lack of identification of the types of proteoglycan makes it difficult to compare these studies.

In solution, the large aggregating proteoglycans are effective mineralization inhibitors [1,2,5,23], while the small biglycan, a major proteoglycan constituent of the hypertrophic cell matrix [60] is a promoter of mineralization [24]. Decorin, in contrast has little effect on in vitro mineralization [24]. The effects of the other small proteoglycans on mineralization have not yet been reported. Knockout and transgenic animals and genetic analyses of human diseases provide additional insight into the functions of the proteoglycans. Brachymorphic mice with undersulfated PGs are dwarfed [5]. The biglycan knockout mouse contains less mineral in their bones, while the mineral crystals are smaller, in agreement with the view that biglycan is one of the apatite nucleators [61]. Turner's syndrome and Klinefelder's syndromes are associated with over- and under-expression- of biglycan, and respectively show premature- and delayed- growth plate closure, and excessively short and long limbs [62]. The perlecan knockout [57] has chondrodysplasia, with shortened collagen fibrils, elevated expression of other cartilage matrix protein genes, and excessive degradation, implying perlecan, predominately a basement membrane protein, protects cartilage from break down.

Increases in enzymes associated with proteoglycan degradation have been demonstrated in the lower half of the growth plate [15,63,64], implying that there is modification of the larger proteoglycans and other matrix components coincident with chondrocyte hypertrophy and mineralization. The growth plate proteinases include MMP-1 (interstitial collagenase), MMP-2 (gelatinase), MMP-3 (stromelysin-1), MMP-7 (matrilysin), MMP-8 (gelatinase), MMP-9, and MMP-13, as well as the cathepsins, lysosomal proteinases that degrade extracellular matrix proteins [65]. Addition of stromelysin (MMP-3) to micro-mass chondrocyte cultures derived from chick limb-buds accelerated calcification, as did degradation of proteoglycans by addition of other enzymes [58]. Similarly, doxycycline, which inhibits several of the metalloproteinases, prevented calcification in organ culture [66].

Fibronectin

Fibronectin is one of the first proteins produced by mesenchymal cells in culture, and its functions are known to include organization of the collagen matrix, and cell binding [67]. In vitro, soluble fibronectin retards apatite crystal growth, but when immobilized, it promotes apatite formation [30]. Mutations in fibronectin genes have not been reported to cause abnormalities in growth plate development or mineralization, however fibronectin fragments have been associated with activation of stromelysin, implying that this molecule may also be important for matrix turnover [68].

Matrix gla protein and osteocalcin

The gamma carboxy glutamate containing proteins found in the epiphysis include the more abundant matrix gla protein (MGP) and low levels of bone gla protein (osteocalcin) [5]. In vitro effects of MGP on hydroxyapatite formation and growth are unknown, probably because of the difficulty in isolating the protein and maintaining it in solution due to its high hydrophobicity. In contrast, osteocalcin is an effective mineralization inhibitor [32,33]. Osteocalcin forms complexes with several other matrix proteins, including osteopontin, and osteocalcin-osteopontin complexes can facilitate mineralization in vitro in the absence of cells, while the individual proteins have the opposite effect (Fig. 2). The precise function of osteocalcin in the growth plate is not known, and no growth plate abnormalities were noted in the osteocalcin knockout [69,70]. The osteocalcin knockout has thickened bones, but only shows an altered phenotype when ovariectomized. The mineral in the bones of the osteocalcin knockout fails to mature, suggesting a role for this protein in the regulation of mineral turnover. The MGP knockout mouse, in contrast, shows extensive aberrant calcification [71,72], with excessively large crystals, pointing to the role of this protein as a mineralization inhibitor. Viral-induced overexpression of MGP in a chondrocyte cell line also resulted in excessive calcification [73], and Keutel's syndrome, a human disease associated with abnormal cartilage calcification, has been linked to abnormalities in the MGP gene [74].

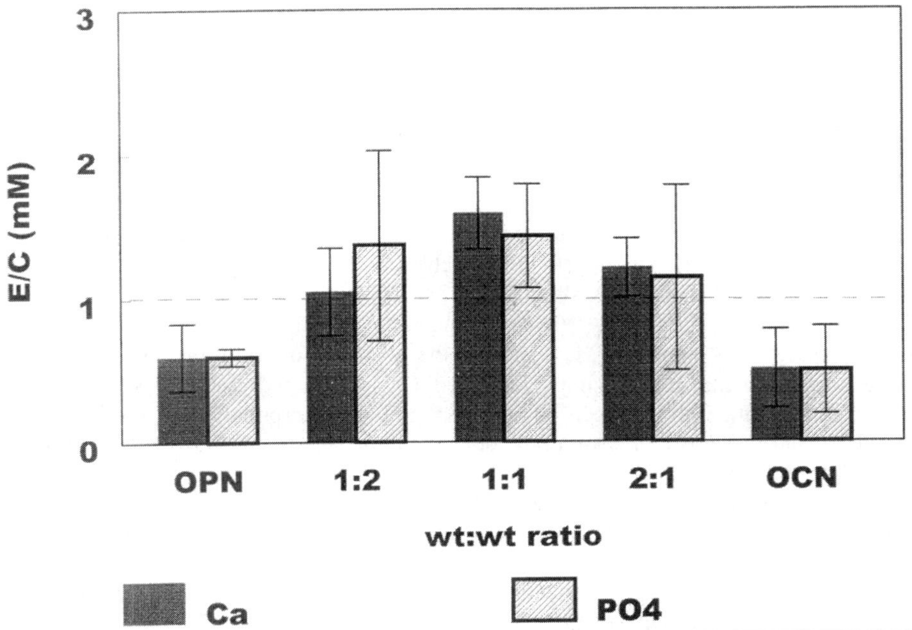

Figure 2. Effects of osteopontin-osteocalcin complexes on HA formation in a gelatin gel system [21] Bars show mineral ion accumulation when calcium and phosphate diffused into a 10% gelatin gel containing the indicated concentrations of Osteocalcin (OCN) and Osteopontin (OPN) in a 100 μl band. Data presented as experimental/control (E/C) for each experiment, shows mean +/- SD for n=3 independent gels.

Matrilins and CASP

The matrilins originally named cartilage oligomeric matrix protein (COMP) or CMP (cartilage matrix proteins) are products of all chondrocytes [75,76]. Matrilin-1, which is more abundant in the lower half of the epiphysis, associates with collagen and proteoglycans, and is thought to be important for regulation of collagen organization [75], and it appears to play a role in regulating the organization of the collagen fibrils. However, while matrilin-1 knockout animals show no abnormalities in their skeletons [77], some variants of human pseudoachondroplasia have been linked to altered matrilin genes [78]. Direct effects on mineralization in vitro have not yet been reported, although it is likely that similar to type X collagen, this protein will not have a direct effect. CRTAP (cartilage associated protein) is a unique product of hypertrophic chondrocytes in the chick [79]. It has also been identified in mouse [80] and man [81]. Its effect, if any, on the mineralization process has not yet been determined.

Phosphorylated Proteins

The phosphorylated proteins synthesized by mature chondrocytes and present in the extracellular matrix of the growth plate include bone sialoprotein (BSP), osteopontin

(OPN), and osteonectin (ON) [82,83]. In the pig, OPN and ON are found throughout the cartilage and bone, while BSP is restricted to the calcifying cartilage and bone [84] implying different functions. Each of these proteins in solution have well documented effects on apatite formation; In vitro BSP is a potent hydroxyapatite nucleator at low concentrations [20,25] and an inhibitor at higher concentrations [25,85]. OPN, as extracted from bone, is an inhibitor of apatite formation (Fig. 2) and growth [20,26,5], although the more highly phosphorylated milk protein is an apatite nucleator [2]. ON is both an apatite inhibitor and a nucleator [33,86]. Mice that lack each of these proteins have been generated and they have distinct bone phenotypes, but their calcified cartilage appears normal. The BSP-knockout has diminished mechanical strength [Aubin, personal communication]. The OPN knockout has increased mineral content with larger than normal mineral crystals [2] suggestive of an impaired remodeling process. The OPN knockout also does not remodel bone properly [87]. The ON knockout is mechanically weaker, and has slightly modified mineral properties [88], but more significantly has increased collagen maturity, indicative of impaired matrix degradation [Boskey, unpublished data]. These proteins are phosphorylated by specific kinases prior to the initiation of calcification, and blocking kinase activity in the mineralizing micro-mass chick limb bud culture system decreases mineralization [89], without effects on cell proliferation or morphology.

There are a variety of naturally occurring animal models which provide additional insight into the importance of the phosphorylated proteins in cartilage calcification. In a turkey model of tibial dyschondroplasia, in which fibrous non-calcified lesions develop, chondrocytes in the normal animals expressed type II collagen, OPN, and type X collagen, whereas the chondrocytes in the lesions expressed neither type X collagen nor OPN [90], implying a role for OPN in mineralization. Rats that lack the ability to synthesize their own ascorbic acid (a requirement for collagen synthesis), and therefore exhibit defective fracture healing in the absence of dietary ascorbate supplementation, provide an interesting insight into the functions of the phosphorylated cartilage proteins. During fracture healing, chondrocytes in the callus of ascorbate deficient animals expressed MGP and ON, but no OPN [91], while all three proteins were expressed by hypertrophic chondrocytes in the ascorbate supplemented rats, again demonstrating the importance of OPN. Hypophosphatemia and other models of nickets are also associated with decreased matrix protein phosphorylation and impaired minerilation [5].

Conclusions

Mineralization of the calcified cartilage within the growth plate is a critical step in the development of bone. The mineral stabilizes the calcified cartilage anlage, and is important for the vascular invasion and remodeling of the cartilaginous tissue into bone. The sequential steps which enable that process to occur are still under investigation. Studies with cell and organ cultures, and studies of transgenic animals are providing insights. From this review it should be apparent that there are proteins in the non-mineralizing areas of the growth plate (e.g., MGP, aggrecan) that are effective inhibitors of mineralization, and that their removal makes the physical chemical process of nucleation easier. Matrix vesicles, because they accumulate ions [92], facilitate nucleation [13], and also provide enzymes that degrade the matrix [14], aid in the initiation of the mineralization process. Within the hypertrophic cell zone, specific proteins facilitate the formation, growth and proliferation of the apatite crystals. The functions of these proteins most likely are redundant, as suggested by the failure of any of the knockout mice studied to date to reveal a total cessation of calcification, as is seen in the parathyroid hormone related peptide (PTHrP) knockout [93] and the core binding factor a (cbfa-1) knockout [94]. This redundance is critical because of the essential nature of the mineralization process.

Acknowledgements

Supported by NIH grants AR037661 and DE04141.

References

[1] Boskey AL 1989 Noncollagenous Matrix Proteins and their role in Mineralization. Bone and Mineral 6:111-123.

[2] Boskey AL, Paschalis EP 2000 Matrix Proteins and Biomineralization. In: Davis JE (ed.). Bone Engineering. emsquared, Inc , Toronto, Canada, pp 44-62.

[3] Glimcher MJ 1989 Mechanisms of calcification in bone: Role of collagen fibrils and collagen-phosphoprotein complexes in vitro and in vivo. Anat Rec 224:139-153.

[4] Glimcher MJ 1996 The nature of the mineral phase in bone: biological and clinical implications. In: Avioli LV, Krane SM (eds.). Metabolic Bone Disease. Academic Press, New York, NY pp. 23-52.

[5] Robey PG, Boskey AL 1996 The Biochemistry of Bone, In: Marcus R, Feldman D, Kelsey J (eds.). Osteoporosis. Academic Press, New York, NY pp. 95-184.

[6] Boskey AL. 2001 Bone Mineralization. In: Cowin SC (ed.). Bone Mechanics Handbook, second edition, CRC Press, Boca Raton, Fla. pp. 5.1-5.33.

[7] Belcher AM, Gnooch S 2000 Protein components and inorganic structure in shell nacre. In: Baeuerlein E (ed.). Biomineralization. Wiley-VCH, Weinheim, Germany pp. 223-250.

[8] Akisaka T, Nakayama M, Yoshida H, Inoue M 1998 Ultrastructural modifications of the extracellular matrix upon calcification of growth plate cartilage as revealed by quick-freeze deep etching technique. Calcif Tissue Int 63:47-56

[9] Anderson HC 1995 Molecular biology of extracellular matrix vesicles. Clin Orthopaed 314:266-280.

[10] Kirsch T, Nah HD, Shapiro IM, Pacifici M 1997 Regulated production of mineralization-competent matrix vesicles in hypertrophic chondrocytes. J Cell Biol 137:1149-60, 1997.

[11] Kirsch T, Claassen H. 2000 Matrix vesicles mediate mineralization of human thyroid cartilage. Calcif Tissue Int. 66:292-7.

[12] Wu LN, Ishikawa Y, Sauer GR, Genge BR, Mwale F, Mishima H, Wuthier RE 1995 Morphological and biochemical characterization of mineralizing primary cultures of avian growth plate chondrocytes: Evidence for cellular processing of Ca^{2+} and Pi prior to matrix mineralization. J Cell Biochem 57: 218-237.

[13] Wu LN, Genge BR, Dunkelberger DG, LeGeros RZ, Concannon B, Wuthier RE 1997 Physicochemical characterization of the nucleational core of matrix vesicles. J Biol Chem 272:4404-11.

[14] Dean DD, Boyan BD, Muniz OE, Howell DS, Schwartz Z 1996 Vitamin D metabolites regulate matrix vesicle metalloproteinase content in a cell maturation-dependent manner. Calcif Tissue Int. 59:109-16.

[15] Ehrlich MG, Armstrong AL 1997 Aggregate degradation by growth plate proteases. Clin Orthop. 334:298-304.

[16] Golub EE, Harrison G, Taylor AG, Camper S, Shapiro IM 1992 The role of alkaline phosphatase in cartilage mineralization. Bone Mineral 17:273-278.

[17] D'Angelo M, Yan Z, Nooreyazdan M, Pacifici M, Sarment DS, Billings PC, Leboy PS 2000 MMP-13 is induced during chondrocyte hypertrophy. J Cell Biochem. 77:678-93.

[18] Loong CK, Rey C, Kuhn LT, Combes C, Wu Y, Chen S, Glimcher MJ 2000 Evidence of hydroxyl-ion deficiency in bone apatites: an inelastic neutron-scattering study. Bone. 26: 599-602.

[19] Sandell LJ, Adler P 1999 Developmental patterns of cartilage. Front Biosci. D731-42.

[20] Boskey AL 1995 Osteopontin and related phosphorylated sialoproteins: Effects on mineralization. Ann NY Acad Sci 760:249-256.

[21] Boskey, AL, Maresca M, Appel J 1989 The Effects of Noncollagenous Matrix Proteins on Hydroxyapatite formation and proliferation in a collagen gel system, Conn Tissue Res, 21: 171-178.

[22] Boskey AL, Maresca M, Ullrich W, Doty SB, Butler WT, Prince CW: Osteopontin-hydroxyapatite interactions in vitro: inhibition of hydroxyapatite formation and growth in a gelatin-gel. Bone and Mineral 22:147-159.

[23] Chen CC, Boskey AL 1986 The effects of proteoglycans from different cartilage types on in vitro hydroxyapatite proliferation. Calcif Tissue Int. 39:324-7.

[24] Boskey AL, Spevak L, Doty SB, Rosenberg L 1997 Effects of bone CS-proteoglycans, DS-decorin, and DS-biglycan on hydroxyapatite formation in a gelatin gel. Calcif Tissue Int. 61:298-305.

[25] Hunter GK, Goldberg HA 1993 Nucleation of hydroxyapatite by bone sialoprotein. PNAS US 90:8562-8565.

[26] Hunter GK, Kyle CL, Goldberg HA 1994 Modulation of crystal formation by bone phosphoproteins: structural specificity of the osteopontin-mediated inhibition of hydroxyapatite formation. Biochem J 300:723-728.

[27] Hunter GK, Hauschka PV, Poole AR, Rosenberg LC, Goldberg HA 1996 Nucleation and inhibition of hydroxyapatite formation by mineralized tissue proteins. Biochem J 317:59-64.

[28] Hunter GK, Szigety SK 1992 Effects of proteoglycan on hydroxyapatite formation under non-steady-state and pseudo-steady-state conditions. Matrix. 12:362-8.

[29] Schinke T, Amendt C, Trindl A, Poschke O, Muller-Esterl W, Jahnen-Dechent W 1996 The serum protein alpha2-HS glycoprotein/fetuin inhibits apatite formation in vitro and in mineralizing calvaria cells. A possible role in mineralization and calcium homeostasis J Biol Chem. 271:20789-96.

[30] Couchourel D, Escoffier C, Rohanizadeh R, Bohic S, Daculsi G, Fortun Y, Padrines M 1999 Effects of fibronectin on hydroxyapatite formation. J Inorg Biochem. 73:129-36.

[31] Tsortos A, Ohki S, Zieba A, Baier RE, Nancollas GH 1996 The Dual Role of Fibrinogen as Inhibitor and Nucleator of Calcium Phosphate Phases: The Importance of Structure. J Colloid Interface Sci. 177:257-262.

[32] Boskey AL, Wians FH Jr, Hauschka PV 1985 The effect of osteocalcin on in vitro lipid- induced hydroxyapatite formation and seeded hydroxyapatite growth. Calcif Tissue Int. 37:57-62.

[33] Romberg RW, Werness PG, Riggs BL, Mann KG 1986 Inhibition of hydroxyapatite crystal growth by bone-specific and other calcium-binding proteins. Biochemistry. 25:1176-80.

[34] Ferguson C, Alpern E, Miclau T, Helms JA 1999 Does adult fracture repair recapitulate embryonic skeletal formation? Mech Dev 87:57-66.

[35] Kosher RA, Kulyk WM, Gay SW 1986 Collagen gene expression during limb cartilage differentiation. J Cell Biol 102:1151-6.

[36] Balmain N, Leguellec D, Elkak A, Nars G, Toury R, Schoevaert D 1995 Zonal variations of types II, IX and XI collagen mRNAs in rat epiphyseal cartilage chondrocytes: quantitative evaluation of in situ hybridization by image analysis of radioautography. Cell Mol Biol (Noisy-le-grand) 41:197-212.

[37] Alini M, Matsuui Y, Dodge GR, Poole AR 1992 The extracellular matrix of cartilage in the growth plate before and during calcification: changes in composition and degradation of type II collagen. Calcif Tiss Int 50:327-335.

[38] Hiltunen A, Metsaranta M, Virolainen P, Aro HT, Vuorio E 1994 Retarded chondrogenesis in transgenic mice with a type II collagen defect results in fracture healing abnormalities.. Dev Dyn. 200:340-9.

[39] Fernandes RJ, Wilkin DJ, Weis MA, Wilcox WR, Cohn DH, Rimoin DL, Eyre DR 1998 Incorporation of structurally defective type II collagen into cartilage matrix in kniest chondrodysplasia. Arch Biochem Biophys. 355(2):282-90.

[40] Weis MA, Wilkin DJ, Kim HJ, Wilcox WR, Lachman RS, Rimoin DL, Cohn DH, Eyre DR 1998 Structurally abnormal type II collagen in a severe form of Kniest dysplasia caused by an exon 24 skipping mutation J Biol Chem. 273(8):4761-8.

[41] Mundlos S, Chan D, McGill J, Bateman JF 1996 An alpha 1(II) Gly913 to Cys substitution prevents the matrix incorporation of type II collagen which is replaced with type I and III collagens in cartilage from a patient with hypochondrogenesis. Am J Med Genet. 63:129-36.

[42] Maddox BK, Garofalo S, Keene DR, Smith C, Horton WA 1997 Type II collagen pro-alpha-chains containing a Gly574 Ser mutation are not incorporated into the cartilage matrix of transgenic mice. Matrix Biol. 16:93-103.

[43] Tiller GE, Polumbo PA, Weis MA, Bogaert R, Lachman RS, Cohn DH, Rimoin DL, Eyre DR 1995 Dominant mutations in the type II collagen gene, COL2A1, produce spondyloepimetaphyseal dysplasia, Strudwick type. Nat Genet. 11:87-9.

[44] Mortier GR, Weis M, Nuytinck L, King LM, Wilkin DJ, De Paepe A, Lachman RS, Rimoin DL, Eyre DR, Cohn DH 2000 Report of five novel and one recurrent COL2A1 mutations with analysis of genotype-phenotype correlation in patients with a lethal type II collagen disorder.J Med Genet. 37:263-71.

[45] Boskey AL, Stiner D, Binderman I, Doty SB 2000 Type I collagen influences cartilage calcification: an immunoblocking study in differentiating chick limb-bud mesenchymal cell cultures. J Cell Biochem. 79:89-102.

[46] Bianco P, Cancedda FD, Riminucci M, Cancedda R 1998 Bone formation via cartilage models: the "borderline" chondrocyte. Matrix Biol. 17:185-92.

[47] Yagami K, Kakuta S, Tachibana H, Kimura Y, Nagumo M 2000 Establishment of a cell line with phenotypes of chondrocyte from a human osteogenic sarcoma of the mandible. J Oral Pathol Med. 29:321-30.

[48] Hagg R, Hedbom E, Mollers U, Aszodi A, Fassler R, Bruckner P 1997 Absence of the alpha1(IX) chain leads to a functional knock-out of the entire collagen IX protein in mice. J Biol Chem. 272:20650-4.

[49] Fassler R, Schnegelsberg PN, Dausman J, Shinya T, Muragaki Y, McCarthy MT, Olsen BR, Jaenisch R 1994 Mice lacking alpha 1 (IX) collagen develop noninflammatory degenerative joint disease. Proc Natl Acad Sci U S A 91:5070-4.

[50] Jacenko O 2000 Genetic-engineered models of skeletal diseases. I. Collagen type X. Methods Mol Biol. 137:471-90.

[51] Sasano Y, Takahashi I, Mizoguchi I, Kagayama M, Takita H, Kuboki Y 1998 Type X collagen is not localized in hypertrophic or calcified cartilage in the developing rat trachea. Anat Embryol (Berl). 197:399-403.

[52] Paschalis EP, Jacenko O, Olsen B, deCrombrugghe B, Boskey AL 1996 The role of type X collagen in endochondral ossification as deduced by Fourier transform infrared microscopy analysis. Connect Tissue Res 35:371-7.

[53] Paschalis EP, Jacenko O, Olsen B, Mendelsohn R and Boskey AL 1996 FT-IR microscopic analysis identified alterations in mineral properties in bones from mice transgenic for type X collagen, Bone, 18:151-156.

[54] van Osch GJ, van der Veen SW, Buma P, Verwoerd-Verhoef HL 1998 Effect of transforming growth factor-beta on proteoglycan synthesis by chondrocytes in relation to differentiation stage and the presence of pericellular matrix. Matrix Biol 17:413-24.

[55] Knudson CB, Knudson W 2001 Cartilage proteoglycans. Semin Cell Dev Biol. 12:69-78.

[56] Matsushima N, Ohyanagi T, Tanaka T, Kretsinger RH 2000 Super-motifs and evolution of tandem leucine-rich repeats within the small proteoglycans--biglycan, decorin, lumican, fibromodulin, PRELP, keratocan, osteoadherin, epiphycan, and osteoglycin. Proteins. 38:210-25.

[57] Costell M, Gustafsson E, Aszodi A, Morgelin M, Bloch W, Hunziker E, Addicks K, Timpl R, Fassler R 1999 Perlecan maintains the integrity of cartilage and some basement membranes. J Cell Biol. 47:1109-22.

[58] Boskey AL, Stiner D, Binderman I, Doty SB 1997 Effects of proteoglycan modification on mineral formation in a differentiating chick limb-bud mesenchymal cell culture system. J Cell Biochem 64:632-43.

[59] Hunter GK, Weinert CA 1996 Inhibition of proteoglycan biosynthesis decreases the calcification of chondrocyte cultures. Connect Tissue Res. 35:379-84.

[60] Takagi M, Kamiya N, Urushizaki T, Tada Y, Tanaka H 2000 Gene expression and immunohistochemical localization of biglycan in association with mineralization in the matrix of epiphyseal cartilage. Histochem J. 32:175-86.

[61] Xu T, Bianco P, Fisher LW, Longenecker G, Smith E, Goldstein S, Bonadio J, Boskey A, Heegaard AM, Sommer B, Satomura K, Dominguez P, Zhao C, Kulkarni AB, Robey PG, Young MF 1998 Targeted disruption of the biglycan gene leads to an osteoporosis-like phenotype in mice Nat Genet. 20:78-82.

[62] Geerkens C, Vetter U, Just W, Fedarko NS, Fisher LW, Young MF, Termine JD, Robey PG, Wohrle D, Vogel W 1995 The X-chromosomal human biglycan gene BGN is subject to X inactivation but is transcribed like an X-Y homologous gene. Hum Genet. 96:44-52.

[63] Cawston T, Billington C, Cleaver C, Elliott S, Hui W, Koshy P, Shingleton B, Rowan A1996 The regulation of MMPs and TIMPS in cartilage turnover, Annals NY Acad Sci, 120-129.

[64] Ehrlich MG, Armstrong AL 1997 Aggregate degradation by growth plate proteases. Clin Orthop. 334:298-304.

[65] Soderstrom M, Salminen H, Glumoff V, Kirschke H, Aro H, Vuorio E 1999 Cathepsin expression during skeletal development. Biochim Biophys Acta 1446:35-46.

[66] Cole AA, Chubinskaya S, Luchene LJ, Chlebek K, Orth MW, Greenwald RA, Kuettner KE, Schmid TM 1994 Doxycycline disrupts chondrocyte differentiation and inhibits cartilage matrix degradation. Arth Rheum 37:1727-34.

[67] Heinegard D, Oldberg A 1989 Structure and biology of cartilage and bone matrix noncollagenous macromolecules. FASEB J. 3:2042-51.

[68] Bewsey KE, Wen C, Purple C, Homandberg GA 1996 Fibronectin fragments induce the expression of stromelysin-1 mRNA and protein in bovine chondrocytes in monolayer culture. Biochim Biophys Acta. 1317:55-64.
[69] Ducy P, Desbois C, Boyce B, Pinero G, Story B, Dunstan C, Smith E, Bonadio J, Goldstein S, Gundberg C, Bradley A, Karsenty G 1996 Increased bone formation in osteocalcin-deficient mice. Nature 382:448-52.
[70] Boskey AL, Gadaleta S, Gundberg C, Doty SB, Ducy P, Karsenty G 1998 Fourier transform infrared microspectroscopic analysis of bones of osteocalcin-deficient mice provides insight into the function of osteocalcin. Bone. 23:187-96.
[71] Luo G, Ducy P, McKee MD, Pinero GJ, Loyer E, Behringer RR, Karsenty G 1997 Spontaneous calcification of arteries and cartilage in mice lacking matrix GLA protein. Nature. 386:78-81.
[72] Boskey AL, Karsenty G, McKee MD 2000 Mineral characterization of bones and soft tissues in matrix gla protein deficient mice. In: Goldber M, Boskey A, Robinson C (eds.). Chemistry and Biology of Mineralized Tissues. Am Acad Orthopaed Surgeons, Chicago, Il, pp. 63-67.
[73] Yagami K, Suh JY, Enomoto-Iwamoto M, Koyama E, Abrams WR, Shapiro IM, Pacifici M, Iwamoto M 1999 Matrix GLA protein is a developmental regulator of chondrocyte mineralization and, when constitutively expressed, blocks endochondral and intramembranous ossification in the limb. J Cell Biol. 147:1097-108.
[74] Munroe PB, Olgunturk RO, Fryns JP, Van Maldergem L, Ziereisen F, Yuksel B, Gardiner RM, Chung E 1999 Mutations in the gene encoding the human matrix Gla protein cause Keutel syndrome. Nat Genet. 21:142-4, 1999.
[75] Thur J, Rosenberg K, Nitsche DP, Pihlajamaa T, Ala-Kokko L, Heinegard D, Paulsson M, Maurer P 2001 Mutations in cartilage oligomeric matrix protein causing pseudoachondroplasia and multiple epiphyseal dysplasia affect binding of calcium and collagen I, II, and IX. J Biol Chem. 276:6083-92.
[76] Deak F, Wagener R, Kiss I, Paulsson M 1999 The matrilins: a novel family of oligomeric extracellular matrix proteins. Matrix Biol. 18:55-64.
[77] Stevens JW 1999 Pseudoachondroplastic dysplasia: an Iowa review from human to mouse. Iowa Orthop J. 19:53-65.
[78] Aszodi A, Bateman JF, Gustafsson E, Boot-Handford R, Fassler R 2000 Mammalian skeletogenesis and extracellular matrix: what can we learn from knockout mice? Cell Struct Funct. 25:73-84.
[79] Castagnola P, Gennari M, Morello R, Tonachini L, Marin O, Gaggero A, Cancedda R 1997 Cartilage associated protein (CASP) is a novel developmentally regulated chick embryo protein. J Cell Sci. 110:1351-9.
[80] Tonachini L, Morello R, Monticone M, Skaug J, Scherer SW, Cancedda R, Castagnola P 1999 cDNA cloning, characterization and chromosome mapping of the gene encoding human cartilage associated protein (CRTAP). Cytogenet Cell Genet. 87:191-4.
[81] Morello R, Tonachini L, Monticone M, Viggiano L, Rocchi M, Cancedda R, Castagnola P 1999 cDNA cloning, characterization and chromosome mapping of CRTAP encoding the mouse cartilage associated protein. Matrix Biol. 18:319-24.
[82] Barak-Shalom T, Schickler M, Knopov V, Shapira R, Hurwitz S, Pines M 1995 Synthesis and phosphorylation of osteopontin by avian epiphyseal growth-plate chondrocytes as affected by differentiation. Comp Biochem Physiol C Pharmacol Toxicol Endocrinol 111:49-59.
[83] Castagnola P, Bet P, Quarto R, Gennari M, Cancedda R 1991 cDNA cloning and gene expression of chicken osteopontin. Expression of osteopontin mRNA in chondrocytes is enhanced by trypsin treatment of cells. J Biol Chem. 266:9944-9.
[84] Chen J, Zhang Q, McCulloch CA, Sodek J 1991 Immunohistochemical localization of bone sialoprotein in foetal porcine bone tissues: comparisons with secreted phosphoprotein 1 (SPP-1, osteopontin) and SPARC (osteonectin). Histochem J. 23:281-9.
[85] Stubbs JT 3rd, Mintz KP, Eanes ED, Torchia DA, Fisher LW 1997 Characterization of native and recombinant bone sialoprotein: delineation of the mineral-binding and cell adhesion domains and structural analysis of the RGD domain. J Bone Miner Res. 12:1210-22.
[86] Termine JD, Kleinman HK, Whitson SW, Conn KM, McGarvey ML, Martin GR 1981 Osteonectin, a bone-specific protein linking mineral to collagen. Cell. 26:99-105.
[87] Yoshitake H, Rittling SR, Denhardt DT, Noda M 1999 Osteopontin-deficient mice are resistant to ovariectomy-induced bone resorption. Proc Natl Acad Sci U S A. 96:8156-60.
[88] Delany AM, Amling M, Priemel M, Howe C, Baron R, Canalis E 2000 Osteopenia and decreased bone formation in osteonectin-deficient mice. J Clin Invest. 105:915-23.
[89] Boskey AL, Guidon P, Doty SB, Binderman I, Stiner D, Dow D and Leboy P 1996 The Mechanism of ß-glycerophosphate action in mineralizing chick limb-bud mesenchymal cells. J Bone Min Res, 11:1694-1702.

[90] Knopov V, Hadash D, Hurwitz S, Leach RM, Pines M 1997 Gene expression during cartilage differentiation in turkey tibial dyschondroplasia, evaluated by in situ hybridization. Avian Dis. 41:62-72.

[91] Sugimoto M, Hirota S, Sato M, Kawahata H, Tsukamoto I, Yasui N, Kitamura Y, Ochi T, Nomura S 1998 Impaired expression of noncollagenous bone matrix protein mRNAs during fracture healing in ascorbic acid-deficient rats. J Bone Miner Res. 13:271-8.

[92] Kirsch T, Harrison G, Worch KP, Golub EE 2000 Regulatory roles of zinc in matrix vesicle- mediated mineralization of growth plate cartilage. J Bone Miner Res. 15:261-70.

[93] Amizuka N, Warshawsky H, Henderson JE, Goltzman D, Karaplis AC 1994 Parathyroid hormone-related peptide-depleted mice show abnormal epiphyseal cartilage development and altered endochondral bone formation. J Cell Biol. 126:1611-23.

[94] Hoshi K, Komori T, Ozawa H 1999 Morphological characterization of skeletal cells in Cbfa1-deficient mice. Bone. 25:639-51.

The Growth Plate
I.M. Shapiro et al. (Eds.)
IOS Press, 2002

Regulated Production of Mineralization-Competent Matrix Vesicles by Terminally Differentiated Chondrocytes

Wei Wang and Thorsten Kirsch
Department of Orthopaedics & Rehabilitation, Hershey Medical Center, Penn State College of Medicine, Hershey, Pennsylvania, U.S.A.

Abstract. Biomineralization is a highly regulated process which is under strict cellular control. In this study, we show that treatment of hypertrophic chondrocytes with retinoic acid (RA) led to terminal differentiation of these cells and the release of matrix vesicles (MV), which initiate the mineralization process. These vesicles contain high amounts of annexins II, V and VI, and alkaline phosphatase activity. The annexins form Ca^{2+} channels in these vesicles, enabling the influx of Ca^{2+} into the vesicles and the formation of the first crystal phase inside the vesicle lumen. RA-treatment of chondrocyte cultures led to a 3-fold increase in the cytosolic calcium concentration, followed by a relocation of annexins II, V, and VI, which require Ca^{2+} to bind to phospholipids, from the cytoplasm to the plasma membrane, and the release of annexin-containing MV. Chelation of cytosolic calcium with BAPTA-2AM led to significant decrease of mineralization in RA-treated cultures, and to a reduction of the amount of annexins and alkaline phosphatase activity in MV. In addition, these vesicles were not able to take up Ca^{2+}. In conclusion, changes in the concentration of cytosolic calcium regulate the release of mineralization-competent MV from the plasma membrane of terminally differentiated chondrocytes and subsequent mineralization.

Introduction

Chondrocyte differentiation involves complex processes, such as cell proliferation, hypertrophy, terminal differentiation, mineralization and cell death. [1-3] In growth plate cartilage, these events are necessary to ensure normal growth and development of the skeleton. If the same hypertrophic, mineralization, and terminal differentiation events, however, are activated in articular chondrocytes, as shown by our and other laboratories, articular cartilage will experience devastating destructive changes, as seen in osteoarthritis. [4,5] Despite the obvious importance of terminal differentiation and mineralization of chondrocytes during endochondral ossification, very little is known about the regulation of these processes.

Mineralization of growth plate cartilage is under control of the maturing chondrocyte. Thus, the mineralization process in growth plate cartilage and other skeletal tissues must be highly regulated and restricted to sites where mineral formation is required for proper tissue function. Indeed, uncontrolled mineralization can have severe consequences. For example, mineral depositions in articular cartilage leads to inflammation and cartilage destruction. [5,6] Calcification of vascular tissue, including arteries, heart valves, and cardiac muscle, contributes both to morbidity and mortality.

We and others have demonstrated that MV initiate the mineralization process in growth plate cartilage. [3,7] MV are membrane-enclosed particles which are released from the plasma membrane of mineralizing chondrocytes. [7] Recently, we have demonstrated that treatment of hypertrophic growth plate chondrocytes with vitamin C stimulated the release of mineralization-competent MV and subsequent mineralization of these cultures. [3] Thus, it is reasonable to hypothesize that chondrocytes regulate the mineralization process by controlling the release of mineralization-competent MV, and only terminally differentiated chondrocytes release mineralization-competent MV. However, very little is known about the mechanisms involved in regulating the release of mineralization-competent MV from the plasma membrane of terminally differentiated chondrocytes. Several studies have provided evidence that growth plate chondrocytes accumulate large amounts of cytosolic calcium just before the initiation of mineralization. [8,9,10] Thus, it is possible that changes in Ca^{2+} homeostasis in growth plate chondrocytes may play a crucial role in regulating the release of mineralization-competent MV from the plasma membrane of terminally differentiated chondrocytes. To address this question, we isolated hypertrophic chondrocytes from 19 day fetal chick growth plate cartilage and cultured these cells after they have reached confluency, in the absence or presence of retinoic acid (RA). Previously, it has been shown that RA drastically increased mineralization in hypertrophic chondrocyte cultures. [11] MV were isolated from these cultures, analyzed for their composition and ability to initiate mineralization. The role of changes in Ca^{2+} homeostasis in regulating the initiation of mineralization in these cultures was tested by culturing chondrocytes in the presence of the cell-permeable Ca^{2+}-chelating agent, BAPTA-2AM.

Materials and Methods

Chondrocyte Culture

Chondrocytes were isolated from the hypertrophic zone of 19 day chick embryonic tibial growth plate cartilage as described previously.[3] Cells were plated at a density of 3 x 10^6 cells/10cm tissue culture dish and grown in monolayer cultures in Dulbecco's modified Eagle's medium (DMEM; GibcoBRL, Gaithersburg, MD, USA) containing 5% fetal calf serum (FCS; Hyclone, Logan, UT, USA), 2mM L-glutamine, and 50 U/ml penicillin and streptomycin (complete medium). After the cultures have reached confluency, chondrocytes were cultured in the presence of phosphate (1.5mM), and in the absence or presence of (a) RA (35nM; Sigma, St. Louis, MO, USA) or (b) RA (35nM) and 10μM BAPTA-2AM (Molecular Probes Inc., Eugene, OR, USA) for a maximum of 6 days. Cytosolic calcium, $[Ca^{2+}]_i$, in these cultures was measured after 1 day treatment, MV were isolated after 3 day treatment, and the degree of mineralization in these cultures was measured after 6 day treatment.

Measurement of Cytosolic Calcium Concentration $[Ca^{2+}]_i$

Cells were trypsinized and then 2 x 10^6 cells were incubated with 4μM of fura-2AM in complete medium at 37°C. The concentration of cytosolic calcium was measured as described previously.[10] Briefly, labeled cells were resuspended in measuring buffer (140mM NaCl, 5mM KCl, 1mM $CaCl_2$, 20mM (N-[2-hydroxyethyl}-piperazine-N-[2-ethanesulfonic acid] (HEPES), 1mM NaH_2PO_4, 5.5mM glucose pH 7.4), transferred to a cuvette (magnetically stirred) and fluorescence was measured in a fluorimeter (Photon Technology Instruments) using the excitation wavelength of 340nm (Ca^{2+}-bound fura) and the emission wavelength of 510nm. The fluorescence maximum (Fmax) was determined by addition of ionomycin (2pmol; Calbiochem), and the fluorescence minimum (Fmin) was determined in the presence of 1mM EDTA/10mM Tris, pH7.4. $[Ca^{2+}]_i$ was calculated according the following equation; $[Ca^{2+}]_i$ =Kd [(F-Fmin)/(Fmax-F)] with K_d=224nM.

Isolation of Cytosolic and Membrane Fractions

Cells were trypsinized, pelleted and resuspended in phosphate-buffered saline (PBS, pH 7.4). The cell suspension was frozen in liquid nitrogen and thawn three times. After removing cell debris at low speed centrifugation, the membrane fraction was obtained by ultracentrifugation at 100,000 x g for 1h. The supernatant contained the cytosolic fraction.

Isolation of MV

MV were isolated from chondrocyte cultures as described previously.[3] Briefly, the cell layers were incubated with crude collagenase (500U/ml; Sigma, St. Louis, MO, USA) in Hank's balanced salt solution at 37°C for 3h. MV were harvested by differential ultracentrifugation.

Ca^{2+} Uptake by MV

After 24h incubation of MV aliquots (100µg of protein) in 1 ml of synthetic cartilage lymph (SCL) at 37°C in the absence or presence of 200nM of annexin II, V, or VI, MV were pelleted, washed twice in 150mM NaCl, 10mM TES (pH7.4) and 200µM EDTA (buffer 1), and resuspended in 1ml of buffer 1 containing 1µM fura-2 (Molecular Probes Inc.). MV suspensions were then incubated with Triton X-100 to burst MV and to release intralumenal Ca^{2+} (blast method). Changes in the fluorescence ratio, 340:380nm, were measured. 340nm is the excitation wavelength of Ca^{2+}-bound fura, while 380nm is the excitation wavelength of Ca^{2+}-free fura-2.

Measurement of Alkaline Phosphatase (APase) Activity, Calcium (Ca^{2+}), Phosphate (P_i) and Protein Content

The measurement of APase activity, protein content and Ca^{2+} and Pi content in the cell layer and in MV was measured as described previously.[3]

Recombinant Annexin Proteins and Antibodies

Recombinant annexin II, V, or VI were prepared using the pGEX expression vector (Pharmacia, Piscataway,NJ) as described previously.[12] The preparation of antibodies specific for annexin II, V or VI were also described elsewhere.[12]

SDS-Polyacrylamide Gel Electrophoresis and Immunoblotting

Samples were dissolved in 3% SDS sample buffer with dithiothreitol, denatured at 100°C for 3min, and analyzed by electrophoresis in 10% or 12% (w:v) SDS-polyacrylamide gels. Samples were electroblotted onto nitrocellulose filters after electrophoresis. After blocking with a solution of low fat milk protein, blotted proteins were immunostained with primary antibodies followed by peroxidase-conjugated secondary antibodies.

Results

Treatment of cells with RA for 1 day led to an approximately 3-fold increase in $[Ca^{2+}]_i$ compared to untreated cells. Thus, $[Ca^{2+}]_i$ of untreated chondrocytes was 765nM; $[Ca^{2+}]_i$ of RA-treated chondrocytes was 2211nM. Treatment with RA also led to a significant relocation of annexin II, V and VI from the cytoplasm to the plasma membrane (Fig.1). Annexin II, V and VI are cytoplasmic proteins, which in the presence of Ca^{2+} bind to membranes.[13] Co-treatment with RA and BAPTA significantly reduced the percentage of annexin II, V and VI bound to the plasma membrane to levels similar to untreated cells (Fig. 1).

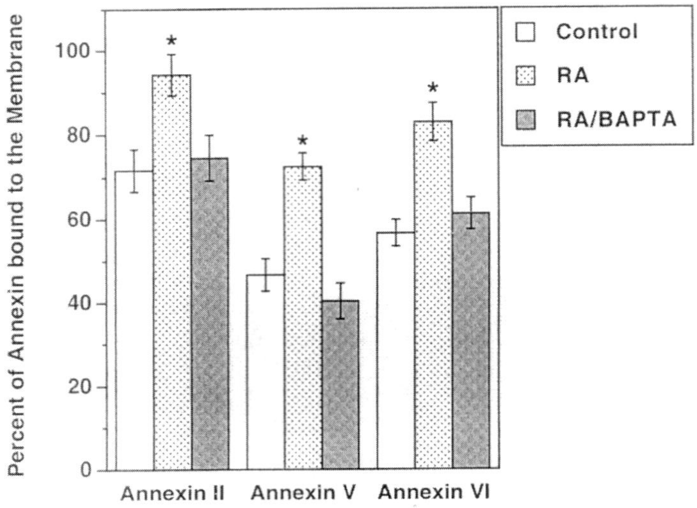

Fig. 1: Percentage of annexin II, V and VI bound to the plasma membrane of untreated (Control), RA-treated, or RA/BAPTA-treated growth plate chondrocytes.
After treatment with RA or RA/BAPTA for 3 days, the membrane and cytosolic fractions were isolated and analyzed by SDS PAGE and immunostaining with antibodies specific for annexin II, V or VI. The densities of the annexin bands in the membrane and cytosolic fractions were quantitated by densitometry. Data were obtained from three different experiments and are expressed as means ± SD; *$p<0.01$.

We next isolated MV from untreated, RA-treated, and RA/BAPTA-treated cultures. Specific APase activity was significantly increased in MV isolated from RA-treated chondrocytes (11,997 ± 2350 nmol/min hydrolyzed pNPP/mg protein) compared to the APase activity in MV isolated from untreated cultures (828 ± 341 nmol/min hydrolyzed pNPP/mg protein). Cotreatment with BAPTA and RA reduced the level of APase activit y in MV to levels similar in MV isolated from untreated cultures (648 ± 292 nmol/min hydrolyzed pNPP/mg protein). Furthermore, only vesicles isolated from RA-treated cultures were able to take up Ca^{2+}, while the other two MV fractions isolated from untreated and RA/BAPTA-treated cells did not take up Ca^{2+}. Immunoblot analysis using antibodies specific for annexin II, V or VI revealed that only MV isolated from RA-treated cultures contained significant amounts of annexin II, V and VI, while MV isolated from untreated cultures or RA/BAPTA co-treated cultures contained only little amounts of these annexins (Fig.2).

We and others have previously demonstrated that annexin II, V and VI form Ca^{2+} channels when inserted into lipid bilayers and mediate Ca^{2+} influx into phosphatidylserine-rich liposomes.[12,14] In addition, annexin II, V or VI-mediated Ca^{2+} influx into liposomes can be inhibited by K-201, a compound which specifically binds to these annexins.[12,15] More importantly, K-201 also inhibited Ca^{2+} uptake and mineralization by authentic MV isolated from chicken growth plate cartilage, suggesting that annexin II, V and VI mediate Ca^{2+} influx into MV and the formation of the first crystal phase inside the vesicle lumen.[12] Here we extend these findings and show that exogenous annexin II, V or VI can restore the ability of Ca^{2+} uptake by MV isolated from untreated chondrocyte cultures. While MV isolated from RA-treated cultures, which contained significant amounts of annexin II, V and VI (see Fig. 2), were able to take up Ca^{2+}, vesicles isolated from untreated cultures showed no significant Ca^{2+} uptake. However, the addition of either 200nM of annexin II, V or VI led to Ca^{2+} influx into these vesicles (Fig. 3).

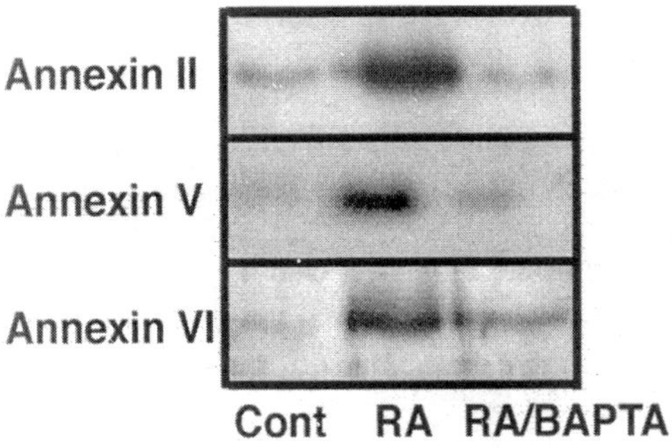

Fig. 2: Immunoblot analysis of MV isolated from untreated, RA-treated and RA/BAPTA-treated chondrocyte cultures with antibodies specific for annexin II, V or VI.
MV were isolated from untreated (Cont), RA-treated (RA), and RA/BAPTA-treated (RA/BAPTA) chondrocyte cultures after 3 day treatment. 50μg of total MV protein were separated by SDS gel electrophoresis. Electrotransferred proteins were immunostained with antibodies specific for annexin II, V or VI.

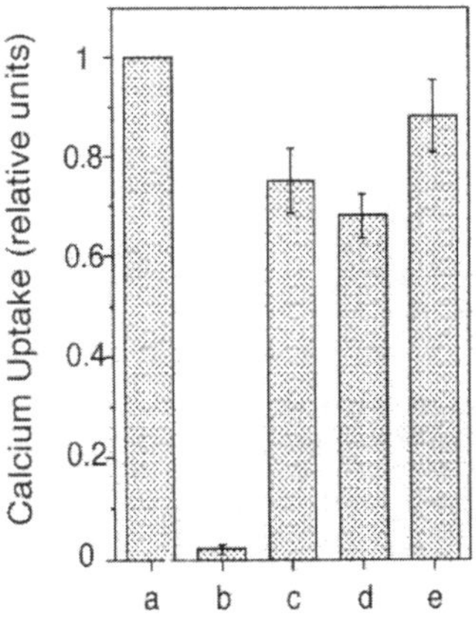

Fig. 3: Ca^{2+} uptake by MV isolated from untreated and RA-treated chondrocyte cultures in the absence or presence of exogenous annexin II, V, or VI.
MV isolated from RA-treated cultures were able to take up Ca^{2+} (a), while MV isolated from untreated cultures showed no significant Ca^{2+} uptake (b). However, addition of 200nM annexin II (c), V (d), or VI (e) to vesicles from untreated cultures restored their abilty to take up Ca^{2+}.

RA treatment led to a more than 5-fold increase in Ca^{2+} and P_i content of the cell layer compared to Ca^{2+} and P_i content of untreated cells. Co-treatment with RA and BAPTA significantly reduced the accumulation of calcium and phosphate in the cell layer to levels similar to levels in untreated cells.

Discussion

In this study we provide evidence that the mineralization process in growth plate cartilage is under strict cellular control, and only mineralization-competent, maturing chondrocytes release MV which initiate the mineralization process. These vesicles contain alkaline phosphatase, an enzyme which cleaves organic phosphate compounds, and thus provides inorganic phosphate.[7] Furthermore, these vesicles contain annexin II, V and VI. As shown in this and our previous studies, annexin II, V and VI are able to mediate Ca^{2+} influx into MV. [12] Ca^{2+} influx into the vesicles is required for the formation of the first crystal phase inside the vesicle lumen.[7,12] RA induces a cascade of events finally leading to matrix mineralization. RA treatment leads to an increase in the cytosolic calcium concentration. Interfering with this increase inhibits mineralization potential of terminally differentiated chondrocytes.

An increase in the cytosolic calcium concentration controls a diverse range of cell functions, including adhesion, motility, gene expression, proliferation and differentiation. Interestingly, many studies have demonstrated that growth plate chondrocytes, especially hypertrophic cells, can accumulate large amounts of cytosolic calcium, suggesting that these increases might regulate chondrocyte differentiation.[8,9,10] Here we provide direct evidence that an increase in the cytosolic calcium concentration in hypertrophic chondrocytes leads to the release of mineralization-competent MV and the initiation of mineralization. A direct consequence of the increased level of cytosolic calcium is the relocation of annexin II, V and VI from the cytoplasm to the plasma membrane. Annexins are proteins which require a certain Ca^{2+} concentration to bind to the membrane.[13] Interestingly, both calcium and annexins have been shown to be involved in vesiculation and membrane budding processes.[16,17] Thus, it is possible that high concentrations of membrane-associated calcium together with annexins trigger the release of mineralization-competent MV.

Once MV are released into the extracellular matrix, they contain all the components which enable them to initiate the mineralization process. In contrast, hypertrophic, non-mineralizing chondrocytes release vesicles, which do not have these components, including alkaline phosphatase and annexins, and thus these vesicles fail to mineralize. The function of these non-mineralizing vesicles still remains to be elucidated.

In conclusion, our study provides evidence that changes in Ca^{2+} homeostasis in terminally differentiated chondrocytes play a crucial role in the regulation of mineralization in growth plate cartilage. Increases in cytosolic calcium trigger the relocation of annexin II, V and VI from the cytoplasm to the plasma membrane and the release of mineralization-competent, annexin-containing MV. Annexin II, V and VI form Ca^{2+} channels in the membrane of the vesicles. In addition, annexin V binds to types II and X collagen. These interactions anchor the vesicles to the extracellular matrix and stimulate Ca^{2+} channels activities of annexin V.[12] Together, the Ca^{2+} channel activities of annexin II and VI, and the stimulated Ca^{2+} channel activities of annexin V lead to the rapid influx of Ca^{2+} into MV required for the formation of the first mineral phase inside the vesicles (Fig. 4) and initiation of mineralization.

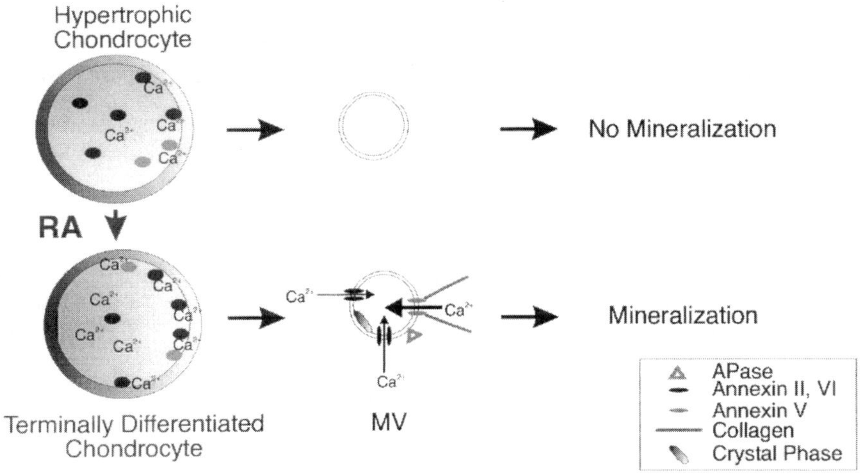

Fig. 4: Model showing the role of an increased cytosolic calcium concentration in terminally differentiated chondrocytes in the release of mineralization-competent MV and the function of annexin II, V and VI in initiation of MV-mediated mineralization.

Acknowledgements

This study was supported by grants from the National Institute of Arthritis and Musculoskeletal and Skin Diseases (AR-43732 and AR-46245 to T.K.)

References

[1] von der Mark K 1986 Differentiation, modulation and dedifferentiation of chondrocytes. Rheumatology **10**:272-315.

[2] Hatori M, Klatte KJ, Teixeira CC, Shapiro IM 1995 End labeling studies of fragmented DNA in the avian growth plate: evidence of apoptosis in terminally differentiated chondrocytes. J Bone Miner Res **10**:1960-1968.

[3] Kirsch T, Nah HD, Shapiro IM, Pacifici M 1997 Regulated production of mineralization-competent matrix vesicles in hypertrophic chondrocytes. J Cell Biol **137**:1149-1160.

[4] Hashimoto S, Ochs RL, Rosen F, Quach J, Mccabe G, Solan J, Seegmiller JE, Terkeltaub R, Lotz M 1998 Chondrocyte-derived apoptotic bodies and calcification of articular cartilage. Proc Natl Acad Sci USA **95**:3094-3099.

[5] Kirsch T, Swoboda B, Nah H-D 2000 Activation of annexin II and V expression, terminal differentiation, mineralization and apoptosis in human osteoarthritic cartilage. Osteoarthritis&Cartilage **8**:294-302.

[6] Doyle DV 1982 Tissue calcification and inflammation in osteoarthritis. J Path **136**:199-216.

[7] Anderson HC 1995 Molecular biology of matrix vesicles. Clin Orthop Rel Res **314**:266-

[8] Gunter TE, Zuscik MJ, Puzas JE, Gunter KK, Rosier RN 1990 Cytosolic free calcium concentrations in avian growth plate chondrocytes. Cell Calcium **11**:445-457.

[9] Iannotti JP,Brighton CT 1989 Cytosolic ionized calcium concentration in isolated chondrocytes from each zone of the growth plate. J Orth Res **7**:511-518.

[10] Kirsch T, Swoboda B, von der Mark K 1992 Ascorbate independent differentiation of human chondrocytes in vitro: simultaneous expression of types I and X collagen and matrix mineralization. Differentiation **52**:89-100.

[11] Iwamoto M, Shapiro IM, Yagami K, Boskey AL, Leboy PS, Adams SL, Pacifici M 1993 Retinoic acid induces rapid mineralization and expression of mineralization-related genes in chondrocytes. Exp Cell Res **207**:413-420.

[12] Kirsch T, Harrison G, Golub EE, Nah H-D 2000 The roles of annexins and types II and X collagen in matrix vesicle-mediated mineralization of growth plate cartilage. J Biol Chem **275**:35577-35583.

[13] Geisow MJ, Walker JH, Boustead C,Taylor W 1988 Annexins - a new family of Ca^{2+} regulated phospholipid-binding proteins. In: Thorn NA, Traiman M, Peterson OH (eds) Molecular Mechanisms in Secretion. Munksgaard, Copenhagen, pp 598-608.

[14] Hofmann A, Escherich A, Lewit-Bentley A, Benz J, Raguenes-Nicol C, Russo-Marie F, Gerke V, Moroder L, Huber R 1998 Interactions of benzodiazepine derivatives with annexins. J Biol Chem **273**:2885-2894.

[15] Kaneko N 1994 New 1,4-benzothiazepine derivative, K201, demonstrates cardioprotective effects against sudden cardiac cell death and intracellular blocking action. Drug Dev Res **33**:429-438.

[16] Terasaki M, Miyake K, Mcneil PL 1997 Large plasma membrane disruptions are rapidly resealed by Ca^{2+}- dependent vesicle-vesicle fusion events. J Cell Biol **139**:63-74.

[17] Creutz CE 1992 The annexins and exocytosis. Science **258**:924-931.

The Growth Plate
I.M. Shapiro et al. (Eds.)
IOS Press, 2002

Linking Endochondral Ossification to Hematopoiesis

O. Jacenko, M. R. Campbell and D. W. Roberts
University of Pennsylvania School of Veterinary Medicine, Department of Animal Biology, Philadelphia, PA

Abstract. Each skeletal element where marrow develops is first defined by a hypertrophic cartilage blueprint. Through endochondral ossification (EO), hypertrophic cartilage is replaced by trabecular bone and marrow, with accompanying tissue growth. Here we summarize how the key features of the disease phenotypes seen in the collagen X transgenic (Tg) and null mice support an intimate link between EO and marrow establishment. Specifically, Tg mice with dominant interference mutations for collagen X, a unique hypertrophic cartilage matrix protein, develop skeleto-hematopoietic abnormalities manifested as growth plate compressions, reduced trabecular bone, marrow hypoplasia, and impaired immunity. Similar hematopoietic defects also occur in the collagen X null mice [1]. Acute growth plate defects and depletion of marrow hematopoiesis are prominent in a subset of both the Tg (~25%) and null (~10%) mice that exhibit perinatal lethality. These mice display reduced thymuses with a paucity of cortical T cells, and diminished spleens with indistinct lymphatic nodules and an altered B and T cell content; such lymphatic organ atrophy is consistent with the hypoplastic marrow's inability to replenish maturing lymphocytes. The surviving subset of Tg (~75%) and null (~90%) mice have a reduced number of marrow lymphocytes throughout life. Moreover, all collagen X mice have elevated serum cytokines, suggestive of an impaired or inappropriate immune response. In Tg growth plates, the hypertrophic cartilage pericellular matrix, typically composed of a lattice-like network, is disrupted. Moreover, the growth plate distribution of glycosaminoglycans (GAGs) and proteoglycans (PGs) is altered in both the Tg and null mice, as indicated by a paucity of staining for hyaluronan (HA) and heparan sulfate PGs (HSPGs). A provocative hypothesis links the disruption of the collagen X pericellular network and GAG/PG decompartmentalization to the locus for hematopoietic failure in the collagen X mice.

Introduction

In vertebrates, the process of endochondral ossification (EO) is responsible for the formation of the axial and appendicular skeleton, as well as of certain cranial bones. This multi-step sequence of events initiates within the cartilagenous "blueprint" of the skeleton, and involves the replacement of the hypertrophic cartilage matrix by trabecular bone and marrow (for review see [2]). A distinctive feature of this process comprises the emergence of hypertrophic cartilage, which heralds the morphogenetic events of EO. Moreover, the appearance of hypertrophic cartilage and the initiation of EO define each skeletal element where future marrow develops.

Specifically, chondrocyte hypertrophy is accompanied by a transformation from a non-calcified avascular cartilage matrix to a calcifiable one that is permissive to vascularization. As invading blood vessels import mesenchymal cells, hematopoietc precursors, and osteoclasts/chondroclasts, the hypertrophic cartilage matrix begins to be degraded, while differentiating osteoblasts line the hypertrophic cartilage cores and deposit osteoid (for review see [2]). The resultant network of trabecular bony spicules protrudes into the marrow (Fig. 1), and likely provides hematopoietic niches. Moreover, blood cells colonize the spaces carved out from the hypertrophic cartilage, and hematopoiesis ensues almost exclusively within the endochondral bone. Thus, the outcome of EO is two-fold. First, trabecular (e.g. "spongy", "cancellous") bone is formed upon hypertrophic cartilage cores. Unlike the dense, cortical bone that results from intramembranous ossification and is ideally suited for weight bearing, the endochondrally-derived trabecular bone is porous and enveloped by the marrow (Fig. 1A&B). Second, the marrow becomes established providing the definitive microenvironment for blood cell differentiation.

The marrow environment is critical for promoting hematopoietic progenitor cell proliferation, differentiation, and controlled egress into the lymphatics or systemic circulation. The hematopoietic niches in the marrow are not precisely defined, but are likely both spatial and chemical in nature [3, 4]. For example, the trabecular spicule network protruding into the marrow (Fig. 1), the endosteal network lining the inner surfaces of the periosteal and trabecular bone that interfaces with marrow, as well as the stromal cells, vascular cells, and various matrix components (e.g. collagens, PGs, glycoproteins) may physically compartmentalize the marrow space. Likewise, associated bioactive components may include growth factors and cytokines, synthesized by resident cells or made elsewhere, and sequestered by the matrix. Moreover, interactions between the physical and diffusable factors may contribute towards the establishment of a marrow microenvironment supportive for hematopoiesis. One example of such interactions likely involves HSPGs, which have been proposed to orchestrate hematopoietic niches by binding and immobilizing cytokines, and then presenting them to the marrow stromal and hematopoietic progenitor cells [5, 6, 7, 8].

Since the marrow provides niches for blood cell differentiation, it is conceivable that alterations in the multi-step process of EO, involving the transition from hypertrophic cartilage to trabecular bone and marrow, may affect both the physical and chemical constituents of the marrow and thus alter its stromal and/or hematopoietic residents. We have proposed this hypothesis based on the skeleto-hematopoietic disease phenotype of our transgenic (Tg) mice with altered type X collagen [9, 1, 10]. These mice have alterations in hypertrophic cartilage and in EO, and exhibit abnormalities in their growth plates, trabecular bone, marrow, and lymphatic organs. Our data imply that these defects arise as a consequence of the initial changes in hypertrophic cartilage, and support the possibility that EO may contribute towards the establishment of an appropriate environment in the marrow for subsequent hematopoiesis. Below we summarize the key features of the disease phenotype seen in the collagen X mice, and discuss how these changes support an intimate link between EO and marrow establishment.

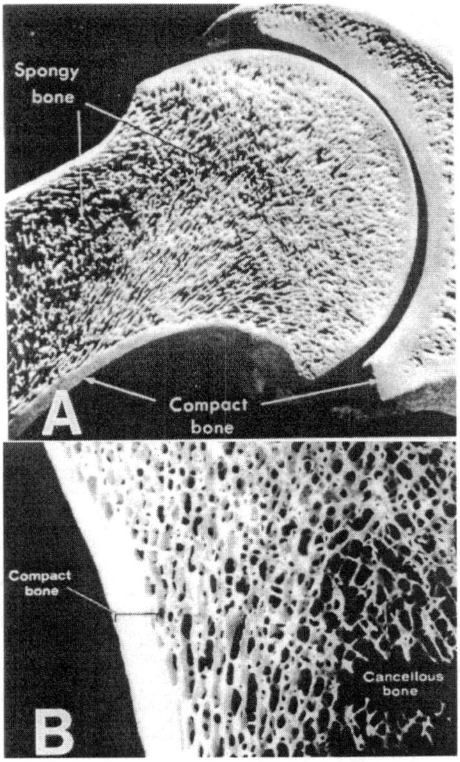

Figure 1. Architecture of a spongy (e.g. trabecular, cancellous) bone derived through endochondral ossification is compared to that of a compact (e.g. cortical, dense) bone derived through intramembranous ossification. **A.** A saggital section of the proximal end of the humerus in relation to the glenoid fossa of the scapula at the shoulder joint is shown. **B.** A thick ground section of the tibia is depicted, illustrating the cortical compact bone and the lattice of trabeculae of cancellous bone. (Figures reproduced from Bloom and Fawcett, A Textbook of Histology; D.W. Fawcett, ed.)

Material and Methods

Breeding and maintenance of mice, DNA and RNA analysis, conventional histology, immunohistochemistry, electron microscopy, and flow cytometry for lymphatic organs has been described [9, 1, 10]. Below we summarize protocols introduced in this chapter.

Fluoresence-Activated Cell Sorting (FACS) of Marrow Cells
Tibiae and femurs, obtained from various ages of wild type control and collagen X Tg and KO mice with mild and perinatal lethal phenotypes, were flushed with calcium-magnesium-free Hank's Balanced Salt Solution (HBSS) using a $26^{5/8}$G needle, and the marrow was suspended by pipeting and gentle vortexing. After erythrocyte lysis (0.17M Tris, 0.16M NH_4CL; 2ml/5ml suspension; 2.5 min, 24^0C), cells were pelleted, resuspended, and counted by hemacytometer. Cells (10^6/100μl HBSS) were single labeled with 20μl (10μg/ml) of the following antibodies: CD138/Syndecan-1, CD106, CD11b, IgM, and IgD (BD Pharmingen). Double labeling was performed using IgM/B220 and IgD/B220. For single labeling, the cells were incubated with antibodies (30 min, 4°C), washed, resuspended in

100μl FACS buffer (0.4% BSA/PBS), and incubated with FITC conjugated anti-rat IgG (30 min, 4°C). For double labeling, PE conjugated B220 was added with FITC anti-rat IgG. Cells were then washed, and resuspended in 0.5ml FACS buffer. Propidium Iodide (PI; BD Pharmingen) was added 5 min prior to analysis for dead cell exclusion. Data acquisition and analysis were performed as described previously [1].

Analysis of Serum Cytokine Levels by ELISA
Peripheral blood, obtained by heart puncture, wasa allowed to settle (30 min, 4^0C). Samples were then centrifuged (6,000 rpms, 10 min, 4^0C), and serum was sequestered. A two site enzyme-linked immunoadsorbant assay was performed to detect IL-12 p40, IL-6, and INF-γ levels as described [11]. Antibodies [12] were generously provided by Dr. C. Hunter, Univ. Penn. Sch. Vet. Med.

Results and Discussion

A Variable Disease Phenotype in Mice with Altered Collagen X

Collagen X is a predominant biosynthetic product of hypertrophic cartilage during the fundamental events of EO under both normal and pathologic conditions (for review see [2]). This short chain collagen is non-fibril-forming, but has been proposed to contribute structurally to a hexagonal lattice-like array, which may be key to its function [13, 14, 2, 15]. To understand its role in hypertrophic cartilage and to define its involvement in EO, we generated a dominant interference phenotype for collagen X [9]. Specifically, we introduced transgene constructs with either a 4.7 or a 1.6 Kb promoter region of the chicken α1(X) gene, followed by a chicken type X collagen cDNA which contained deletions encoding either 21 or 293 amino acid residues within the central, triple-helical domain. The premise for this design was that expression of truncated, partially functional chick collagen X peptides would compete with full length mouse collagen X peptides for association at the carboxyl domain; however, the triple-helical deletions would prevent subsequent trimerization of hybrid trimers, leading to dominant interference [9]. This scenario is supported by our demonstration that co-expression of the truncated chicken collagen X transgenes with full-length mouse collagen X molecules in a cell-free translation system yielded chick-mouse hybrid trimers and truncated chick homotrimers; this indicated that the mutant transgene product could assemble with endogenous mouse collagen X, and thus has potential for dominant interference [10].

We have characterized 14 Tg mouse lines with random transgene insertion sites, and have established that all Tg lines exhibited a similar disease phenotype that was variable in severity [9, 1]. Specifically, by weeks 2-3 after birth, ~25% of the Tg mice developed a perinatal lethal phenotype. The visible onset of the disease manifested as neck lordosis and hunching of the back, wasting, low weight (Table 1A), lethargy, and death within 24 hours. The other ~75% of the mice showed varying degrees of transient dwarfism, ranging from severe to barely distinguishable (Table 1A). While the survivors had normal life-spans, they were prone to developing skeletal deformities, chronic hyperplastic dermatitis, aggressive lymphosarcomas, as well as additional suppressed immunity-related changes [1]; (Jacenko, O., Roberts, D.W., Gress, C.J., Tao, Z., and Campbell, (2002), Linking hematopoiesis to endochondral skeletogenesis through analysis of mice transgenic for collagen X, Am. J. Pathol., 162, In Press). We have also observed that mice with inactivated collagen X [16] (generously provided by Drs. B. de Crombrugghe and R. Behringer, M.D. Anderson Cancer Center, Univ. Texas) also exhibited a variable disease phenotype [1]. Specifically, ~10% of the collagen X null or knock-out (KO) mice developed the acute perinatal-lethal defects around week-3 after birth; these abnormalities mirrored those seen in the Tg mice. However, the majority of the KO mice did not display any readily identifiable gross skeletal or hematopoietic defects.

Skeletal Defects in Mice and Humans with Altered Collagen X

In the collagen X Tg mice, similar histological defects were detected in EO-derived tissues, with the severity of the defects correlating with the outward murine phenotype [1]. These abnormalities involved the chondro-osseous junction, as well as the marrow and lymphatic organs. Skeletal defects included mild metaphyseal flaring [2], growth plate compressions, diminished hypertrophy, and a reduction in trabecular bony spicules [9, 1]. In all Tg mice, the thinning of the growth plates became apparent between weeks 2-3 after birth, which correlated with the pup's increased mobility, as well as with the establishment and maturation of the secondary centers of ossification and the growth plate. While these compressions involved all growth plate zones, the one most effected was the hypertrophic cartilage, where both the endogenous collagen X and the transgene product were co-expressed [1, 10]. In mice with the most severe perinatal-lethal phenotype, the hypertrophic chondrocytes often appeared smaller, displayed picnotic nuclei, and were occasionally flattened, as if the matrix surrounding them had collapsed [9]. This was contrary to the defects seen in the surviving subset of KO mice. While the collagen X null mice with perinatal lethality mirrored the defects seen in the perinatal-lethal Tg mouse subset, the surviving KO mice exhibited only subtle growth plate changes [1]. Specifically, the growth plates were slightly thinner than those of controls and were wavy; moreover, the zone most affected comprised the proliferative chondrocytes [17, 1] rather than the hypertrophic cells.

These subtle phenotypic discrepancies in growth plate defects seen in the collagen X Tg and null mice may provide insights into the underlying pathogenic mechanisms. It is not unexpected for the dominant interference collagen X mutations to yield a stronger disease phenotype than one resulting from gene inactivation. The competition of the transgene products for trimerization with normal mouse collagen chains could lead either to a depletion of the hybrid trimers and a loss-of-function, or to a persistence of these abnormal molecules, leading to a gain-of-function. It is likely that both of these scenarios contributed to the growth plate abnormalities in the Tg mice. Specifically, the proliferative cartilage zones appeared to be similarly affected in both sets of mice, and likely arose as a consequence of loss of collagen X function. In contrast, the changes in hypertrophic cartilage in the Tg mice may have resulted from gain-of-function due to the presence of defective collagen X molecules. These scenarios have recently been confirmed by interbreeding the Tg and KO mice, and ascertaining the contribution of the transgene on a null background towards the emergence of a more acute disease phenotype (unpublished data).

In addition to the most acute growth plate changes described above, the subset of Tg and KO mice exhibiting perinatal lethality also had the most dramatic reduction of trabecular bony spicules. In these mice, generalized osteopenia also affected the periosteal and cranial intramembranously-derived bones. Moreover, the perinatal-lethal mutants displayed easily discernable red marrows due to erythrocyte predominance and leukocyte depletion, which indicated an alteration of marrow hematopoiesis [1]; (Jacenko, O., Roberts, D.W., Gress, C.J., Tao, Z., and Campbell, (2002), Linking hematopoiesis to endochondral skeletogenesis through analysis of mice transgenic for collagen X, Am. J. Pathol., 162, In Press).

The predominant skeletal disease phenotype of mild dwarfism with metaphyseal involvement seen in the collagen X Tg mice hinted that mutations in the COL10A1 gene would most likely be responsible for certain human chondrodysplasias. Consequently, over 30 different collagen X mutations have been identified in patients with the autosomal dominant disorders of Schmid metaphyseal chondrodysplasia (SMCD) [2] and the "Japanese" type spondylometaphyseal dysplasia (SMD) [18]. Interestingly, most mutations were localized to microdomains within the carboxyl NC1 domain. For SMCD, haploinsufficiency was proposed as a disease mechanism, although for certain mutations, dominant interference remains a possibility [2, 19]. The skeletal defects in affected individuals are intermediate between those seen in the Tg and KO mice. Specifically, the onset of SMCD correlates with weight bearing and affects skeletal elements under the greatest mechanical stress; patients exhibit mild dwarfism, coxa vara and a waddling gait [20]. Moreover, a recent clinical re-evaluation of SMCD has identified vertebral changes as a variable component of SCMD, indicating that SMCD and the "Japanese" type SMD are identical collagen X disorders [21]. Although the SMCD/SMD patients shared specific skeletal defects and dwarfism seen in either the Tg or KO mice, hematopoietic or immune function abnormalities were not described in the affected individuals. We find it noteworthy that to date, no triple-helical collagen X mutations have yet been described in any disorders. This implies that a spectrum of abnormalities may ensue from mutations in different domains of collagen X, as is the case with type I collagen and osteogenesis imperfecta [22]. Moreover, certain triple-helical mutations would likely yield a distinct, possibly more severe phenotype that may include hematopoietic defects. The association of skeletal defects and dysfunction of the immune system has been recognized for a while, with a number of immuno-osseous dysplasias resembling the murine skeleto-hematopoietic disease phenotype [23, 24, 25, 26, 27, 28].

Hematopoietic Defects in Collagen X Tg and Null Mice

Hematopoietic defects were obvious morphologically in ~25% of Tg and ~10% of KO mice with the most severe perinatal-lethal phenotype, although flow cytometry confirmed altered hematopoiesis in all collagen X mice (e.g. Table 1). In the perinatal-lethal mutants, these defects were manifest as marrow hypoplasia and lymphatic organ atrophy [1]. Specifically, upon dissection of endochondrally-derived skeletal elements, the marrows were conspicuously red in the severely-affected mice. Subsequent histology demonstrated in the marrows an erythrocyte predominance and a notable decrease of other hematopoietic cells. Likewise, gross examination revealed substantially reduced thymuses and spleens (Table 1A), and undetectable lymph nodes. Moreover, histology confirmed that spleens were pale due to red pulp depletion [1]. It is possible that the prominent marrow erythrocytes had infiltrated from the splenic red pulp to compensate for the depleting hematopoietic compartment; this is consistent with the spleen's ability in certain species to contract and release red blood cells into the circulation upon stress [29].

The hematopoietic changes involving the marrow and lymphatic organs were unpredicted. We have recently confirmed, using reverse transcriptase-polymerase chain reaction, northern blot analysis, in-situ hybridization (Jacenko, O., Roberts, D.W., Gress, C.J., Tao, Z., and Campbell, (2002), Linking hematopoiesis to endochondral skeletogenesis through analysis of mice transgenic for collagen X, Am. J. Pathol., 162, In Press), as well as immunohistochemistry [10], that the transgene was co-expressed temporo-spatially with endogenous collagen X in the endochondral skeleton of Tg mice. Furthermore, we have established that endogenous collagen X was not expressed in any of the affected extraskeletal sites. This raised the intriguing possibility that the changes in hypertrophic cartilage have led to the hematopoietic abnormalities. Moreover, this implied that EO, where hypertrophic cartilage is replaced by bone and marrow, may establish the marrow microenvironment prerequisite for blood cell differentiation, and that changes in the lymphatics are secondary to changes in the marrow.

To explore this hypothesis, we studied the nature of the changes in the marrow, thymus, and spleen by histology, immunohistochemistry, and flow cytometry [1]; (Jacenko, O., Roberts, D.W., Gress, C.J., Tao, Z., and Campbell, (2002), Linking hematopoiesis to endochondral skeletogenesis through analysis of mice transgenic for collagen X, Am. J. Pathol., 162, In Press). Our approaches have confirmed variations in the lymphoid lineages in these organs in all collagen X Tg mice, as well as in the KO mice exhibiting perinatal lethality. Moreover, in the surviving subset of KO mice, several subtle changes were also detected (Table 1).

In marrows, differential analysis of smears confirmed a reduction in lymphocytes in both the Tg and KO mice at week-3 after birth, and especially in the perinatal-lethal mutants [1]; (Jacenko, O., Roberts, D.W., Gress, C.J., Tao, Z., and Campbell, (2002), Linking hematopoiesis to endochondral skeletogenesis through analysis of mice transgenic for collagen X, Am. J. Pathol., 162, In Press). Likewise, flow cytometry using antibodies against cell surface markers representing B lineage cells at different stages of maturation (e.g. anti-B220, CD138, IgM, and IgD), showed reductions in all B lymphocytes in marrows from week-3 Tg, and Tg and KO perinatal-lethal mutants (Table 1B). In the Tg survivors, B cell reductions generally persisted throughout life (Jacenko, O., Roberts, D.W., Gress, C.J., Tao, Z., and Campbell, (2002), Linking hematopoiesis to endochondral skeletogenesis through analysis of mice transgenic for collagen X, Am. J. Pathol., 162, In Press). In KO survivors, B cells appeared slightly elevated at week-3, but then were slightly decreased throughout life (unpublished observations). These subtle hematopoietic changes are consistent with the observation that the KO mice are not as profoundly affected as the Tg mice. Moreover, slight elevations in B cells may suggest that the KO mice are more capable of coping with various assaults (e.g. opportunistic infections) on their immune system than the Tg mice, and thus display a lower ratio of perinatal lethality.

FACS analysis of additional marrow cell surface markers included CD11b (for macrophages, dendritic cells, granulocytes), and CD106 (for marrow stromal cells), c-kit (for hematopoietic progenitors), and Ter119 (for erythroid lineage). Of these, neither CD11b nor CD 106 showed significant changes throughout life in all the collagen X mice. However, preliminary analysis indicated that Tg and KO perinatal-lethal mutants have a reduction in c-kit-positive cells (unpublished data), and an increase in Ter119-positive erythroblasts (Table 1B).

Histology and immunohistochemistry of thymuses from week-3 collagen X Tg and KO mice revealed a diminished cortex and a paucity of cortical T lymphocytes in the perinatal-lethal mutants [1]. The virtual depletion of the immature T cell population, distinguished by expression of both CD4 and CD8 cell surface markers, was quantitated by single and double label FACS (Table 1C). For example, a 7.5 and 16- fold reduction of immature double-positive T cells was observed in thymuses of KO and Tg mice exhibiting perinatal lethality, respectively (e.g. ~5-11%), when compared to the wild type controls (e.g. ~86%) (Table 1C). This immature T cell population derives from the marrow, and emigrates to the thymic cortex; as these cells mature and become single-positive for CD4 or CD8, they progress to the medulla. A depletion of immature cortical T cells, coupled with an increased number of single-positive cells in the medulla (Table 1C), is consistent with the premise that the onset of marrow hypoplasia would impair the marrow's ability to replenish the maturing cortical T lymphocytes.

The reduced spleens of week-3 collagen X Tg and KO mice exhibiting perinatal lethality displayed an altered architecture that was characterized by poorly distinguishable lymphatic nodules and a diminished red pulp [1]. Moreover, flow cytometry indicated an

A. Organ Weights

	Wild Type	Col X Tg	Col X Tg Mutant	Col X KO	Col X KO Mutant
Body (gm)	9.24 ± 0.22 (15)	7.34 ± 0.31 (7)	4.74 ± 0.14 (45)	8.41 ± 0.17 (33)	5.40 ± 0.27 (10)
Kidney (mg)	72.10 ± 1.76 (16)	56.09 ± 2.32 (7)	38.49 ± 1.05 (46)	61.66 ± 1.60 (41)	45.69 ± 2.35 (13)
Kidney : Body	128 : 1 (15)	131 : 1 (7)	123 : 1 (45)	141:1 (33)	118:1 (10)
Spleen (mg)	79.93 ± 7.44 (10)	31.38 ± 2.63 (7)	9.06 ± 0.50 (47)	58.73 ± 3.71 (44)	10.12 ± 1.12 (10)
Thymus (mg)	37.31 ± 1.46 (15)	37.06 ± 1.89 (7)	8.50 ± 0.68 (47)	48.58 ± 2.23 (40)	8.52 ± 1.35 (12)
Spleen : Kidney	0.90 : 1 (10)	0.57 : 1 (7)	0.23 : 1 (46)	0.95:1 (41)	0.22:1 (10)
Thymus : Kidney	0.52 : 1 (15)	0.66 : 1 (7)	0.22 : 1 (46)	0.79:1 (40)	0.19:1 (12)

B. Bone Marrow - Single Label Flow Cytometry

	Wild Type	Col X Tg	Col X Tg Mutant	Col X KO	Col X KO Mutant
# Live Nucleated Cells	8.62E6±0.60E6 (5)	4.70E6±0.54E6 (15)	4.73E6±0.24E6 (3)	7.43E6 ± 0.46E6 (24)	3.52E6 ± 0.93 (4)
% CD138 + Cells	31.02 ±1.30 (17)	21.19 ± 0.86 (8)	2.5 ±0.47 (21)	30.69 ± 1.66 (8)	3.74 ± 1.23 (2)
% B220 + Cells	34.01 ± 0.76 (11)	26.92 ± 0.58 (9)	12.75 ± 1.33 (23)	38.63 ± 1.05 (8)	18.98 (1)
% IgM + Cells	6.65 ± 0.56 (8)	4.42 ± 0.44 (7)	1.26 ± 0.37 (5)	8.88 ± 0.45 (4)	3.72 (1)
% IgD + Cells	2.59 ± 0.52 (7)	1.50 ± 0.18 (7)	1.76 ± 0.30 (11)	4.42 ± 0.67 (6)	3.01 (1)
% Ter + Cells	21.65 ± 2.65 (14)	29.66 ± 3.36 (10)	35.63 ± 2.56 (12)	18.16 ± 1.30 (6)	29.10 ± 1.05 (2)

C. Thymus - Single Label Flow Cytometry

	Wild Type	Col X Tg	Col X Tg Mutant	Col X KO	Col X KO Mutant
# Live Cells	7.97E7 ± 0.37E7 (30)	4.00E7 ± 0.55E7 (20)	7.27E6 ± 1.72E6 (23)	8.46E7 ±0.72 (25)	1.90E7 ± 0.20 (6)
% Live Lymph.	94.01 ± 0.58 (21)	95.82 ± 0.64 (12)	80.15 ± 1.82 (14)	91.51 ± 1.70 (25)	76.04 ± 6.26 (10)
% CD4 + Lymph.	91.78 ± 0.79 (19)	91.67 ± 0.52 (21)	61.94 ± 2.70 (21)	91.99 ± 0.43 (25)	58.71 ± 4.60 (10)
% CD8 + Lymph.	88.37 ± 0.73 (13)	86.80 ± 1.03 (19)	21.09 ± 1.53 (20)	89.51 ± 0.60 (25)	31.72 ± 5.59 (10)
% CD4/CD8 Lymph.	85.57 ± 0.82 (5)	84.52 ± 1.10 (9)	5.27 ± 1.49 (14)	87.61 ± 1.71 (6)	11.34 ± 3.68 (7)

D. Spleen - Single Label Flow Cytometry

	Wild Type	Col X Tg	Col X Tg Mutant	Col X KO	Col X KO Mutant
# Live Cells	9.87E7 ± 0.84E7 (31)	3.72E7 ± 0.44E7 (17)	1.14E7 ± 0.15E7 (23)	9.70E7 ± 1.04E7 (23)	1.30E7 ± 0.24 (7)
% Live Lymph.	33.91 ± 1.95 (21)	30.79 ± 1.77 (12)	10.34 ± 1.13 (20)	36.28 ± 1.75 (24)	30.24 ± 4.36 (10)
% CD4 + Lymph.	5.56 ± 0.20 (16)	10.10 ± 0.60 (12)	21.81 ± 1.94 (20)	8.71 ± 0.54 (16)	15.99 ± 2.22 (8)
% CD8 + Lymph.	2.50 ± 0.16 (18)	3.37 ± 0.41 (15)	8.5 ± 0.85 (19)	3.42 ± 0.21 (20)	4.82 ± 0.54 (8)
% B220 + Lymph.	55.55 ± 1.56 (16)	35.94 ± 2.07 (16)	28.83 ± 2.42 (12)	59.04 ± 1.40 (12)	30.90 ± 3.68 (9)
% Ter + Erythroblasts	46.31 ± 2.29 (18)	70.38 ± 1.62 (15)	40.15 ± 3.94 (12)	55.10 ± 2.45 (14)	22.30 ± 4.39 (8)

Table 1. Comparison of organ weights, and flow cytometry of marrow, thymus, and spleen from wild type control mice (Wild Type), mice transgenic for collagen X with mild (Col X Tg) or perinatal-lethal phenotype (Col X Tg Mutant), and collagen X null mice with mild (Col X KO) or perinatal-lethal phenotype (Col X KO Mutant) at week-3 after birth. Numbers in parentheses indicate number of animals tested. For organ weights, 1 kidney and 2 thymic lobes are recorded.

altered percentage of both T and B lymphocytes in spleens of all subsets of collagen X Tg and KO mice (Table 1D). Specifically, CD4 and CD8 single-positive T cells were elevated relative to the rest of the lymphocyte population in all mice at week-3 (Table 1D), as well as throughout life (Jacenko, O., Roberts, D.W., Gress, C.J., Tao, Z., and Campbell, (2002), Linking hematopoiesis to endochondral skeletogenesis through analysis of mice transgenic for collagen X, Am. J. Pathol., 162, In Press). These cells likely represent the mature T cell population that comes to the spleen from the thymic medulla, and thus escapes the effects of marrow hypoplasia. In contrast, splenic B cells were particularly effected. Flow cytometry for B220-positive cells showed a reduction of B cells in the KO perinatal-lethal mutants, and in all Tg mice; in the latter, this reduction persisted throughout life (Jacenko, O., Roberts, D.W., Gress, C.J., Tao, Z., and Campbell, (2002), Linking hematopoiesis to endochondral skeletogenesis through analysis of mice transgenic for collagen X, Am. J. Pathol., 162, In Press). In KO survivors, the splenic B cell profile was similar to that described above for marrow B cells, since this population was relatively unaltered at week-3 (Table 1D), but then was slightly decreased throughout life (unpublished observations). Since B cell differentiation and maturation take place in the marrow, a decrease in splenic B cells may reflect the limited ability of the hypoplastic marrow to provide an appropriate environment for B cell maturation, causing a diminished number of B cells being generated. Furthermore, a life-long diminution of B cells in the marrow and spleens may reflect a persistent lowering of the humoral immune response in the surviving subset of Tg and KO mice, and thus could contribute to an environment permissive to tumor formation.

Taken together, characterization of the marrow, thymuses, and spleens from the collagen X mice indicates that in addition to the skeletal defects, all mice with altered collagen X have to some extent an alteration in hematopoiesis, which likely ensues from an impaired marrow environment. In an effort to understand how the skeletal changes involving hypertrophic cartilage in growth plates could translate to an alteration of the marrow niches, we next analyzed the chondro-osseous junctions of the collagen X Tg and KO mice.

Altered Chondro-Osseous Environment in Collagen X Tg and KO Mice

Our ultrastructural data suggest that the likely morphological consequence of collagen X perturbation by dominant interference in Tg mice is the disruption of the pericellular lattice surrounding hypertrophic chondrocytes [10]. A similar lattice was first observed by Dr. C. Farnum in hypertrophic cartilage of Yucatan swine [13]. Grant and co-workers have demonstrated that collagen X could spontaneously assemble into a comparable hexagonal array in vitro [14], and Lunstrum and co-workers have visualized this supramolecular aggregation in hypertrophic chondrocyte cultures [15]. Our ability to visualize the extensive, hexagonal lattice-like pericellular matrix in control mice was contingent on the inclusion of ruthenium hexamine trichloride (RHT) in the fixative [10]; RHT precipitates GAGs and PGs, thereby stabilizing the pericellular matrix. In Tg mice, the pericellular matrix was reduced or lacking, and the lattice-like network was no longer apparent; rather, RHT-positive aggregates were seen accumulating near the hypertrophic chondrocyte cell surfaces. The disruption of this collagen X-containing network, which may normally be stabilized by GAGs and PGs, is thought to represent the primary defect in our Tg mice. Moreover, in these mice, RHT-positive aggregates were distributed throughout the growth plate, and collagen fibrils in the proliferative cartilage were not readily apparent [15]. These changes were consistent with those described for the collagen X KO mice by Cheah and co-workers [17], who proposed that the inactivation of collagen X results in growth plate decompartmentalization. Our results extend this hypothesis in that this altered partitioning of RHT-positive aggregates, likely GAGs and PGs, may be a consequence of a disruption in the hexagonal lattice-like network containing type X collagen in the pericellular matrix around hypertrophic chondrocytes.

These observations lead to the screening of GAGs and PGs for possible altered distribution in the collagen X Tg and KO mice. To date, two such differences have been identified [10]. Free HA has been proposed be involved in the assembly and stabilization of the pericellular matrix and in the interstitial enlargement of growth plates, and has been localized in the zone between the pericellular region of hypertrophic chondrocytes and their lacunae [30, 31]. Immunohistochemistry revealed intense staining for HA around hypertrophic chondrocytes of wild type mice, whereas in the collagen X Tg and KO sections, this staining was often faint or undetected [10]. The variability in staining intensity of HA in Tg and KO samples implies that HA may be leached out from tissues during fixation because of a lack of retention by another matrix component. Alternatively, the reduction of free HA may globally impact growth plate integrity and function, since HA provides the backbone for aggrecan supramolecular assembly in cartilage and thus represents an essential component for its compressibility [32]. It may also be noteworthy that CD44, a HA receptor, has been implicated in lymphocyte homing and adhesion during hematopoiesis [33]; this may represent a direct link between the growth plate defects and altered hematopoiesis in the marrow.

The second molecule to display an altered distribution in the growth plates of the collagen X Tg and KO mice was heparan sulfate [10]. This is noteworthy primarily since HSPGs have been implicated as key components of marrow niches required for hematopoiesis [5, 7, 8]. Immunostaining for heparan sulfate-related epitopes, present in HSPGs, was seen in proliferative and hypertrophic cartilage, as well as trabecular bony spicules in wild type controls (Fig. 2A). In contrast, Tg mice with mild phenotypes showed HSPG localization as pericellular in proliferative cartilage, and as diffuse in the hypertrophic zone (Fig. 2B). In KO mice with mild phenotypes (Fig. 2E), as well as in the Tg and KO mice exhibiting perinatal lethality (Fig. 2C&F, respectively), limited growth plate staining for HSPG was restricted to the pericellular region of proliferative chondrocytes; hypertrophic cartilage was virtually negative [10]. To identify which HSPG(s) is affected by collagen X disruption, it would be necessary to localize HSPG members in the chondro-osseous junction. Obvious candidates include perlecan, glypicans, and syndecans. Of these, perlecan and glypican are found in cartilage undergoing EO [34]; moreover, recent inactivation of the perlecan gene has resulted in an acute chondrodysplasia with variable disease phenotypes [35]. Glypican is also present in trabecular bone, and perlecan, glypican and syndecan were all shown to be either made by marrow stromal cells, or deposited as marrow matrix [7, 8, 36]. It is particularly noteworthy that HSPGs are proposed to orchestrate local microenvironmental niches in the marrow by sequestering cytokines, and by juxtaposing hematopoietic stem cells with those cytokines and the stroma to which HSPGs bind [7]. A provocative possibility links the disruption of a collagen X-containing matrix at the hypertrophic cartilage/marrow interface, to an altered GAG/PG distribution and a potential locus for hematopoietic failure in Tg and KO mice.

Altered Cytokine Profile in Collagen X Tg and KO Mice

Regulation of cytokine bioactivity and bioavailability by PGs may represent an important mechanism for locally controlling hematopoietic cell development. Moreover, fluctuations in cytokine levels may represent the cause or the consequence of impaired hematopoiesis and immune function. Several of our preliminary studies suggested that both the collagen X Tg and KO mice may be prone to opportunistic infections, and that perinatal lethality may ensue from an inappropriate and overwhelming inflammatory response to a challenge (unpublished data). To understand the mechanism of defective immunity leading to cachexia or wasting in the perinatal lethal mutants, serum levels of selected cytokines were assessed [37]. Specifically, interleukin-6 (IL-6) was shown to mediate cachexia in several murine models [38]. Likewise, interferon gamma (INF-(γ), whose expression is induced by IL-12 has been reported as one of several inducers of cachexia [39]. It was thus noteworthy that IL-12 p40, IL-6, and INF-(γ) serum levels were aberrant in all the collagen X Tg and KO mice throughout most of their lives (unpublished data). Moreover, Tg and KO mice exhibiting perinatal lethality, IL-6, IL-12, and INF-(γ) serum levels were consistently elevated when compared to controls (Fig. 3). We are currently investigating the basis for the acute and inappropriate cytokine production, whether it may lead to perinatal lethality, and if this can be linked to the altered HA and HSPG distribution in the chondro-osseous junction.

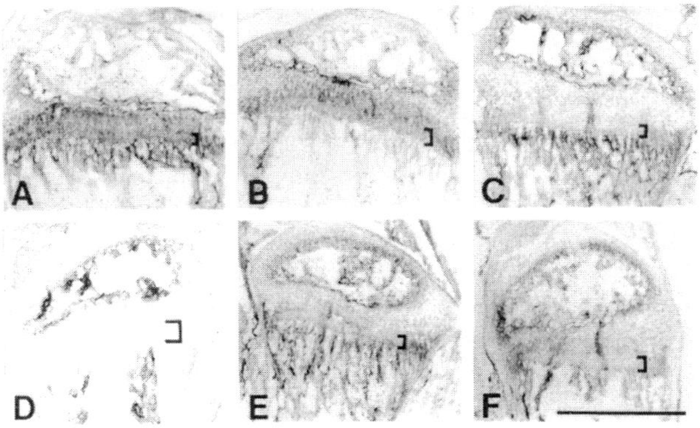

Figure 2. Immunohistochemical localization of HSPG in growth plates of wild-type and collagen X Tg and KO mice. Longitudinal sections are shown of week-3 tibiae from wild type controls (WT; **A,D)**, Tg mice (TG; **B**), KO mice (KO; **E**), and Tg (TG-Mut; **C**) or KO (KO-Mut; **F**) mice exhibiting perinatal lethality. Note HSPG localization to proliferative and hypertrophic cartilage, and to trabecular bone in **A**. In **B**, staining is pericellular in proliferative chondrocytes, and intensity is reduced in hypertrophic cartilage. In **C**, **E**, and **F**, staining is either faint or absent in the growth plate. **D**: control where heparitinase was omitted. Brackets approximate width of hypertrophic cartilage. Figure is modified from [10].

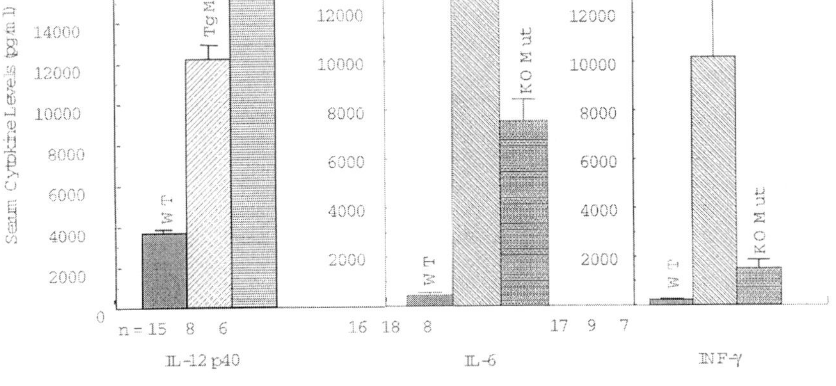

Figure 3. Comparison of serum IL-12, IL-6, and INF-γ levels by ELISA from week-3 wild type (WT), collagen X Tg (Tg Mut) or null (KO Mut) mice exhibiting perinatal lethality. Error bars = standard error; animal number is indicated (n).

Linking Hypertrophic Cartilage, Endochondral Ossification, and Marrow Establishment
Taken together, our data indicate that all mice with altered collagen X, generated either by transgenesis or through gene inactivation, to some extent have both skeletal as well as hematopoietic defects. This in turn highlights a previously unforeseen yet intimate link between hypertrophic cartilage, EO, and marrow establishment. An interdependence between endochondrally-derived bone and hematopoiesis had long been apparent [40], however, it has not included hypertrophic cartilage as a potential contributor in hematopoieitc interactions. It is conceivable that hematopoietic niches in the marrow may be compartmentalized both spatially, via porous trabecular bone projections (e.g. Fig. 1) or various matrix components, and chemically, via bioactive molecules such as growth factors and cytokines. Defects in hypertrophic cartilage may thus contribute either alone or in concert towards skeletal defects and marrow alterations. Specifically, we revealed a disruption of the pericellular lattice network, likely composed of collagen X, around hypertrophic chondrocytes in Tg mice, which we believe represents the primary locus of the collagen X defect [10]. The consequent loss of structural integrity in the pericellular matrix may result in a re-distribution of molecules such as the GAGs/PGs, which likely associate with and stabilize the lattice, and may thus impact the microenvironment of the chondro-osseous and marrow junction. Towards this end, we detected an altered distribution for free HA and HSPG in growth plates of both the collagen X Tg and KO mice [10]. It is noteworthy that both HA and HSPG have been implicated as key components of marrow niches required for hematopoiesis, and likely mediate their affects by regulating cytokine bioactivity and availability [5, 6, 7, 8]. We are currently testing whether collagen X supramolecular aggregates require associations with specific classes of GAGs/PGs for its stability and function, and if these interactions are prerequisite for hematopoiesis.

A second scenario of how an alteration in the chondro-osseous junction may impact the marrow environment involves the trabecular bone. Our histology indicated a reduction in the length and/or amount of trabecular bony spicules, with the least trabecular bone present in mice with the most acute phenotype [1]; (Jacenko, O., Roberts, D.W., Gress, C.J., Tao, Z., and Campbell, (2002), Linking hematopoiesis to endochondral skeletogenesis through analysis of mice transgenic for collagen X, Am. J. Pathol., 162, In Press). It is still unresolved whether the reduction of trabecular spicules stems from an altered hypertrophic cartilage matrix, which provides the scaffold upon which bone is deposited. Alternatively, osteopenia may ensue from the action of osteolytic cytokines, which may be elevated in the collagen X mice. Our data favor the first possibility since TRAP assays suggest that osteoclast activity is reduced, rather than increased (unpublished data). Regardless, a decrease in trabecular spicules may alter the spatial arrangement of marrow into hematopoietic niches, as well as decrease the surface area of bone-hematopoietic stem cell interactions [41]. Specifically, both the trabecular and periosteal bone surfaces are lined with a continuous membrane of bone-lining (endosteal) cells. These endosteal cells not only provide a physical separation between the osseous and extracellular fluid compartments, but have been implicated in regulating hematopoiesis [42]. In particular, several reports described an intimate association between proliferative (generative) hematopoietic stem cells and endosteal surfaces [42].

The third scenario that could lead to a skeleto-hematopoietic phenotype involves a chemical imbalance, and may provide an alternate explanation for trabecular spicule reduction and generalized osteopenia, which is most pronounced in the perinatal-lethal mutants. Our preliminary data suggest that interactions of hematopoietic cytokines with hypertrophic cartilage and/or trabecular bone may be impaired. Specifically, we have detected elevated serum cytokine levels in both the collagen X Tg and KO mice, and in particular, in the perinatal-lethal mutants (Fig. 3). These osteolytic cytokines may directly induce osteopenia and inhibit hematopoiesis. Moreover, cytokine overproduction may be indicative of an inappropriate and acute immune response to opportunistic infections, and may contribute towards phenotypic variability and the acute phenotype in the collagen X mice. A modifier gene [43] whose actions may be linked to collagen X may also contribute to phenotypic variability; however, persistence of the variable phenotype after inbreeding the collagen X mice over 10 generations into several pure mouse strains argues against this (unpublished data).

In summary, each of these three scenarios, either alone or in conjunction, could compromise marrow-derived hematopoietic precursors, leading to impaired blood cell maturation, immune dysfunction, and in extreme cases, marrow aplasia. Interestingly, links between altered hematopoiesis and skeletal defects were observed in multiple mouse models (see references within [1]). Likewise, in humans, the association of inborn defects of skeletal development and dysfunction of the immune system has been recognized, and the number of characterized immuno-osseous disorders is increasing. A few of potential relevance to our murine phenotype include Spondylo-mesomelic-acrodysplasia with severe combined immunodeficiency [23], Schimke dysplasia [24], Cartilage Hair Hypoplasia [25], Dubowitz [26], Kyphomelic [27] and Kostmann's [28] syndromes.

Despite these examples of immuno-osseous defects, it cannot be overlooked that several murine models have altered EO, hypertrophic cartilage, and perhaps collagen X, but no hematopoietic defects have been reported. One explanation would be that since a direct causal link between the endochondral skeleton and hematopoiesis has not yet been established, until now there has been no precedence to look for such changes. We are analyzing selected models for hematopoietic defects resembling those in collagen X mice, as well as generating additional Tg mice with altered hypertrophic cartilage. In particular, through a collaboration with P. LuValle and co-workers, we had recently mis-expressed the proto-oncogene *c-myc* in hypertrophic cartilage of mice (unpublished data). Histology of the *c-myc* Tg mice revealed both skeletal and hematopoietic changes involving the growth plate, trabecular bone, and marrow. Specifically, chondrocytes were densely packed and aligned in vertical columns, trabecular bone was extensive and obscured the hypertrophic cartilage-marrow interface, and marrows were hypercellular. Based on our observations of the collagen X and the *c-myc* mice, we propose that alterations in either hypertrophic cartilage, or in the molecules regulating the transition process of EO, would result at least to a certain extent, in associated marrow and hematopoietic changes.

Acknowledgments

We are extremely grateful to Dr. C. Hunter (Univ. Penn. Vet. Sch.) and his laboratory for assistance with cytokine assays, and Dr. J. San Antonio (Thomas Jefferson Univ.) for critical manuscript review. Support included NIH grants AR43362 and DK57904, and funding from the Mari Lowe Center for Comparative Oncology and Univ. Penn. Research Foundation to OJ.

References

[1] Gress CJ, Jacenko O 2000 Growth plate compressions and altered hematopoiesis in collagen X null mice. J. Cell Biol. **149:**983-993.

[2] Chan D, Jacenko O 1998 Phenotypic and biochemical consequences of collagen X mutations in mice and humans. Matrix Biol. **17:**1169-1184.

[3] Charbord P, Tavian M, Humeau L, Peault B 1996 Early ontogeny of the human bone marrow from long bones: an immunohistochemical study of hematopoiesis and its microenvironment. Blood **87:**4109-19.

[4] Kincade P 1992 Cell interaction molecules and cytokines which participate in B lymphopoiesis. Baillieres Clin. Haematol. **5:**575-98.

[5] Bruno E, Luikart SD, Long MW, Hoffman R 1995 Marrow-derived heparan sulfate proteoglycan mediates the adhesion of hematopoietic progenitor cells to cytokines. Exp. Hematol. **23:**1212-7.

[6] Gupta P, McCarthy JB, Verafaillie CM 1996 Stromal fibroblasts heparan sulfate is required for cytokine-mediated ex vivo maintenance of human long-term culture-initiating cells. Blood **87:**3229-36.

[7] Gupta P, Oegama TRJ, Brazil JJ, Dudek AZ, Slungaard A, Verfaille CM 1998 Structurally specific heparan sulfate supports primitive human hematopoiesis by formation of a multimolecular stem cell niche. Blood **92:**4641-4651.

[8] Siebertz B, Stocker G, Drzeniek Z, Handt S, Just U, Haubeck HD 1999 Expression of glipican-4 in haematopoietic progenitor and bone marrow stromal cells. Biochem. J. **344:**937-943.

[9] Jacenko O, LuValle P, Olsen BR 1993 Spondlylometaphyseal dysplasia in mice carrying a dominant negative mutation in a matrix protein specific for cartilage-to-bone transition. Nature **365:**56-61.

[10] Jacenko O, Chan, D., Franklin, A., Ito, S., Bateman, J.F., Campbell, M.R. 2001 A dominant interference collagen X mutation disrupts hypertrophic chondrocyte pericellular matrix and glycosaminoglycan and proteoglycan distribution in transgenic mice. Am. J. Pathol. **159:**2257-2269.

[11] Sander B, Hoiden, I., Andersson, U., Moller, E., Abrams, J.S. 1993. Similar frequencies and kinetics of cytokine producing cells in murine peripheral blood and spleen. Cytokine detection by immunoassay and intracellular immunostaining. J. Immunol. Methods **3:**201-214.

[12] D'Andrea A, Rengaraju, M., Valiente, N.M., Chehimi, J., Kubin, M., Aste, M., Chan, S.H., Kobayashi, M., Young, D., Nickbarg, E., Chizzonite, R., Wolf, S.F., Trinchieri, F. 1992. Production of natural killer cell stimulatory factor (interleukin 12) by peripheral blood mononuclear cells. J. Exp. Med. **176:**1387.

[13] Jacenko O, Olsen BR, LuValle P 1991 Organization and regulation of collagen genes. Critical Reviews in Eukaryotic Gene Expression., vol. 1. CRC Press, Boca Raton, FL, pp 327-353.

[14] Kwan APL, Cummings CE, Chapman JA, Grant ME 1991 Macromolecular organization of chicken type X collagen in vitro. J. Cell Biol. **114:**597-604.

[15] Lunstrum GP, Keene DR, Weksler NB, Cho YJ, Cornwall M, Horton WA 1999 Chondrocyte differentiation in a rat mesenchymal cell line. J. Histochem. Cytochem. **47:**1-6.

[16] Rosati R, Horan GS, Pinero GJ, Garofalo S, Keene DR, Horton WA, Vuorio E, de Crombrugghe B, Behringer RR 1994 Normal long bone growth and development in type X collagen null mice. Nature Gen. **8:**129-135.

[17] Kwan KM, Pang MK, Zhou S, Cowen SK, Kong RY, Pfordte T, Olsen BR, Sillence DO, Tam PP, Cheah KS 1997 Abnormal compartmentalization of cartilage matrix components in mice lacking collagen X: implications for function. J. Cell Biol. **136:**459-71.

[18] Ikegawa S, Nishimura G, Nagai T, Hasegawa T, Ohashi H, Nakamura Y 1998 Mutation in the type X collagen gene (COL10A1) causes spondylometaphyseal dysplasia. Am. J. Hum. Genet. **63:**1659-1662.

[19] McLaughlin SH, Conn SN, Bulleid NJ 1999 Folding and assembly of type X collagen mutations that cause metaphyseal chondrodysplasia-type Schmid. Evidence for co-assembly of the mutant and wild-type chains and binding to molecular chaperones. J. Biol. Chem. **274:**7570-5.

[20] Lachman RS, Rimoin DL, Spranger J 1988 Metaphyseal chondrodysplasia, Schmid type. Clinical and radiographic delineation with a review of the literature. Pediat. Radiol. **18:**930102.

[21] Savarirayan R, Cormier-Daire V, Lachman RS, Rimoin DL 2000 Schmid type metaphyseal chondrodysplasia: a spondylometaphyseal dysplasia identical to the "Japanese" type. Pediatr. Radiol. **30:**460-463.

[22] Kuivaniemi H, Tromp G, Prockop DJ 1997 Mutations in fibrillar collagens (types I, II, III, and XI), fibril-associated collagen (type IX), and network-forming collagen (type X) cause a spectrum of diseases of bone, cartilage, and blood vessels. Human Mutation **9:**300-15.

[23] Castriota-Scanderbeg A, Mingarelli R, Caramia G, Osimani P, Lachman RS, Rimoin DL, Wilcox WR, Dallapiccola B 1997 Spondylo-mesomelic-acrodysplasia with joint dislocations and severe combined immunodeficiency: a newly recognized immuno-osseous dysplasia. J. Med. Genet. **34:**854-856.

[24] Spranger J, Hinkel GK, Stoss H, Thoenes W, Wargowski D, Zepp F 1991 Schimke immuno-osseous dysplasia: a newly recognized multisystem disease. J. Pediatrics **119:**64-72.

[25] Polmar SH, Pierce GF 1986 Cartilage hair hypoplasia: immunological aspects and their clinical implications. Clin. Immun. Immunopath. **40:**87-93.

[26] Winter RM 1995 Dubowitz Syndrome. In: Congenital Malformation Syndromes. **D. Donnai and R.M. Winter, eds.:**133-6.

[27] Turnpenny PD, Dakwar RA, Boulos FN 1995 Kyphomelic dysplasia. In: Congenital Malformation Syndromes. **D. Donnai and R.M. Winter, eds.:**199-205.

[28] Guba SC, Sartor CA, Hutchinson R, Boxer LA, Emerson SG 1994 Granulocyte colony-stimulating factor (G-CSF) production and G-CSF receptor structure in patients with congenital neutropenia. Blood **83:**1486-92.

[29] Kuwahira I, Kamiya, U., Moue, Y., Urano, T., Ohta, Y., Gonzalez, N.C. 1999 Splenic contraction-induced reversible increase in hemaglobin cencentration in intermittent hypoxia. J. Appl. Physiol. **86:**181-187.

[30] Pavasant P, Shizari TM, Underhill CB 1994 Distribution of hyaluronan in the epiphyseal growth plate: turnover by CD44-expressing osteoprogenitor cells. J. Cell Sci. **107:**2669-2677.

[31] Pavasant P, Shizari T, Underhill CB 1996 Hyaluronan contributes to the enlargement of hypertrophic lacunae in the growth plate. J. Cell Sci. **109:**327-334.

[32] Comper WD 1991 Physicochemical aspects of cartilage extracellular matrix. In: Cartilage: Molecular Aspects. Vol. 3. B. Hall and S.A. Newman, editors. CRC Press, Boca Raton.:59-78.

[33] Gunthert U 1993 CD44: a multitude of isoforms with diverse functions. Curr. Topics microbiol. Immunol. **184:**47.

[34] Handler M, Yurchenco PD, Iozzo RV 1997 Developmental expression of perlecan during murine embryogenesis. Devel. Dynamics **210:**130-145.

[35] Arikawa-Hirasawa E, Watanabe H, Takami H, Hassell JR, Yamada Y 1999 Perlecan is essential for cartilage and cephalic development. Nature Genet. **23:**354-358.

[36] Kincade PW, Yamashita Y, Borghesi L, Medina K, Oritani K 1998 Blood cell precursors in context. Composition of the bone marrow microenvironment that supports B lymphopoiesis. Vox Sanguis **74 Suppl.:**265-268.

[37] Argiles JM, Lopez-Soriano, F.J. 2000 The role of cytokines in cancer cachexia. In: Kirsner JB (ed.) Inflammatory Bowl Disease. W.B. Saunders, pp 223-248.

[38] Tanaka Y, Eda, H., Udagwa, T., Ishikawa, T., Hordii, I., Ishitsuka, H., Kataoka, T., Taguchi, T. 1990. Experimental cancer cachexia induced by transplantable colon 26 adenocarcinoma in mice. Cancer Res. **50:**2290-2295.

[39] Matthys P, Dijkmas, R., Proost, P., Van Damme, J., Heremans, H., Sobis, H., Billiau, A. 1991 Severe cachexia in mice inoculated with interferon-gamma-producing tumor cells. Int. J. Cancer **49:**77-82.

[40] Taichman RS, Emerson SG 1994 Human osteoblasts support hematopoiesis through the production of granulocyte colony-stimulating factor. J. Exp. Med. **179:**1677-1682.

[41] Taichman RS, Reilly MJ, Emerson SG 1996 Human osteoblasts support human hematopoietic progenitor cells in in vitro bone marrow cultures. Blood **87:**518-24.

[42] Islam A, Glomski, C., Henderson, E.S. 1990. Bone lining (endosteal) cell and hematopoiesis: a light microscopic study of normal and pathologic human bone marrow in plastic-embedded sections. Anat. Rec. **227:**300-306.

[43] Erickson RP 1996 Mouse models of human genetic disease: which mouse is more like a man? BioEssays **18:**993-998.

The Growth Plate
I.M. Shapiro et al. (Eds.)
IOS Press, 2002

Fibroblast Growth Factor Receptor (FGFR) Mutations in Achondroplasia and Related Skeletal Dysplasias

Melissa A. Rasar, Jae Cho, Gregory P. Lunstrum and William A. Horton
Research Center, Shriners Hospital for Children, Portland, OR, and Department of Molecular and Medical Genetics
Oregon Health Sciences University, Portland, OR

Review of FGFR3

The emergence of FGFR3 as an important local regulator of linear bone growth is historically tied to the search for the "achondroplasia gene" and its mutations. Achondroplasia is by far the most common human chondrodysplasia. It is the prototype of a group of disorders that range from the much more severe thanatophoric dysplasia types I and II (TDI, TDII) to the less severe hypochondroplasia [1]. Achondroplasia was genetically mapped to chromosome 4p in 1994, and heterozygous mutations of *FGFR3* were identified shortly afterwards [2-4]. *FGFR3* mutations were subsequently discovered for the TDs and hypochondroplasia (Fig. 1) [5-7]. Remarkable degrees of genetic homogeneity and genotype:phenotype correlation soon became apparent as virtually all patients with classic achondroplasia were found to have the same G380R mutation in the transmembrane domain of this tyrosine kinase receptor [1]. Similarly, all infants with TDII had the identical K650E mutation in the distal kinase domain, whereas a N540K mutation in the proximal kinase domain was detected in most patients with hypochondroplasia. Moreover, almost all infants with TDI had mutations that introduced free cysteine residues in the proximal extracellular ligand-binding domain of the receptor. Interestingly, the vast majority of these mutations arise as new mutations during spermatogenesis making *FGFR3* one of the most mutable genes in the human genome, especially in the regions of recurrent mutation [8]. *FGFR3* encodes one of four closely related FGF receptors (FGFR1-4) in mammals. Comparable mutations have been observed in *FGFR1* and *FGFR2* in human craniosynostosis syndromes [9].

After initial speculation that achondroplasia mutations lead to loss of receptor function, the generation of transgenic and knockout mice as well as experiments in which achondroplasia and TD mutations were expressed in cultured cells revealed that they result in gain of FGFR3 function [10-13]. Ornitz subsequently showed a correlation between the extent of this gain and the severity of human clinical phenotypes associated with the different mutations [13,14]. Of note is that the function(s) that is (are) gained differ depending on the cell in which the FGFR3 is expressed. It promotes mitosis and blocks differentiation in several non-chondrocytic cells. In fact, activating TD mutations have been found in colon and bladder carcinoma and multiple myeloma [15-17]. It is not clear if this reflects the mutability of FGFR3, a role for FGFR3 in tumorogenesis or probably both.

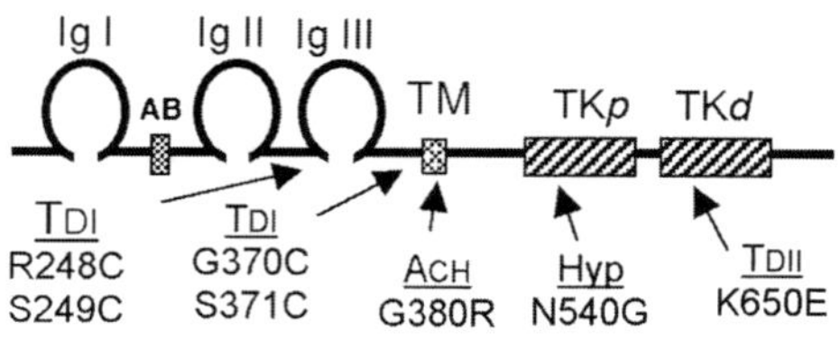

Figure 1. Domain structure of FGFR3 and major sites of mutations. Ig – immunoglobulin, AB – acid box, TM – transmembrane, TK*p*/*d* – proximal and distal tyrosine kinase domains, ACH – achondroplasia, HYP – hypochondroplasia.

In contrast, activation of FGFR3 in differentiated chondrocytes inhibits proliferation and terminal differentiation [18,19]. This has been best demonstrated in mice in whom achondroplasia and TD mutations or a ligand that activates FGFR3 is misexpressed in cartilage under the *Col2* promoter/enhancer, mutant *FGFR3* is expressed under its own promoter or the activating mutations are knocked into the endogenous *FGFR3* by gene targeting [12,20-22]. These mice consistently show postnatal shortening of bones associated with a reduction of proliferating and terminally differentiating chondrocytes in the growth plate. Figs. 2 and 3 show skeletal Xrays and growth plate morphology from mice in which FGF9 was overexpressed in cartilage to activate the endogenous FGFR3 [22].

Current dogma holds that wild type FGFR3 monomers dimerize - more precisely, the equilibrium between monomers and dimers shifts toward dimers - in response to binding ligand(s) [23,24]. Which of the 23 known FGFs is (are) the physiologic ligand(s) for FGFR3 is (are) not known, although FGFs 2, 4 and 9 are candidates because they are detected in and around cartilage, have high affinity for FGFR3 in binding assays and activate the receptor in functional assays [14,25-28]. FGFR3 may utilize different ligands at different developmental stages and in different regions of the growth plate. Heparin sulfate-bearing proteoglycans on the cell surface, such as syndecans and perlecan as well as alternative splicing of ligand binding subdomains influence binding specificity [29]. Dimerization activates intrinsic receptor kinase activity and promotes transphosphorylation of key tyrosine residues in the cytoplasmic domain that serve as docking sites for adaptor proteins and signal effectors containing SH2 and PTB binding domains that are recruited to the activated receptors and which propagate FGFR3 signals [23,24,30]. FGFR3 signals influence a variety of cellular events and processes largely through inducing or repressing expression of target genes in a cell-specific context. Four signaling pathways have been identified to date to propagate FGFR3 signals: STAT, MAPK-ERK, PLCγ and PI3K-AKT [18,19,28,31-34]. The target genes for FGFR3 are not well defined, although there is evidence for induction of p21, p16, p18, p19, Ink4, osteocalcin, BMP4, Ihh and Sox 9 [12,31,35-37].

Several mutation-specific mechanisms have been proposed to explain how the activating mutations of *FGFR3* enhance FGFR3 signals [1]. The transmembrane achondroplasia mutation is thought to stabilize FGFR3 dimers following ligand-induced dimerization [38]. Monsonego-Ornan et al have suggested that this mutation slows receptor internalization leaving it on the surface to signal [39]. The free cysteine residues introduced by the TDI mutations are believed to form disulfide bonds resulting in dimerization [40]. The TDII mutation alters the conformation of the kinase domain constitutively activating enzyme activity [38,41]. It is not clear if receptors carrying the TD mutations must reach the cell surface to become activated. Kinase is also activated by the common hypochondroplasia mutation but to a lesser extent.

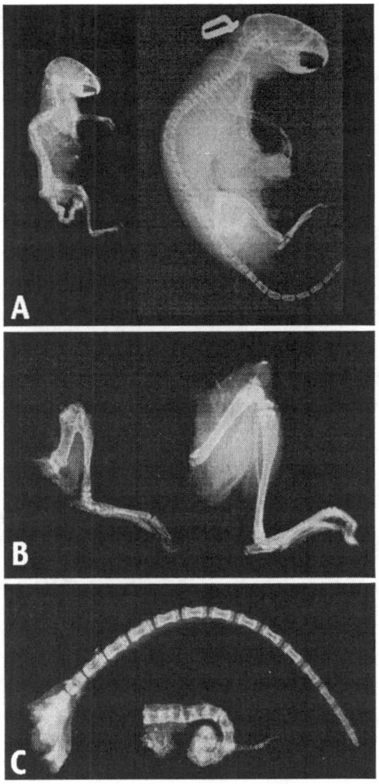

Figure 2. Xrays of 6 wk old *Col2-FGF9* transgenic mouse (left) compared to wild type littermate (right). A. lateral view of spine. B lower extremity. C tail. The shortening is greater for proximal than for distal bones from the transgenic mice (B). *From Garofalo et al, J Bone Min Res 14:1909, 1999.*

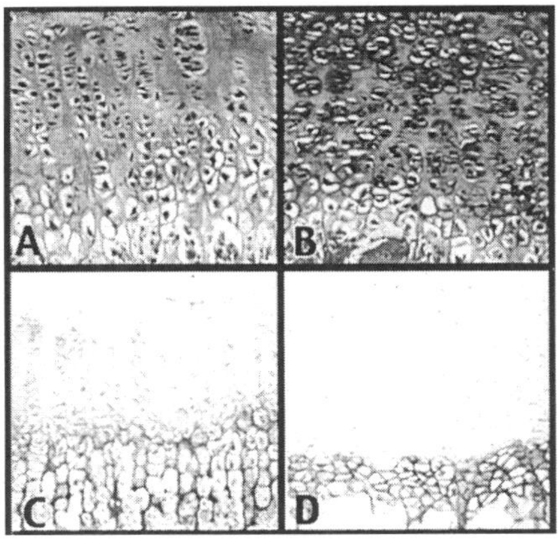

Figure 3. Morphology (A, B) and immunostaining for type X collagen (C, D) of growth plate from wild type (A, C) and *Col2-FGF9* transgenic (B, D) mice. Note reduction of proliferative and hypertrophic chondrocytes, size of hypertrophic chondrocytes and in distribution of type X collagen in transgenic growth plate. *From Garofalo et al, J Bone Min Res 14:1909, 1999.*

Signaling Experiments

We have previously shown that FGF2 and FGF9 differentially affect differentiation of the prechondrocytic RCJ3.1C5.18 cells through activation of FGF receptors [19,42]. FGF2 promotes proliferation and blocks all differentiation, whereas FGF9 inhibits only terminal differentiation (Fig. 4). However, since FGFR1, FGFR2 and FGFR3 were all activated, we were unable to determine which receptor(s) was (were) responsible for these phenotypic effects. To shed light on the issue, we introduced the equivalent of the TDII (kinase activating) mutation into FGFR1 and FGFR2 to generate constitutively activated (ca) forms of these receptors. cDNAs encoding all three caFGFRs were cloned into retroviral expression vectors and subsequently transduced into RCJ3.1C5.18 cells. The cells were cultured under conditions that support chondrocyte differentiation. All three promoted mitosis and inhibited chondrocyte differentiation and subsequent terminal differentiation, although there was minimal differentiation of cells transduced with FGFR2. Results for cells expressing caFGFR1 and caFGFR3 are shown in Fig. 5. Moreover, when receptors were assayed by immunoprecipitation followed by western blotting, only the FGFR that had been transduced was detected suggesting that expression of the engineered receptors suppressed expression endogenous receptors in these cells.

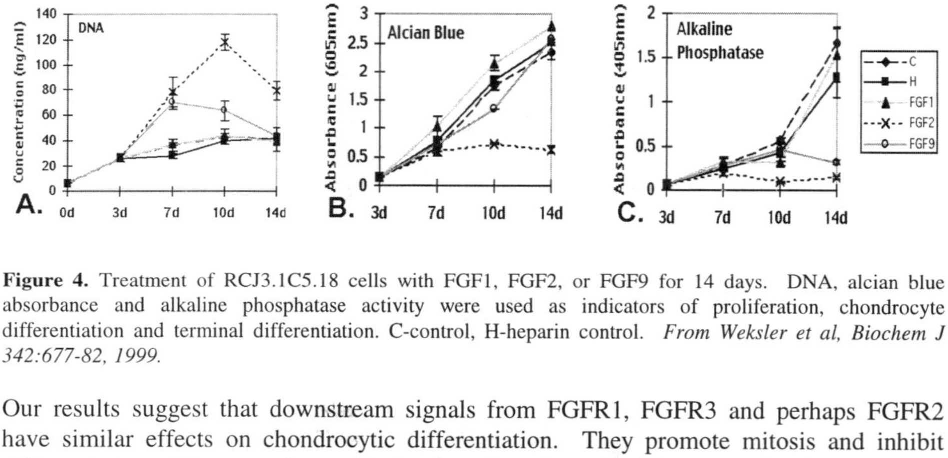

Figure 4. Treatment of RCJ3.1C5.18 cells with FGF1, FGF2, or FGF9 for 14 days. DNA, alcian blue absorbance and alkaline phosphatase activity were used as indicators of proliferation, chondrocyte differentiation and terminal differentiation. C-control, H-heparin control. *From Weksler et al, Biochem J 342:677-82, 1999.*

Our results suggest that downstream signals from FGFR1, FGFR3 and perhaps FGFR2 have similar effects on chondrocytic differentiation. They promote mitosis and inhibit differentiation. These results could simply reflect the artificial nature of cell culture or a behavior unique to the RCJ3.1C5.18 cells. However, the detection of activating FGFR3 mutations, which are antimitotic in growth plate chondrocytes, in rapidly proliferating cells in several types of tumors suggests that the cellular responses to FGFR signals may depend on the type and stage of differentiation of cells. These findings are very consistent with the emerging view noted above that cell context is very important for determining propagation pathways and specific responses to extracellular signals [24].

Acknowledgements

This work was supported by Shriners Hospital for Children.

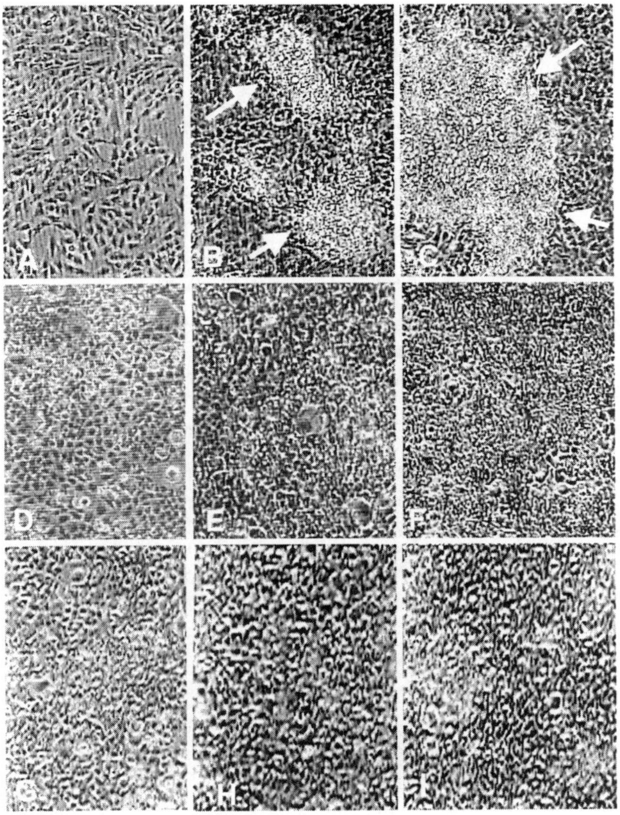

Figure 5. RCJ3.1C5.18 cells at 3 (A, D, G), 7 (B, E, H) and 10 (C, F, I) days after plating and under conditions that promote differentiation. A-C shows chondrocytic differentiation of non-transduced cells. Arrows indicate cartilage nodules. Cells in D-F were transduced with caFGFR1 and in H-J with caFGFR3. The cartilage nodules present in B and C do not form in the presence of FGFR1 or FGFR3 signals. Cells transduced with caFGFR2 had minimal nodule formation (not shown).

Acknowledgements

This work was supported by Shriners Hospital for Children.

References

[1] Horton WA 1997 Fibroblast growth factor receptor 3 and the human chondrodysplasias. Curr Opin Pediatr **9**(4)**:**437-42.

[2] Shiang R, Thompson LM, Zhu Y-Z, Church DM, Fielder TJ, Bocian M, Winokur ST, Wasmuth JJ 1994 Mutations in the transmembrane domain of FGFR3 cause the most common genetic form of dwarfism, achondroplasia. Cell **78:**335-342.

[3] Rousseau F, Bonaventure J, Legeal-Mallet L, Pelet A, Rozet J, Maroteaux P, Le Merrer M, Munnich A 1995 Mutations in the gene encoding fibroblast growth factor recetpor-3 in achondroplasia. Nature **371:**252-254.

[4] Bellus G, Hefferon T, Ortiz de Luna R, Hecht J, Horton W, Machado M, Kaitila I, McIntosh I, Francomano C 1995 Achondroplasia is defined by recurrent G380R mutations in FGFR3. Am J Hum Genet **56:**368-373.

[5] Tavormina P, Shiang R, Thompson L, Zhu Y, Wilkin D, Lachman R, Wilcox W, Rimoin D, Cohn D, Wasmuth J 1995 Thanatophoric dysplasia (types I and II) caused by distinct mutations in fibroblast growth factor receptor 3. Nat Genet **9:**321-328.

[6] Rousseau F, El Ghouzzi V, Delezoide AL, Legeai-Mallet L, Le Merrer M, Munnich A, Bonaventure J 1996 Missense FGFR3 mutations create cysteine residues in thanatophoric dwarfism type I (TD1). Hum Molec Genet **5**(4)**:**509-512.

[7] Bellus GA, McIntosh I, Smith EA, Aylsworth A.S., Kaitila I, Horton WA, Greenhaw GA, Hecht JT, Francomano CA 1995 A recurrent mutation in the tyrosine kinase domain of fibroblast growth factor receptor 3 causes hypochondroplasia. Nat Genet **10:**357-359.

[8] Wilkin DJ, Szabo JK, Cameron R, Henderson S, Bellus GA, Mack ML, Kaitila I, Loughlin J, Munnich A, Sykes B, Bonaventure J, Francomano CA 1998 Mutations in fibroblast growth-factor receptor 3 in sporadic cases of achondroplasia occur exclusively on the paternally derived chromosome. Am J Hum Genet **63**(3)**:**711-6.

[9] Hehr U, Muenke M 1999 Craniosynostosis syndromes: from genes to premature fusion of skull bones. Mol Genet Metab **68**(2)**:**139-51.

[10] Deng C, Wynshaw-Boris A, Zhou F, Kuo A, Leder P 1996 Fibroblast growth factor receptor 3 is a negative regulator of bone growth. Cell **84**(6)**:**911-21.

[11] Colvin JS, Bohne BA, Harding GW, McEwen DG, Ornitz DM 1996 Skeletal overgrowth and deafness in mice lacking fibroblast growth factor receptor 3. Nat Genet **12**(4)**:**390-397.

[12] Naski MC, Colvin JS, Coffin JD, Ornitz DM 1998 Repression of hedgehog signaling and BMP4 expression in growth plate cartilage by fibroblast growth factor receptor 3. Development **125**(24)**:**4977-88.

[13] Naski MC, Wang Q, Xu J, Ornitz DM 1996 Graded activation of fibroblast growth factor receptor 3 by mutations causing achondroplasia and thanatophoric dysplasia. Nat Genet **13**(2)**:**233-7.

[14] Ornitz DM, Xu J, Colvin JS, McEwen DG, MacArthur CA, Coulier F, Gao G, Goldfarb M 1996 Receptor specificity of the fibroblast growth factor family. J Biol Chem **271**(25)**:**15292-7.

[15] Jang JH, Shin KH, Park JG 2001 Mutations in fibroblast growth factor receptor 2 and fibroblast growth factor receptor 3 genes associated with human gastric and colorectal cancers. Cancer Res **61**(9)**:**3541-3.

[16] van Rhijn BW, Lurkin I, Radvanyi F, Kirkels WJ, van der Kwast TH, Zwarthoff EC 2001 The fibroblast growth factor receptor 3 (FGFR3) mutation is a strong indicator of superficial bladder cancer with low recurrence rate. Cancer Res **61**(4)**:**1265-8.

[17] Chesi M, Brents LA, Ely SA, Bais C, Robbiani DF, Mesri EA, Kuehl WM, Bergsagel PL 2001 Activated fibroblast growth factor receptor 3 is an oncogene that contributes to tumor progression in multiple myeloma. Blood **97**(3)**:**729-36.

[18] Sahni M, Ambrosetti DC, Mansukhani A, Gertner R, Levy D, Basilico C 1999 FGF signaling inhibits chondrocyte proliferation and regulates bone development through the STAT-1 pathway. Genes Dev **13**(11)**:**1361-6.

[19] Weksler NB, Lunstrum GP, Reid ES, Horton WA 1999 Differential effects of fibroblast growth factor (FGF) 9 and FGF2 on proliferation, differentiation and terminal differentiation of chondrocytic cells in vitro. Biochem J **342 Pt 3**(11)**:**677-82.

[20] Chen L, Adar R, Yang X, Monsonego EO, Li C, Hauschka PV, Yayon A, Deng CX 1999 Gly369Cys mutation in mouse FGFR3 causes achondroplasia by affecting both chondrogenesis and osteogenesis. J Clin Invest **104**(11)**:**1517-25.

[21] Wang Y, Spatz MK, Kannan K, Hayk H, Avivi A, Gorivodsky M, Pines M, Yayon A, Lonai P, Givol D 1999 A mouse model for achondroplasia produced by targeting fibroblast growth factor receptor 3. Proc Natl Acad Sci U S A **96**(8)**:**4455-60.

[22] Garofalo S, Kliger-Spatz M, Cooke JL, Wolstin O, Lunstrum GP, Moshkovitz SM, Horton WA, Yayon A 1999 Skeletal dysplasia and defective chondrocyte differentiation by targeted overexpression of fibroblast growth factor 9 in transgenic mice. J Bone Miner Res **14**(11)**:**1909-15.

[23] Schlessinger J, Plotnikov AN, Ibrahimi OA, Eliseenkova AV, Yeh BK, Yayon A, Linhardt RJ, Mohammadi M 2000 Crystal structure of a ternary FGF-FGFR-heparin complex reveals a dual role for heparin in FGFR binding and dimerization. Mol Cell **6**(3)**:**743-50.

[24] Schlessinger J 2000 Cell signaling by receptor tyrosine kinases. Cell **103**(2)**:**211-25.

[25] Chellaiah AT, McEwen DG, Werner S, Xu J, Ornitz DM 1994 Fibroblast growth factor receptor (FGFR) 3. Alternative splicing in immunoglobulin-like domain III creates a receptor highly specific for acidic FGF/FGF-1. J Biol Chem **269**(15)**:**11620-11627.

[26] Hecht D, Zimmerman N, Bedford M, Avivi A, Yayon A 1995 Identification of fibroblast growth factor 9 (FGF9) as a high affinity, heparin dependent ligand for FGR receptors 3 and 2 but not for FGF receptors 1 and 4. Growth Factors **12:**223-233.

[27] Santos-Ocampo S, Colvin JS, Chellaiah A, Ornitz DM 1996 Expression and biological activity of mouse fibroblast growth factor-9. J Biol Chem **271**(3)**:**1726-1731.

[28] Kanai M, Göke M, Tsunekawa S, Podolsky DK 1997 Signal transduction pathway of human fibroblast growth factor receptor 3. J Biol Chem **272**(10)**:**6621-6628.

[29] Ornitz DM 2000 FGFs, heparan sulfate and FGFRs: complex interactions essential for development. Bioessays **22**(2)**:**108-12.

[30] Mohammadi M, Dikic I, Sorokin A, Burgess WH, Jaye M, Schlessinger J 1996 Identification of six novel autophosphorylation sites on fibroblast growth factor receptor 1 and elucidation of their importance in receptor activation and signal transduction. Mol Cell Biol **16**(3)**:**977-89.

[31] Su W-CS, Kitagawa M, Xue N, Xie B, Garofalo S, Cho J, Deng C, Horton WA, Fu X-Y 1997 Activation of Stat1 by mutant fibroblast growth-factor receptor in thanatophoric dysplasia type II dwarfism. Nature **386:**288-291.

[32] Legeai-Mallet L, Benoist-Lasselin C, Delezoide AL, Munnich A, Bonaventure J 1998 Fibroblast growth factor receptor 3 mutations promote apoptosis but Do not alter chondrocyte proliferation in thanatophoric dysplasia [In Process Citation]. J Biol Chem **273**(21)**:**13007-14.

[33] Hart KC, Robertson SC, Kanemitsu MY, Meyer AN, Tynan JA, Donoghue DJ 2000 Transformation and Stat activation by derivatives of FGFR1, FGFR3, and FGFR4. Oncogene **19**(29)**:**3309-20.

[34] Choi DY, Toledo-Aral JJ, Lin HY, Ischenko I, Medina L, Safo P, Mandel G, Levinson SR, Halegoua S, Hayman MJ 2000 Fibroblast Growth Factor Receptor 3 induces gene expression primarily through Ras-independent signal transduction pathways. J Biol Chem **14:**11.

[35] Li C, Chen L, Iwata T, Kitagawa M, Fu XY, Deng CX 1999 A Lys644Glu substitution in fibroblast growth factor receptor 3 (FGFR3) causes dwarfism in mice by activation of STATs and ink4 cell cycle inhibitors. Hum Mol Genet **8**(1)**:**35-44.

[36] McEwen DG, Green RP, Naski MC, Towler DA, Ornitz DM 1999 Fibroblast growth factor receptor 3 gene transcription is suppressed by cyclic adenosine 3',5'-monophosphate. Identification of a chondrocytic regulatory element. J Biol Chem **274**(43)**:**30934-42.

[37] Murakami S, Kan M, McKeehan WL, de Crombrugghe B 2000 Up-regulation of the chondrogenic Sox9 gene by fibroblast growth factors is mediated by the mitogen-activated protein kinase pathway. Proc Natl Acad Sci U S A **97**(3)**:**1113-8.

[38] Webster MK, Donoghue DJ 1996 Constitutive activation of fibroblast growth factor receptor 3 by the transmembrane domain point mutation found in achondroplasia. EMBO J **15**(3)**:**520-527.

[39] Monsonego-Ornan E, Adar R, Feferman T, Segev O, Yayon A 2000 The transmembrane mutation G380R in fibroblast growth factor receptor 3 uncouples ligand-mediated receptor activation from down-regulation. Mol Cell Biol **20**(2)**:**516-22.

[40] Sorokin A, Lemmon MA, Ullrich A, Schlessinger, J. 1994 Stabilization of an active dimeric form of the epidermal growth factdor receptor by introduction of an inter-receptor disulfide bond. J. Biol. Chem.**:**9752-9759.

[41] Mohammadi M, Schlessinger J, Hubbard SR 1996 Structure of the FGF receptor tyrosine kinase domain reveals a novel autoinhibitory mechanism. Cell **86:**577-587.

[42] Lunstrum GP, Keene DR, Weksler NB, Cho YJ, Cornwall M, Horton WA 1999 Chondrocyte differentiation in a rat mesenchymal cell line. J Histochem Cytochem **47**(1)**:**1-6.

Fibrodysplasia Ossificans Progressiva: Evolving Insights from a Rare Disease

Frederick S. Kaplan, M.D.[1,2], Jaimo Ahn, Ph.D.[1], Eileen M. Shore, Ph.D.[1,3]
Departments of Orthopaedic Surgery[1], Medicine, [2] and Genetics, [3], The University of Pennsylvania Medical Center, Philadelphia, PA 19104

Abstract. Fibrodysplasia ossificans progressiva (FOP), an autosomal dominant genetic disorder is the most disabling form of heterotopic ossification known to mankind. FOP is characterized by anterior-patterning malformations of the great toes and by progressive induction of endochondral osteogenesis at ectopic sites. Recombinant human bone morphogenetic protein 4 (BMP4), a potent osteogenic morphogen, can induce the entire developmental program of endochondral osteogenesis at an ectopic site in a manner identical to that seen in FOP. BMP4 mRNA and protein are uniquely over-expressed in lymphocytes and lesional cells from patients who have FOP, and preliminary findings suggest that a defect in the BMP4 pathway in FOP cells severely limits their ability to express BMP antagonists in response to a BMP4 stimulus. The BMP4 gene is not mutated in FOP, and the BMP4 locus has recently been excluded from linkage to the condition. These data strongly suggest that the primary defect in FOP is not in the BMP4 gene, but in a component of the BMP4 pathway or interacting pathway that leads to over-expression of BMP4 and misregulation of BMP4 antagonist expression. A more complete understanding of the molecular genetics and pathophysiology of FOP will provide valuable insight into the molecular mechanisms that regulate the induction of osteogenesis and will permit a more rational approach to devising effective treatments for FOP and related disorders.

Numerous genes are involved in bone formation, but few have been implicated in bone induction. A novel approach to the isolation, detection, and control of genes responsible for the induction and regulation of skeletogenesis involves the identification and study of genetic diseases in which skeletal induction is specifically dysregulated. [1-2]

Clinical Features of Fibrodysplasia Ossificans Progressiva

Fibrodysplasia ossificans progressiva (FOP) is a rare genetic disorder of connective tissue characterized by progressive, disabling heterotopic osteogenesis. [3-6] Congenital malformation of the great toes is the earliest phenotypic feature of FOP, and is present in nearly all affected individuals. [2] Progressive heterotopic ossification usually begins in the first five years of life. [7-8] Impending heterotopic ossification is heralded by the appearance of large painful swellings of highly vascular fibroproliferative tissue, involving tendons, ligaments, and skeletal muscle.[9-10] These pre-osseous swellings progress along a pathway of endochondral ossification to form mature ossicles of heterotopic bone.[9] The anatomic progression of heterotopic ossification in FOP occurs in specific patterns over time. Involvement typically is seen earliest in dorsal, axial, cranial, and proximal regions of the body and later in ventral, appendicular, caudal, and distal regions. [7,8,11] These developmental patterns are similar to the patterns and progression of embryonic skeletal formation.

Progressive episodes of heterotopic ossification lead typically to ankylosis of all major joints of the axial and appendicular skeleton, rendering movement impossible. [3-6,12-14] Although the rate of disease progression is variable, most patients are confined to a wheelchair by their early twenties and require lifelong assistance in performing activities of daily living. [7,8] People with fibrodysplasia ossificans progressiva have markedly reduced reproductive fitness, and usually succumb later in adulthood to cardiorespiratory complications secondary to severe restrictive disease of the chest wall.[15]

Falls, soft tissue injury, surgical trauma, mandibular blocks, and routine intramuscular immunizations lead to exacerbation of local ossification. At present, there is no effective prevention or treatment.[4, 6,16-19]

FOP is an autosomal dominant disorder, although most cases are attributable to spontaneous new mutations in previously unaffected families.[3, 20] The genetic defect and pathophysiology of the disorder are not known, however, the bone morphogenetic protein genes and other genes in the BMP pathway have been implicated as plausible candidate genes.[11] Studies to identify the cause of FOP have focused on the candidate gene approach. Definitive linkage analysis and positional cloning is difficult, because only four small families with inheritance of FOP have been identified worldwide. Karyotype abnormalities have not been detected in patients with the disorder and lesional tissue is not readily available for study.[9, 20]

Pathologic Features of Fibrodysplasia Ossificans Progressiva

Histologic examination of early FOP lesions reveals an intense perivascular lymphocytic infiltration followed by lymphocyte-associated death of skeletal muscle and robust development of fibroproliferative tissue with extensive neovascularity and mast cell infiltration.[21,22] Tissue from FOP lesions at a later stage of maturation shows characteristic features of endochondral ossification including chondrocyte hypertrophy, calcification of cartilage, and formation of woven bone with marrow elements.[9] Ectopic bone in patients with FOP characteristically develops along an endochondral pathway.[9] Fractures through heterotopic bone appear to heal normally.[23]

The endochondral ossification that occurs in an FOP lesion is similar to that seen in the growth plate or in fracture healing. However, unlike the evolution of bone formation in either of those two processes, the FOP lesion arises at a heterotopic site following lymphocyte-associated death of normal skeletal muscle. As such, the evolution of an FOP lesion has features that are truly different from the normal processes of endochondral ossification seen in a growth plate. The FOP lesion appears closer in its development to embryonic skeletogenesis or to fracture healing.

The cellular origin of the pre-osseous fibroproliferative tissue in FOP lesions remains unknown, but candidate cell types include satellite cells and perivascular cells that reside in the skeletal muscle, and mesenchymal stem cells that reside in the peripheral tissue or in the bone marrow. Preliminary data indicate an intense immunostaining of the fibroproliferative cells with antibodies to bone morphogenetic protein 4 (BMP4) and numerous smooth muscle proteins.[24]

Bone Morphogenetic Proteins: Candidate Genes for FOP

When the FOP research program was established at The University of Pennsylvania in 1992, the major focus was to identify the genetic basis of FOP in order to develop clinical treatments that would prevent the progressive induction of heterotopic bone in people who have this condition. Today, the focus has not changed.

Considering the rarity of FOP and the very few families who show inheritance of FOP from generation to generation, standard genetic linkage studies were not originally feasible. Therefore, the candidate gene approach was used to identify genes that could have an altered expression and function in FOP. The bone morphogenetic proteins (BMPs) were primary candidates since this family of proteins can induce mesenchymal cells to differentiate to bone through an endochondral pathway, an event that closely parallels the progression of bone formation in an FOP lesion.[11] Examination of the expression levels of several of the BMP genes revealed that only the expression of BMP4 was increased in FOP cells, and that these elevated levels were due to an increased rate of transcription of the BMP4 gene.[25-28] Continuing studies are investigating the nature of this altered transcriptional regulation in FOP cells.

Upon the discovery of altered BMP4 expression in FOP, the BMP4 gene was screened for DNA sequence mutations in people who have FOP.[29] Extensive analysis revealed no mutations, and this finding was supported by the absence of genetic linkage of the BMP4 locus with FOP in four small multi-generational families that were identified through the collaborative efforts of our colleagues, nationally and internationally.[30] A genome-wide genetic linkage study using DNAs from these four families has revealed linkage to the 4q27-31 region (which does not include the BMP4 locus) and has unveiled additional candidate genes within this 36 centimorgan segment of chromosome 4 that are being examined for mutations that cause FOP.[30] To date, no mutations have been found in any of the known 4q candidate gene including the genes encoding SMAD1, bFGF, BMPRIb, NF-kB, and 15-hydroxyprostaglandin dehydrogenase.[31]

Although the BMP4 gene does not harbor the genetic mutations that cause FOP, BMP4 overexpression is almost certainly an inductive event for the formation of bone in this disorder.[32] Currently, differential gene expression approaches and somatic cell mitotic recombination studies are being conducted to identify altered cellular pathways and to identify the mutated gene in FOP that leads to overexpression of BMP4.

BMP Antagonists in Skeletal Induction

Recent studies have indicated that BMP4 is capable of upregulating expression of multiple BMP antagonists, thereby establishing an autoregulatory negative feedback loop that exerts spatio-temporal control over its own activity.[33-60]

Skeletogenesis, bone induction, and joint formation are specified in a dose-dependent fashion by tightly regulated morphogenetic gradients established in part by bone morphogenetic proteins (BMPs) and their secreted antagonists. In embryonic skeletal formation and post-natal bone formation, BMPs restrict their own activity by inducing secreted antagonists in responsive cells. Such a remarkable degree of genetically-programmed self-restraint illustrates a general mechanism whereby signaling cascades regulate critical morphogen gradients in many developmental contexts.

BMP antagonists play distinct roles in regulating the quality as well as the magnitude of BMP-signaling.[33,34] Many recently discovered secreted peptides antagonize BMP function [35] BMP4 gradients are established not only by the diffusion of BMP4 but also by the long-range effects of the BMP4 antagonists noggin, chordin, gremlin, follistatin and tob.[33,36-47] The various BMP antagonists are unique proteins, but share the functional property of binding specifically to BMPs, thus preventing BMPs from interacting with their receptors. Endogenous and ectopic expression of BMP antagonists illustrates a conserved mechanism for regulating BMP function during limb and somite patterning as well as fracture healing, and suggests that the inhibitory interaction between BMPs and their antagonists is a widely used mechanism to modulate BMP-signaling during multiple inductive events in vertebrate embryogenesis and post-natal osteogenesis.[33,48]

A balanced combination of positive and negative factors is required during development to define and shape each skeletal element.[49] Excess BMP activity in the absence of BMP antagonists enhances the recruitment of cells into cartilage.[49] A number of recent studies have analyzed the distribution and function of various BMP antagonists such as noggin, chordin, and gremlin in the developing limb. Noggin, for example, is expressed in the chondrogenic condensations and regulates the shape and size of the cartilaginous skeleton as well as the identity of the diarthrodial joints [48-50]; chordin may in part regulate joint formation and axial skeletal development [51-52]; gremlin, a member of the DAN/cerberus gene family, functions as a BMP antagonist in the control of limb outgrowth, relays the sonic hedgehog (shh) signal from the zone of polarizing activity (ZPA) to the apical ectodermal ridge (AER), establishes the SHH/FGF4 feedback loop in limb morphogenesis, delimits the apoptotic areas, and restricts chondrogenesis to the central core mesenchyme of the limb bud. [53,54] These studies suggest that the functions of BMPs are spatially and temporally regulated by different BMP antagonists that act in a complimentary fashion rather than being redundant signals.

Knowledge is beginning to emerge on the transcriptional regulation of the BMP antagonists. Recent studies indicate that BMP4 upregulates expression of noggin, [39,55,56] while noggin inhibits BMP4 protein activity, thus establishing a stable autoregulatory network of morphogen activity. [42,57,58] Noggin diffuses among cells more rapidly than BMP4, and establishes the temporal and spatial extent of the BMP4 morphogenetic gradient. [59]

Attenuation of BMP-4 Induced Noggin and Gremlin Expression in Fibrodysplasia Ossificans Progressiva

Based upon our findings in FOP and the emerging knowledge of BMP - BMP antagonist interactions, we formulated an hypothesis that a defect in the feedback pathway between BMP4 and one or more of its extracellular antagonists could plausibly contribute to elevated BMP4 activity in fibrodysplasia ossificans progressiva. Therefore, we investigated basal and BMP4-induced expression of noggin, chordin, and gremlin mRNA in control and FOP lymphoblastoid cell lines (LCLs).

We observed that steady-state levels of BMP4 mRNA expression were dramatically elevated in FOP cell lines. In the absence of exogenous BMP4 stimulation (basal state), steady-state levels of all three BMP antagonists were similar in FOP and control cell lines. Upon stimulation with recombinant human BMP4, control LCLs exhibited a marked increase in expression of noggin and gremlin mRNA. FOP cells, however, showed an attenuated response to BMP4 simulation indicating that this response is perturbed in FOP cells compared to control cells. Therefore, the absence of upregulation of noggin and gremlin mRNA in response to BMP4 signaling in FOP cells suggested that the level of antagonist expression observed in the FOP cell lines may be inappropriately low for the elevated levels of BMP4 expression.[60]

Bone Induction in Fibrodysplasia Ossificans Progressiva: A Hypothesis

These preliminary observations on the relationship between BMP4 and its secreted antagonists in FOP cells allow us to propose a disease mechanism based on our evolving insights into FOP. Heterotopic ossification in FOP begins in childhood and can be induced by soft tissue injury. BMP4 is produced by lesional lymphocytes and skeletal muscle and its expression is increased at sites of soft tissue injury. Under normal conditions BMP4 dramatically upregulates the expression of its inhibitors which diffuse more rapidly than BMP4 and other TGF-beta family members.[56,59] A blunted BMP4 antagonist response following soft tissue trauma would permit the rapid expansion of a BMP4 morphogenetic gradient conducive to progressive osteogenesis through an endochondral pathway.[49,61] The growth of highly vascular pre-osseous fibroproliferative tissue seen locally in response to BMP over-expression would be magnified in the setting of a blunted BMP4 antagonist response and could explain the explosive bone induction seen during an FOP flare-up.[24] Lymphocytes arriving at a site of soft tissue injury in patients with FOP exacerbate the process by releasing more BMP4, further expanding the BMP gradient and escalating the inflammatory process at the leading edge of a rapidly expanding lesion.[21] Over time, a large vascular fibroproliferative mass replaces skeletal muscle and matures through an endochondral process into the highly-restrictive extra-articular ribbons, sheets, and plates of bone that ankylose the joints and render movement impossible.

The Future

Our preliminary data suggest the existence of a defect in the BMP4-mediated induction of multiple BMP antagonists in FOP cells as compared to controls, but much more work needs to be done in order to substantiate these findings and to better understand the molecular mechanisms and primary genetic abnormality underlying them.

Such progress in unraveling the molecular pathophysiology of FOP underscores the potential for developing extracellular inhibitor-based strategies in the therapy of FOP. Presently, we are pursuing studies that investigate a gene therapy approach to using noggin in the treatment and prevention of BMP4 induced heterotopic ossification. [62,63]

"With so much being discovered about how the BMPs act," says Brigid Hogan, a developmental geneticist at Vanderbilt University in Nashville, Tennessee, "it might be possible to develop drugs that would block some part of the BMP4 pathway – and therefore prevent the progression of what is a horrible nightmare disease." [64]

Acknowledgements

This work was supported in part by The International FOP Association, The Ian Cali Fund, The Roemex Fellowship, The Grampian Fellowship, The Isaac & Rose Nassau Professorship of Orthopaedic Molecular Medicine, The Allison Weiss Fellowships, and The National Institutes of Health (2RO1-AR-41916-04-ORTH).

References

[1] Connor JM (1996) Fibrodysplasia ossificans progressiva - Lessons from rare maladies N Engl J Med 335 591-593

[2] Kaplan FS (1998) Fibrodysplasia ossificans progressiva, an editorial. Clin Orthop Rel Res 346, 2-4

[3] Connor JM, Evans DAP (1982) Fibrodysplasia ossificans progressiva. The clinical features and natural history of 34 patients. J Bone Joint Surgery - Brit Vol 64, 76-83

[4] Kaplan FS, Delatycki M, Gannon FH, Rogers JG, Smith R, Shore EM (1998) Fibrodysplasia ossificans progressiva. In, Neuromuscular Disorders: Clinical and Molecular Genetics, AEH Emery, ed. (Chichester, England, John Wiley and Sons, Ltd.), pp. 289-321.

[5] Shore EM, Rogers JG, Smith R, Gannon FH, Delatycki M, Urtizberea JA, Triffitt J, LeMerrer M, Kaplan FS (2000) Fibrodysplasia ossificans progressiva. In, The Genetics of Osteoporosis and Metabolic Bone Disease, MJ Econs, ed. (Totowa, NJ, Humana Press)

[6] Kaplan FS, Shore, EM, Whyte MP (1999) Fibrodysplasia ossificans progressiva. pp. 435-437. In Favus MJ (ed). Primer on the Metabolic Bone Diseases and Disorders of Mineral Metabolism - Fourth Edition. The American Society for Bone and Mineral Research, Lippincott-Raven, Philadelphia

[7] Cohen RB, Hahn GV, Tabas JA, Peeper J, Levitz CL, Sando A, Sando N, Zasloff M, Kaplan FS (1993) The natural history of heterotopic ossification in patients who have fibrodysplasia ossificans progressiva. A study of forty-four patients. J Bone Joint Surg - Am Vol 75, 215-219

[8] Rocke DM, Zasloff M, Peeper J, Cohen RB, Kaplan FS (1994) Age- and joint-specific risk of initial heterotopic ossification in patients who have fibrodysplasia ossificans progressiva. Clin Orthop Rel Res 301, 243-248

[9] Kaplan FS, Tabas JA, Gannon FH, Finkel G, Hahn GV, Zasloff MA (1993) The histopathology of fibrodysplasia ossificans progressiva. An endochondral process. J Bone Joint Surg - Am Volume 75, 220-230

[10] Kaplan FS, Sawyer J, Connors S, Keough K, Shore EM, Gannon F, Glaser D, Rocke D, Zasloff M, Folkman J (1998) Urinary basic fibroblast growth factor: a biochemical marker for pre-osseous fibroproliferative lesions in patients who have fibrodysplasia ossificans progressiva. Clin Orthop Rel Res 346, 59-65

[11] Kaplan FS, Tabas JA, Zasloff MA (1990) Fibrodysplasia ossificans progressiva: A clue from the fly? Calcif Tiss Int 47, 117-125

[12] Shah PB, Zasloff MA, Drummond D, Kaplan, FS (1994) Spinal deformity in patients who have fibrodysplasia ossificans progressiva. J Bone Joint Surg - Am Vol 76, 1442-1450

[13] Moriatis JM, Gannon FH, Shore EM, Bilker W, Zasloff MA, Kaplan FS (1997) Limb swelling in patients who have fibrodysplasia ossificans progressiva. Clin Orthop Rel Research 336, 247-253

[14] Kaplan FS, Strear CM, Zasloff MA (1994) Radiographic and scintigraphic features of modeling and remodeling in the heterotopic skeleton of patients who have fibrodysplasia ossificans progressiva. Clin Orthop Rel Res 304, 238-247

[15] Kussmaul WG, Esmail AN, Sagar Y, Ross J, Gregory S, Kaplan FS (1998) Pulmonary and cardiac function in advanced fibrodysplasia ossificans progressiva. Clin Orthop Rel Res 346:104-109

[16] Glaser DL, Rocke DM, Kaplan FS (1998) Catastrophic falls in patients who have fibrodysplasia ossificans progressiva. Clin Orthop Rel Res 346, 110-116

[17] Janoff HB, Zasloff MA, Kaplan FS (1996) Submandibular swelling in patients with fibrodysplasia ossificans progressiva. Otolaryn Head Neck Surg 114, 599-604

[18] Lanchoney TF, Cohen RB, Rocke DM, Zasloff MA, Kaplan FS (1995) Permanent heterotopic ossification at the injection site after diphtheria-tetanus-pertussis immunizations in children who have fibrodysplasia ossificans progressiva. J Ped 126, 762-764

[19] Luchetti W, Cohen RB, Hahn GV, Rocke DM, Helpin M, Zasloff M, Kaplan FS, (1996) Severe restriction in jaw movement after routine injection of local anesthetic in patients who have fibrodysplasia ossificans progressiva. Oral Surg Oral Med Oral Path Oral Radiol Endod 81, 21-25

[20] Delatycki M, Rogers JG (1998) The genetics of fibrodysplasia ossificans progressiva. Clin Orthop Rel Res 346, 15-18

[21] Gannon FH, Valentine BA, Shore EM, Zasloff MA, Kaplan FS (1998) Acute lymphocytic infiltration in extremely early lesions of fibrodysplasia ossificans progressiva. Clin Orthop Rel Res 346, 19-25

[22] Gannon FH, Glaser D, Caron R, Thompson LDR, Shore EM, Kaplan FS (2001) Mast cell involvement in fibrodysplasia ossificans progressiva (FOP). Human Pathol, in press

[23] Einhorn TA, Kaplan FS (1994) Traumatic fractures of heterotopic bone in patients who have fibrodysplasia ossificans progressiva. Clin Orthop Rel Res 308:173-177, 1994

[24] Gannon FH, Kaplan FS, Olmsted E, Finkel GC, Zasloff M, Shore E (1997) Bone morphogenetic protein (BMP) 2/4 in early fibromatous lesions of fibrodysplasia ossificans progressiva. Human Pathol 28, 339-343

[25] Shafritz AB, Shore EM, Gannon FH, Zasloff MA, Taub R, Muenke M, Kaplan FS (1996) Overexpression of an osteogenic morphogen in fibrodysplasia ossificans progressiva. N Engl J Med 335, 555-561

[26] Lanchoney TF, Olmsted EA, Shore EM, Gannon FH, Zasloff MA, Rosen V, Kaplan FS (1998) Characterization of bone morphogenetic protein-4 receptors in fibrodysplasia ossificans progressiva. Clin Orthop Rel Res 346, 38-45
[27] Shore EM Xu M-q, Shah PB, Janoff HB, Hahn GV, Deardorff MA, Sovinsky L, Spinner NB, Zasloff MA, Wozney J, Kaplan FS (1998) The human bone morphogenetic protein (BMP-4) gene: molecular structure and transcriptional regulation. Calcif Tissue Int 63: 221-229
[28] Olmsted E, Shore EM, Kaplan FS (2001) Transcriptional regulation of the BMP4 gene in FOP., Clin Orthop Rel Res, submitted
[29] Xu M-q, Shore EM (1998) Mutational screening of the bone morphogenetic protein 4 gene in a family with fibrodysplasia ossificans progressiva. Clin Orthop Rel Res 346, 53-58
[30] Feldman G, Li M, Martin S, Urbanek M, Urtizberea JA, Fardeau M, LeMerrer M, Connor JM, Triffitt J, Smith R, Muenke M, Kaplan FS, Shore EM (2000) Fibrodysplasia ossificans progressiva, a heritable disorder of severe heterotopic ossification, maps to human chromosome 4q27-31. Am J Human Gen 66, 128-135
[31] Kaplan FS, Shore EM (1998) Encrypted morphogens of skeletogenesis: biological errors and pharmacologic potentials. Biochem Pharm 55, 373-382
[32] Xu M-q, Feldman G, LeMerrer M, Shugart YY, Glaser DL, Urtizberea JA, Fardeau M, Connor JM, Triffit J, Smith R, Shore EM, Kaplan FS (2000) Linkage exclusion and mutational analysis of the noggin gene in patients with fibrodysplasia ossificans progressiva. Clin Genet 58: 291-298
[33] Vogt TF, Duboule D (1999) Antagonists go out on a limb. Cell 99, 563-566
[34] Yu K, Srinivasan A, Shimmi O, Biehs B, Raska KE, Kimelman D, O'Connor MB, Bier E. (2000) Processing of the Drosophila Sog protein creates a novel BMP inhibitory activity. Dev 127, 2143-2154
[35] Massagué J, Chen YG (2000) Controlling TGF-beta signaling. Genes & Dev 14, 627-644
[36] Hemmati-Brivanlou A, Melton D (1997) Vertebrate embryonic cells will become nerve cells unless told otherwise. Cell 88, 13-17
[37] Hogan BL (1996) Bone morphogenetic proteins: Multifunctional regulators of vertebrate development. Genes & Dev 10, 1580-1594
[38] Holley SA, Neul JL, Attisano L, Wrana JL, Sasai Y, O'Connor MB, De Robertis EM, Ferguson EL (1996) The Xenopus dorsalizing factor noggin ventralizes Drosophila embryos by preventing DPP from activating its receptor. Cell 86, 607-17
[39] Nifuji A, Noda, M (1999) Coordinated expression of noggin and bone morphogenetic proteins (BMPs) during early skeletogenesis and induction of noggin expression by BMP-7. J Bone Min Res 14, 2057-2066
[40] Pearce JJ, Penny G, Rossant J (1999) A mouse cerberus/Dan-related gene family. Dev Biol 209, 98-110
[41] Piccolo S, Sasai Y, Lu B, De Robertis, EM (1996) Dorsoventral patterning in Xenopus: Inhibition of ventral signals by direct binding of chordin to BMP-4. Cell 86, 589-98.
[42] Re'em-Kalma Y, Lamb T, Frank D (1995) Competition between noggin and bone morphogenetic protein 4 activities may regulate dorsalization during Xenopus development. Proc Nat Acad Sci United States of America 92, 1214-15.
[43] Smith WC, Harland RM (1992) Expression cloning of noggin, a new dorsalizing factor localized to the Spemann organizer in Xenopus embryos. Cell 70, 829-840
[44] Topol LZ, Bardot B, Zhang Q, Resau J, Huillard E, Marx M, Calothy G, Blair DG, (2000) Biosynthesis, post-translation modification, and functional characterization of Drm/Gremlin. J Biol Chem 275, 8785-8793
[45] Valenzuela DM, Economides AN, Rojas E, Lamb TM, Nunez L, Jones PLNY, Espinosa R, Brannan CI, Gilbert DJ, Copeland NG, Jenkins NA, Le Beau MM, Harland RM, Yancopoulos GD (1995) Identification of mammalian noggin and its expression in the adult nervous system. J Neurosci 15, 6077-84
[46] Yoshida Y, Tanaka S, Umemori H, Minowa O, Usui M, Ikematsu N, Hosoda E Imamura T, Kuno J, Yamashita T, Miyazono K, Noda M, Noda T, Yamamato T. (2000) Negative regulation of BMP/Smad signaling by Tob in osteoblasts. Cell 103, 1085-1097
[47] Zimmerman LB, De Jesus-Escobar JM, Harland RM (1996) The Spemann organizer signal noggin binds and inactivates bone morphogenetic protein 40. Cell 86, 599-606
[48] Perrimon N, McMahon AP (1999) Negative feedback mechanisms and their roles during pattern formation. Cell 86, 13-16
[49] Brunet LJ, McMahon JA, McMahon AP, Harland RM (1998) Noggin, cartilage morphogenesis, and joint formation in the mammalian skeleton. Science 280, 1455-1457

[50] Merino R, Ganan Y, Macias D, Economides AN, Sampath KT, Hurle JM (1998) Morphogenesis of digits in the avian limb is controlled by FGFs, TGFbetas, and noggin through BMP signaling. Develop Biol 200, 35-45
[51] Fisher S, Halpern ME (1999) Patterning the zebrafish axial skeleton requires early chordin function. Nat Genet 23, 442-446
[52] Francis-West PH, Parish J, Lee K, Archer CW (1999) BMP/GDF-signalling interactions during synovial joint development. Cell & Tiss Res 296, 111-119
[53] Merino R, Rodriguez-Leon J, Macias D, Ganan Y, Economides AN, Hurle JM (1999) The BMP antagonist Gremlin regulates outgrowth, chondrogenesis and programmed cell death in the developing limb. Develop 126, 5515-5522
[54] Zuniga A, Haramis AP, McMahon AP, Zeller R (1999) Signal relay by BMP antagonism controls the SHH/FGF4 feedback loop in vertebrate limb buds. Nature 401, 598-602
[55] Amthor H, Christ B, Patel K (1999) A molecular mechanism enabling continuous embryonic muscle growth - a balance between proliferation and differentiation. Dev 126, 1041-1053
[56] Gazzerro E, Gangji V, Canalis E (1998) Bone morphogenetic proteins induce the expression of noggin, which limits their activity in cultured rat osteoblasts. J Clin Invest 102, 2106-2114
[57] Becskei A, Serrano L (2000) Engineering stability in gene networks by autoregulation. Nature 405, 590-593
[58] Gardner TS, Collins JJ (2000) Neutralizing noise in gene networks. Nature 405, 520-521
[59] Jones CM, Smith JC (1998) Establishment of a BMP-4 morphogen gradient by long-range inhibition. Dev Biol 194, 12-17
[60] Ahn J, Shore EM, Kaplan FS (2000) Attenuation of BMP4-induced noggin and gremlin expression in fibrodysplasia ossificans progressiva . Abstr Proc The Third International Symposium on Fibrodysplasia Ossificans Progressiva. University of Pennsylvania School of Medicine, November 2-5, 2000, Philadelphia
[61] Pathi S, Rutenberg JB, Johnson RL, Vortkamp A (1999) Interaction of Ihh and BMP/Noggin signaling during cartilage differentiation. Dev Biol 209, 239-253
[62] Kimble RB, Liu X, Gannon FH, Kaplan FS, Shore EM, Economides AN, Stahl N (1998) Noggin inhibits bone formation in a model of heterotopic ossification. Bone 23, S244
[63] Kaplan FS, Shore EM (2001) Tenth Annual Report of The FOP Collaborative Research Project. The FOP Connection 14,16-31
[64] Roush W (1996) Protein builds second skeleton. Science 273, 1170

Matrix Vesicle Misfunction in Human Hypophosphatasia

H. Clarke Anderson[1], Howard H. Hsu[1], David C. Morris[1], Kenton N. Fedde[2] and Michael P. Whyte [2,3]
[1]*Department of Pathology and Laboratory Medicine, University of Kansas Medical Center, Kansas City, KS 66160.* [2]*Division of Bone and Mineral Diseases, Washington University School of Medicine at the Barnes-Jewish Hospital, St. Louis, MO 63110.* [3]*Metabolic Research Unit, Shriners Hospital for Children, St. Louis, MO 63131.*

Abstract. Hypophosphatasia, a heritable disease characterized by deficient activity of the tissue non-specific isoenzyme of alkaline phosphatase (TNSALP), results in rickets and osteomalacia [1]. Although identification of TNSALP gene defects in hypophosphatasia establishes a role of ALP in skeletal mineralization, the precise function of the enzyme remains unclear. The initial site of skeletal biomineralization (primary mineralization) normally occurs within the lumen of TNSALP-rich matrix vesicles (MVs) of growth cartilage, bone and dentin [2]. A study was carried out to determine whether defective calcification in hypophosphatasia is due to a paucity of MVs versus a functional failure of MVs to initiate calcification, secondary to their TNSALP deficiency. Non-decalcified autopsy bone and growth plate cartilage from six patients with perinatal (lethal) hypophosphatasia were studied by non-decalcified light and transmission electron microscopy to determine whether MVs 1) were present in normal numbers, 2) were of normal size, shape, and ultra-structure, and 3) contained hydroxyapatite mineral, as would be expected if hypophosphatasia MVs retained their ability to concentrate calcium and phosphate internally despite the paucity of TNSALP in their investing membranes. We found that hypophosphatasia MVs are present in normal numbers and distribution, and that they are capable of initiating internal but not external mineralization [3]. Thus, in hypophosphatasia the failure of bones to calcify appears to involve a block of the vectorial spread of mineral from initial nuclei within MVs, outwards, into the matrix. In conclusion, the study showed that hypophosphatasia MVs can concentrate calcium and phosphate internally despite a deficiency of TNSALP activity.

Introduction

Hypophosphatasia is a heritable form of rickets and/or osteomalacia due to defects within the gene that encodes the tissue non-specific (bone/liver/kidney) isoenzyme of alkaline phosphatase, (TNSALP) [1] In bones, dentin and growth plate cartilage, alkaline phosphatase (ALP) has been known for many years to be associated with calcification [2-5] Although the precise role of ALP in promoting mineralization has long been debated, it is widely agreed that the enzymatic action of ALP and of related phosphoesterases (e.g. ATPase, inorganic pyrophosphatase, etc) does function in calcification [6-9] This assertion concerning TNSALP is strongly supported by investigations of human hypophosphatasia, in which diminished TNSALP phosphohydrolase activity in bones and teeth is accompanied by a failure of hard tissue mineralization [10-12] In two strains of TNALP-deficient mice, examined by Fedde, et al [13] a failure of bone matrix mineralization was observed after about the 18th postnatal day, and increased progressively after that time.

Insight into the physiologic role of ALP could come from a better understanding of the mechanism by which ALP-enriched matrix vesicles (MVs) initiate mineralization. MVs are submicroscopic, extracellular, membrane-invested bodies that are selectively located at initial sites of calcification in developing bone, growth plate cartilage, and in the dentin of teeth [6,14] They function in calcification by generating the first crystals of hydroxyapatite (HAP) mineral within the protective confines of their investing membranes [14]. The phosphatases of MVs, including TNSALP, ATPase and nucleoside triphosphate pyrophosphohydrolase (NTPPase), are concentrated in MVs [15] and are attached to the MV membrane's outer surface [4,16,17] There is considerable experimental evidence indicating that these ectophosphatases can stimulate *in vitro* calcification by isolated mammalian MVs [2, 7, 8, 9, 18, 19] The failure of skeletal mineralization observed in hypophosphatasia could be due solely to a lack of intrinsic phosphatase activity within matrix vesicles. Alternative pathogenetic problems in hypophosphatasia could include 1) MV agenesis leading to a failure of mineralization, or 2) other MV defects in addition to low levels of TNSALP activity, that might inhibit initial mineral crystal formation within the MVs, or might prevent the propagation of performed intravesicular mineral into the extracellular matrix during phase 2 of calcification [6].

The detailed ultrastructure of MVs in hypophosphatasia bone has not been extensively studied. A report in 1985 describes the electron microscopic structure of hypophosphatasia bone [20]. Here it was reported that MVs are present in growth plate cartilage in approximately normal numbers. The authors state that "matrix vesicles of growth plate cartilage and of metaphyseal bone do not appear to mineralize" [20]. However, MVs of bone matrix are not illustrated in the report. In a 1996 report, a 21 week gestation fetus with hypophosphatasia was examined by electron microscopy and microprobe element analysis [21]. These authors reported that "there was almost no hydroxyapatite mineralization in the matrix vesicles" of osteoid. Our electron microscopic observation in six cases of infantile hypophosphatasia, described below, differ significantly from these earlier reports [22].

A definitive answer to the question of whether hypophosphatasia MVs can initiate mineralization, despite their presumed lack of ALP, can provide an important insight into the fundamental mechanism of skeletal calcification. The answer to this question relates to whether ALP is essential for initial MV calcification, or whether the function of ALP can be assumed by other phosphatases present in the vesicle membrane, especially ATPase. Recent investigations have suggested that ATP is more active as a substrate than beta-glycerophosphate in supporting *in vitro* MV calcification [19]. Also, metabolic inhibitors of ALP activity only partially inhibit MV calcification *in vitro* when ATP is supplied as substrate [19]. These findings suggest that 1) MV calcification is promoted by several different phosphatases (one of which is TNSALP) all working together, 2) that the specific ATPase of MVs may be important for initiating MV calcification, and 3) that, despite ALP deficiency, MVs remain capable of initiating calcification through the action of ATPase or other non-ALP phosphatases. Our finding, described below, of mineral in MVs in hypophosphatasia, supports the hypothesis that phosphatases other than ALP can initiate MV calcification, and that TNSALP plays an important role in the propagation of HAP crystals into the collagenous extravesicular matrix [22].

To clarify the role of TNSALP in MV calcification we examined hypomineralized matrix zones in growth plate cartilage and in developing bone in severely affected hypophosphatasia patients using non-decalcified tissue preparations and high-resolution transmission electron microscopy. By this approach, we assessed whether MVs were present in approximately normal numbers and in a normal distribution, and whether these vesicles were "unmineralized", as suggested earlier [20,21], or conversely, whether they showed ultrastructural evidence of some initial mineral deposition either within or at their surfaces.

Materials and Methods

Cases
All six patients studied had clinical findings consistent with perinatal (lethal) hypophosphatasia. Profoundly low levels of serum ALP activity and markedly elevated plasma pyridoxal-5-phosphate (PLP) levels were documented in cases 2 through 5.

Case 1. This 24 week gestation fetus was delivered stillborn from a woman whose previous baby lived only one day, had low serum ALP activity, clinical evidence of severe rickets and osteomalacia, and was diagnosed as having hypophosphatasia. X-rays taken during this second pregnancy showed severe hypomineralization of all fetal bones. Pathologic examination of the tibia and femur showed both bones to be essentially non-calcified. MVs were isolated from an unfixed, frozen sample of growth plate by collagenase digestion and differential centrifugation as previously described [9,15]. The specific activity of ALP was measured in the isolated MVs from this patient and from the growth plate of an age-matched still-born fetus. The hypophosphatasia MVs were totally lacking in ALP activity (Table 1).

Table 1. Alkaline phosphatase specific activity in isolated cells and matrix vesicles from growth plate of perinatal hypophosphatasia (Case 1) and from an age-matched normal infant.
*ALP units (ΔOD_{410}/min/mg) per mg protein. **Level of detectability, 0.001 units.

Sample	ALP-Normal	ALP-hypophosphatasia
cell fraction	9.8*	<0.001 units**
Matrix vesicle fraction	55.7	<0.001 units

Case 2. This boy was born two weeks after the expected date-of-confinement. The pregnancy had been uncomplicated, but fetal distress was detected just prior to birth. Physical examination showed deformities of his chest and the long bones of his extremities. Serum ALP was 5 IU/L (normal = 80-342 IU/L) and plasma PLP 29,500 nM (normal = 30-100 nM). Serum calcium and phosphate levels were normal. Artificial ventilation was discontinued at 10 days-of-age. Autopsy showed severe generalized skeletal hypomineralization, limb deformities, and multiple fractures.

Case 3. Radiographs of this full-term, deformed newborn boy were consistent with severe hypophosphatasia. The serum ALP was 4 IU/L and plasma PLP 6,800 nM. He died at 11 days-of-age.

Case 4. At birth, this boy was found to have poor formation of the membranous bones of the skull. Radiographs showed small plates of bone within the calvarium and multiple, non-displaced rib fractures, consistent with severe hypophosphatasia. Hypercalcemia and hyperphosphatemia were also noted. The serum ALP was 5 IU/L and plasma PLP 5,700 nM. He died at 5 1/2 months-of-age.

Case 5. This girl was delivered by Caesarian section at 39 weeks gestation and weighed 2.5 kgs. Fetal ultrasound was abnormal during the third trimester of pregnancy. At birth, the skull was soft with enlarged fontanels. There were chest and limb deformities. Radiographs showed profound skeletal hypomineralization. At the MRU, serum ALP was <7 IU/L and plasma PLP was 22,100 nM. Urine phosphoethanolamine was elevated. She expired at 2 weeks-of-age.

Case 6. This male infant was born at 36 weeks gestation to a 27 year old mother whose first child, a female, died shortly after delivery with a diagnosis of hypophosphatasia. Fetal ultrasound and x-rays in the second child revealed skeletal dysplasia consistent with hypophosphatasia. At birth, the skull sutures were widely separated, the sternum was short and the ribs pliable. Radiograms showed hypoplastic ribs, a shortened sternum, fracture of the left clavicle and shortened extremities with bowing of the femurs. Serum Ca, PO_4, ALP and PAP were not recorded. The patient died at 2 days of age.

Light and Electron Microscopy
Non-decalcified growth plates with adjacent metaphyses and cortical bone were fixed without decalcification in 2.5% glutaraldehyde buffered with cacodylate, washed, dehydrated and embedded in epoxy resin by standard procedures [14]. Light microscopy was performed on plastic-embedded, 1 micron-thick sections after staining with toluidine blue. In case 1, parallel samples of growth plate and long bone were also embedded in paraffin, sectioned at 4 microns thickness and stained with hematoxylin and eosin. Decalcification was not necessary because of the hypomineralized state of the bones, and was not carried out in any of the cases. Ultra thin sections were stained with uranyl acetate and lead citrate [14] and examined using a Phillips EM 300 or a Zeiss EM 10A electron microscope.

Results

Confirmation of the Presence of Hypomineralization in Growth Plates and Bone
Light microscopy showed that mineral was diminished or absent from the growth plate, and markedly reduced in the perichondrial, metaphyseal and cortical bone in all six cases (Fig. 1). The tibial growth plate in cases 1 and 4 was misshapen with irregular projections of uncalcified epiphyseal cartilage extending downward into the metaphysis. The metaphyseal cortex was devoid of mineral in cases 2, 3 and 4. However, in most cases the central bone matrix of medullary bone trabeculae showed some calcification deep to the osteoid layer (Fig. 1). In all cases, bone surfaces were covered by a thick layer of non-mineralized osteoid. Thus, there was a marked mineral deficit in both cartilage and bone in all cases studied.

Electron Microscopy of Growth Plate Cartilage
MVs were identified in approximately normal numbers in all growth plates examined and were normally distributed within the matrix of longitudinal septa (Fig 2). The majority of growth plate MVs in hypophosphatasia were devoid of mineral. However, in all cases, vesicles located in what would have been the normal calcifying zone of the growth plate contained needle-like crystals of apparent HAP (Fig. 2). These crystals often were attached to the inner leaflet of the vesicle membrane. In all the cases, the MV sap also contained a diffuse, apparently non-crystalline, electron-dense material in addition to the needle-like deposits of apatite-like crystals (Figs. 2 and 3).

Electron Microscopy of Bone Matrix
In the hypophosphatasia bone matrix, a calcification front (i.e. a surface layer for the appositional growth of mineral) was either not detected (cases 1 and 3) or was covered by thick layers of non-mineralized osteoid matrix (cases 2, 4, 5 and 6). In all cases, MVs were identified in the interstices between large collagen fibrils (Fig. 3A). In bone matrix beneath osteoblasts, a few MVs were intact and contained needles of apatite-like mineral (Fig 3), while the majority appeared to be in the early propagation phase of mineralization [6] in which MV membranes are flattened and apparently penetrated by small apatite-like crystals

which project into the intercollagenous spaces (Fig. 3B). In the normal growth plate, mineral clusters grow by surface accumulation of apatite-like crystals until they ultimately fuse to form confluent areas of mineral. However, the growth and ultimate fusion of these radial clusters of mineral appeared to be retarded in hypophosphatasia, thus allowing the persistence of large areas of unmineralized bone matrix.

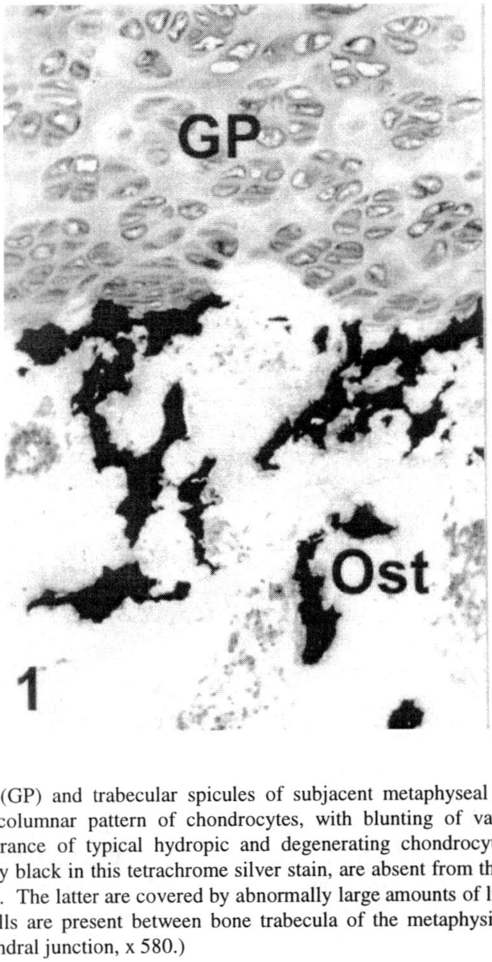

Figure 1. Growth plate (GP) and trabecular spicules of subjacent metaphyseal bone. The growth plate exhibits a disorganized columnar pattern of chondrocytes, with blunting of vascular ingrowth from the metaphysis and disappearance of typical hydropic and degenerating chondrocytes at the base. Mineral deposits, staining intensely black in this tetrachrome silver stain, are absent from the growth plate and greatly reduced in bone trabecula. The latter are covered by abnormally large amounts of light-staining osteoid (Ost). Dark staining marrow cells are present between bone trabecula of the metaphysis in the lower half of the figure. (Case 4, costochondral junction, x 580.)

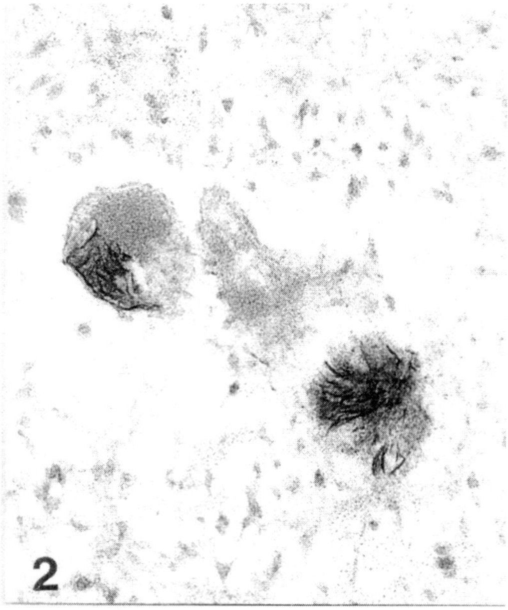

Figure 2. Electron micrograph of matrix vesicles in the lower hypertrophic zone of uncalcified growth plate cartilage matrix in hypophosphatasia. Note the presence of needle-like crystals of apatite within these MVs, but none beyond the confines of the vesicle membrane. Dark-staining small granules of proteoglycan are scattered throughout the surrounding cartilage matrix. (Case 2, x 123,000)

Discussion

Our studies confirm and clarify the association between low TNSALP catalytic activity and the impairment of skeletal mineralization expressed in perinatal (lethal) hypophosphatasia.

Hypophosphatasia is characterized biochemically by a deficiency of TNSALP activity [10,23]. The TNSALP isozyme of ALP is encoded on chromosome 1 p 36.1 - 34 [24]. In hypophosphatasia, inorganic pyrophosphate (PPi) accumulates in the extracellular fluid (ECF) apparently because hydrolysis of PPi is impeded by a deficiency of TNSALP activity. PPi, in supraphysiologic amounts, has been shown to retard HAP crystal growth *in vitro* [25], and thus would be expected to inhibit the propagation of bone mineral, once initial crystal nuclei have formed. As discussed below, accumulation of PPi surrounding ALP deficient MVs in hypophosphatasia may help to explain the rickets and osteomalacia characteristic of this condition.

A notable clinical feature of hypophosphatasia is its extremely variable severity, ranging from death in utero, associated with almost complete failure of skeletal mineralization, to only premature loss of teeth in adults, without bone symptoms [26]. Six clinical phenotypes of hypophosphatasia have been characterized based upon, in part, the age of discovery of skeletal abnormalities [23,26]. A genetic explanation for the variable severity of hypophosphatasia appears to originate from the variety of defects that may occur in the TNSALP gene [28]. Thirteen missense mutations of the TNSALP gene have been reported in severely affected patients [28,30]. Homozygosity or compound heterozygosity of specific TNSALP gene defects can explain the profoundly low ALP activity that has been observed in perinatal (lethal) cases [28,30]. However, as yet, we do not know the precise mutations of the TNSALP gene affecting any of the six cases of perinatal lethal hypophosphatasia reported here (Dr. Paula Henthorn, personal communication). The association of genetic defects that diminish TNSALP activity with secondary failure of skeletal and tooth mineralization in hypophosphatasia patients indicates that at least one biological function of ALP is to promote calcification of hard tissues. As reviewed below, three major hypotheses have been advanced to explain how ALP can promote mineralization [5,25,35].

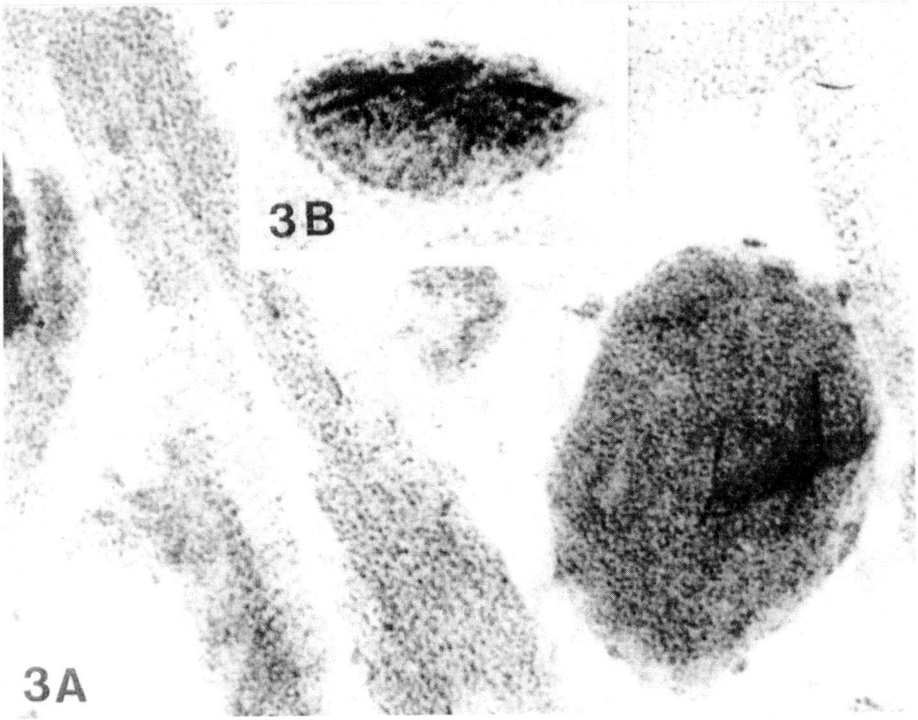

Figure 3. Electron micrographs of MVs in the poorly mineralized bone matrix of hypophosphatasia. Reprinted with permission from the Am. J. Path. **151**:1555,1997.

Figure 3A. MVs appear as rounded electron-dense profiles between broad, banded, type 1 collagen fibrils running diagonally across the field. The MV at lower right is in an early stage of calcification, having acquired electron-dense amorphous content plus a few profiles of apatite-like mineral. Its membrane is still faintly visible. The MV at upper left is heavily mineralized. (Case 2, metaphysis, x 220,000.)

Figure 3B. MV of bone in detail showing an intact membrane and containing rigid, needle-like deposits of apatite-like mineral. (Case 1, tibial metaphysis, x 275,000.)

First, it was proposed by Robison in the 1930's that the phosphoesterase activity of alkaline phosphatase may hydrolyze organic phosphate esters of cartilage and bone matrix, thus releasing orthophosphate (Pi) for incorporation into nascent calcium phosphate mineral in the skeleton [5]. This hypothesis persists despite some problems, including the fact that cells of a variety of tissues, including liver and kidney, also express abundant TNSALP activity but do not mineralize. Despite this apparent paradox, it remains possible that mineralization is augmented in the specific setting of skeletal tissues by the unique presence of ALP-carriers such as MVs, which can concentrate Ca^{2+} and PO_4^{3-} sufficiently to initiate calcification, and are selectively located at the mineralization fronts of cartilage, bone and dentin [6].

Second, it may be that ALP normally hydrolyses and thus eliminates certain organic phosphate ester inhibitors of mineralization [25]. The most frequently cited example, PPi, is normally present at micromolar levels in extracellular fluids, and has been shown to inhibit HAP crystal growth *in vitro* by coating preformed HAP nuclei and thus preventing epitaxial HAP crystal growth [31]. Thus, in hypophosphatasia, as the concentration of endogenous PPi reaches supraphysiologic levels, excess PPi could impede the growth of nascent HAP clusters. Pyrophosphate is generated through the hydrolysis of ATP by the plasma membrane-bound enzyme called nucleoside triphosphate pyrophosphohydrolase (NTPPPH), or PC-1 [32]. NTPPPH is present in the outer membranes of MVs as well as upon osteoblastic cells [34], and its metabolic action of generating PPi is counter balanced by the PPi hydrolytic activity of TNSALP [35]. In hypophosphatasia, the generation of PPi is unopposed due to an insufficient activity on TNSALP. Thus high levels of PPi are generated [10,11,23], and are sufficient to inhibit mineral crystal proliferation [32,34]. The presence of high levels of PPi in the matrix surrounding hypophosphatasia, MVs could explain our finding of a failure of propagation of mineral from within hypophosphatasia MVs, out into the surrounding cartilage or bone matrix.

A third hypothesis relates to the fact that ALP is normally anchored upon the outer surfaces of MV membranes by linkage to phosphatidylinositol [17], and that ALP binds to collagen types II and X [35]. Accordingly, it was suggested that ALP might serve as a bridge-like conduit across which initial mineral, that forms within MVs, may proliferate and spread to involve adjacent collagen fibrils. This hypothesis does not require ALP catalytic activity to explain mineral propagation into the matrix. In hypophosphatasia where the TNSALP molecules may be present at MV surfaces, but rendered inactive by mutations inhibiting catalytic activity, the mineral-bridging activity of ALP could remain intact. However, it remains a possibility that TNSALP gene mutations could also alter the dimeric or tetrameric structure of TNSALP at MV surfaces and thereby impede mineralization by physical disruption of TNSALP bridging to matrix collagen.

Clearly, a combination of more than one of the above pathogenic mechanisms may lead to the defective mineralization that is characteristic of hypophosphatasia. Not only may defective ALP catalytic activity fail to hydrolyze PPi, thus providing insufficient Pi to support mineral growth beyond the confines of MVs, but also the resulting build-up of unhydrolyzed PPi in the extracellular fluid could inhibit the outward proliferation of HAP from initial mineralization sites in MVs.

Regarding the role of MVs in the pathogenesis of hypomineralization in hypophosphatasia, our observations suggest that in this inborn error of metabolism, MVs retain the capability to generate nascent HAP mineral clusters. Thus, the defect in skeletal mineralization in hypophosphatasia appears to occur after crystal initiation, during propagation of HAP from MVs into the surrounding collagenous matrix. Our studies confirm the observation of Ornoy, et al [20] by showing the presence of approximately normal numbers of MVs in hypophosphatasia, in a normal distribution. However, in contrast to the conclusions of Ornoy et al [20] and Terada et al [21], we found that hypophosphatasia MVs are capable of initiating internal mineralization, at least in the near-term or neonatal hypophosphatasia patients we studied. The presence of other non-ALP phosphohydrolases within hypophosphatasia MVs, e.g. ATPase, inorganic PPiase, 5'-AMPase, etc, may account for this ability of hypophosphatasia MVs to initiate mineralization despite a deficiency of ALP catalytic activity.

Acknowledgements

Plasma pyridoxal 5r-phosphate (PLP) was measured courtesy of Dr. Stephen P. Coburn, at the Ft. Wayne Developmental Center, Ft. Wayne, IN. Mr. Paul Moylan was responsible for tissue preparation for transmission electron microscopy.

This study was supported in part by grant #15958 and #15963 from the Shriners Hospitals for Children, and by USPHS grant #DEO5262, NIH, NIDR.

References

[1] Weiss MJ, Henthorn PS, Lafferty MK, Slaughter C, Raducha M, Harris H 1986 Isolation and characterization of a cDNA encoding a human liver/bone/kidney - type alkaline phosphatase. Proc Natl Acad Sci USA **83**:7182-7186

[2] Anderson HC, Sajdera SW 1976 Calcification of rachitic cartilage to study matrix vesicle function. Fed Proc **35**:148-153

[3] Gutman AB, Yu TF 1950 A concept of the role of enzymes in endochondral calcification. Metabolic Interrelations. New York Josiah Macy, Jr Fdn **2**:167-190

[4] Matsuzawa T, Anderson HC 1971 Phosphatases of epiphyseal cartilage studied with electron microscopic cytochemical methods. J Histochem and Cytochem **19**:801-808

[5] Robison R, Rosenheim AH 1934 Calcification of hypertrophic cartilage *in vitro*. Biochem J **28**:684-698

[6] Anderson HC 1995 Molecular biology of matrix vesicles. Clin Orthop Rel Res **314**:266-280

[7] Bellows CG, Rubin JE, Heersch JNM 1991 Initiation and progression of mineralization of bone nodules formed *in vitro*: The role of alkaline phosphatase and organic phosphate. Bone and Min **14**:27-40

[8] Fallon MD, Whyte MP, Teitelbaum SL 1980 Stereospecific inhibition of alkaline phosphatase by L-tetramisole prevents *in vitro* cartilage calcification. Lab Invest **43**:489-494

[9] Hsu HHT, Anderson HC 1977 A simple and defined method to study calcification by isolated matrix vesicles. Effect of ATP and vesicle phosphatase. Biochem Biophys Acta **500**:162-172

[10] Caswell AM, Whyte MP, Russell RGG 1991 Hypophosphatasia and the extracellular metabolism of inorganic pyrophosphate: clinical and laboratory aspects. Crit Rev Clin Lab Sci **28**:175-232

[11] McCance RA, Fairweather DV, Barrett AM, Morrison AB 1956 Genetic, clinical, biochemical and pathological features of hypophosphatasia. Quart J Med **25**:523-538

[12] Whyte M P, Magill HL, Fallon MD, Herrod HG 1986 Infantile hypophosphatasia: normalization of circulating bone alkaline phosphatase activity followed by skeletal remineralization. J Pediatr **108**:82-88

[13] Fedde KN, Blair L, Silverstein J, Coburn SP, Ryan LM, Weinstein RS, Waymire K, Naresawa S, Millan J, McGregor GR, White MP 1999 Alkaline phosphatase knockout mice recapitulate the metabolic and skeletal defects of infantile hypophosphatasia. J Bone Min Res **14**:2015-2026

[14] Anderson HC 1969 Vesicles associated with calcification in the matrix of epiphyseal cartilage. J Cell Biol **41**:59-72

[15] Ali, SY, Sajdera, SW, Anderson HC 1970 Isolation and characterization of calcifying matrix vesicles from epiphyseal cartilage. Proc Nat Acad Sci (USA) **67**:1513-1520.

[16] Akisaka T, Gay CV 1985 Ultrastructural localization of calcium-activated adenosine triphosphatase (Ca^+-ATPase) in growth plate cartilage. J Histochem & Cytochem **33**:925-932

[17] Low JG 1987 Biochemistry of the glycosyl-phosphatidylinositol membrane protein anchors. Biochem J **224**:1-13.

[18] Ali, SY, Evans, L 1973 The uptake of ^{45}Ca Calcium ions by matrix vesicles isolated from calcifying cartilage. Biochem J 647-650.

[19] Hsu HHT, Anderson HC 1995 A role for ATPase in the mechanisms of ATP-dependent Ca and phosphate deposition by isolated rachitic matrix vesicles. Int J Biochem Cell Biol **27**:1349-1356

[20] Ornoy A, Adomian GE, Rimoin DL 1985 Histologic and ultrastructural studies on the mineralization process in hypophosphatasia. Am J Med Genetics **22**:743-758

[21] Terada S, Suzuki N, Ueno H, Uchid K, Kohama T 1996 A congenital lethal form of hypophosphatasia: histologic and ultrastructural study. Acta Obstet Gynecol Scand **75**:502-505.

[22] Anderson HC, Hsu HH, Morris CD, Fedde KN, Whyte MP 1997 Matrix vesicles in osteomalacia hypophosphatasia bone contain apatite-like mineral crystals. Am J Path **151**:1555-1561

[23] Whyte MP 1994 Hypophosphatasia and the role of alkaline phosphatase in skeletal mineralization. Endocr Rev **15**:439-461

[24] Swallow DM, Porvey S, Parkar M 1986 Mapping for the gene coding for the human liver/bone/kidney isozyme of alkaline phosphatase to chromosome 1. Ann Human Genet **50**:229-235

[25] Fleisch H, Bisaz S 1962 Mechanism of calcification: Inhibitory role of pyrophosphate. Nature(London) **195**:911

[26] Whyte MP, Walkenhorst DA, Fedde KN, Henthorn PS, Hill CS 1996 Hypophosphatasia: levels of bone alkaline phosphatase immunoreactivity in serum reflect disease severity. J Clin Endocrinol Metab **81**:2142-2148

[27] Fedde KN, Henthorn PS, Whyte MP 1996 Aberrant properties of alkaline phosphatase in patient fibroblasts correlate with clinical expressivity in severe forms of hypophosphatasia. J Clin Endocrinol Metab **81**:2587-2594

[28] Henthorn PS, Raducha M, Fedde KN, Lafferty MA, Whyte MP 1992 Different missense mutations at the tissue-nonspecific alkaline phosphatase gene locus in autosomal recessively inherited forms of mild and severe hypophosphatasia. Proc Nat Acad Sci (USA) **89**:9924-9928

[29] Ozona K, Yamagata M, Michigami T, Nakaiima S, Sakai N, Satomurak , Yasui N, Okada S, Nakayama M 1996 Identification of novel missense mutations (Phe 310 Leu and Gly 439 Ang) in a neonatal case of hypophosphatasia. J Clin Endocrinol and Metab **81**:4458-4461

[30] Weiss MJ, Cole DEC, Ray K, Whyte MP, Lafferty MS, Mulivor RA, Harris H 1988 A missense mutation in the human liver/bone/kidney alkaline phosphatase gene causing a form of lethal hypophosphatasia. Proc Natl Acad Sci (USA) **85**:7666-7669

[31] Termine JD, Conn KM 1976 Inhibition of apatite formation by phosphorylated metabolites and macromolecules. Calcif Tiss Res **22**:149-157

[32] Johnson K, Moffa A, Chen Y, Pritzger K, Goding J, Terkeltaub R 1999 Matrix vesicle plasma membrane glycoprotein-1 regulates mineralization by murine osteoblastic MC3T3 cells. J Bone Min Res **14**:883-892

[33] Johnson KA, Hessle L, Vaingankar S, Wennberg C, Mauro S, Narisawa S, Goding JW, Sano K, Millan JL, Terkeltaub R 2001 Osteoblast tissue non-specific alkaline phosphatase (TNAP) functions to both antagonize and regulate the mineralization inhibition PC-1. Am J Physiol 2001, (in press).

[34] Wennberg C, Hessle L, Lundberg P, Mauro S, Narisawa S, Lerner VH, Millan JL 2000 Functional characterization of osteoblasts and osteoblasts from alkaline phosphatase knockout mice. J. Bone Min Res **15**:1879-1888

[35] Wu LNY, Genge BR, Lloyd GC, Wuthier RE 1991 Collagen binding proteins in collagenase-released matrix vesicles from cartilage. J Biol Chem **266**:1195-1203

Tibial Dyschondroplasia: A Growth Plate Abnormality Caused by Delayed Terminal Differentiation

Colin Farquharson
Bone Biology Group, Division of Integrative Biology, Roslin Institute, Roslin, EH25 9PS, Scotland, UK.

Abstract. Disruption to the orderly progression of chondrocyte maturation within the growth plate causes many physeal abnormalities resulting in abnormal bone growth. One such disorder is tibial dyschondroplasia (TD), which is a common cause of deformity and lameness in many rapidly growing animals. Cell and matrix components of the growth plate have been studied in order to determine the cause(s) of the premature arrest of chondrocyte differentiation and retention of prehypertrophic chondrocytes observed in TD. Chondrocyte proliferation proceeds normally in TD but markers of the differentiated phenotype, local growth factors and the vitamin D receptor are abnormally expressed within the prehypertrophic chondrocytes above, and within, the lesion. TD is also associated with a reduced incidence of apoptosis suggesting that the lesion contains an accumulation of immature cells that have outlived their normal life span. Immunolocalisation studies of matrix components suggest an abnormal distribution within the TD growth plate that is consistent with a failure of the chondrocytes to fully hypertrophy. In addition, the collagen matrix of the TD lesion is highly crosslinked, which may make the formed lesion more impervious to vascular invasion and osteoclastic resorption. Recent studies have applied the techniques of differential display and semi-quantitative RT-PCR to RNA obtained from discrete populations of growth plate chondrocytes of different maturational phenotype. This strategy has allowed the comparison of phenotypically identical cell fractions from normal and TD growth plates in an attempt to identify possible candidate genes for TD.

Introduction

Retention of articular-epiphyseal cartilage and growth plate cartilage is observed in all domestic animals and is especially common amongst those species selected for rapid growth [1]. The resulting condition is referred to as either dyschondroplasia or osteochondrosis and although it has been proposed that dyschondroplastic lesions are the precursors of osteochondral lesions [2] the terms are often used interchangeably. Dyschondroplasia is most frequently used to describe the growth plate condition in poultry whereas osteochondrosis is more commonly used when describing the condition in mammals such as the pig and horse. Dyschondroplasia has both significant economic and welfare implications to the agricultural industry and much of the research, to date, has been aimed at finding the cell/molecular defect(s) underpinning this disorder. It is also well established that dyschondroplasia has a strong genetic component and, therefore, if the genes responsible can be identified this may lead to marker assisted selection for the elimination of dyschondroplasia from the breeding populations. Much research has been conducted in the chicken, which is particularly susceptible to this disorder and the data reported in this review is predominantly taken from studies on the chicken growth plate. An additional benefit of this research has been the realisation that the dyschondroplastic growth plate also provides the researcher with a useful model to study the key chondrocyte developmental events (proliferation, differentiation, apoptosis, vascularisation and mineralisation) during the endochondral ossification process.

The precise underlying cellular defect that occurs in TD remains unclear. There have been a number of hypotheses, which include a defect in the metaphyseal vasculature such that invasion of the hypertrophic cartilage is impaired [3], abnormal matrix structure leading to an inhibition of vascularisation and mineralisation [4], and a failure in osteoclastic cartilage resorption [5]. It is now generally accepted, however, that TD is a consequence of an inability of the maturing chondrocytes to undergo terminal differentiation that normally leads to vascularisation and mineralisation. In support of this hypothesis, chondrocytes of the TD lesion show reduced ALP and carbonic anhydrase activity and also collagen type X staining all recognised markers of chondrocyte hypertrophy. However, it should be noted that one recognised marker of chondrocyte hypertrophy, extracellular fatty acid binding protein (Ex-FABP) has recently been shown to be upregulated in TD [6] and this may be linked to an acute phase response due to the pathology of the lesion [7]. This series of events would also explain the accumulation of prehypertrophic chondrocytes that are evident in histological sections of TD lesions. This characteristic of TD distinguishes it from the thickened growth plates noted in hypocalcaemic rickets, which is due to an accumulation of randomly orientated proliferating chondrocytes, and also from hypophosphataemic rickets where hypertrophic chondrocytes accumulate.

The regulation and co-ordination of the mechanisms involved in the transition from a proliferative to a terminally differentiated phenotype is clearly central to the development of TD but, to date, this process is not fully understood. This review describes a series of cellular and molecular based studies aimed at obtaining a better understanding of the aetiology of TD.

Gross Pathology

Dyschondroplasia is characterised by a plug of non-calcified avascular opaque cartilage (the lesion) situated below the growth plate and extending into the metaphysis (Fig. 1). It is most commonly found in the proximal tibiotarsus (hence the term tibial dyschondroplasia, TD), but has been reported to occur in most long bones. It was first described by Leach and Nesheim [8] as a spontaneously occurring cartilage abnormality, but it is now known that both nutrition and selective breeding can affect the incidence of TD. The size of the TD lesion is variable. It can exist as a small cartilaginous mass within a discrete area under the growth plate, or can occupy the entire metaphysis. Interestingly, dyschondroplasia within the proximal tibiotarsi appears to be bilateral with the incidence and severity of the TD lesion similar in both legs. The small lesions are likely to be sub-clinical whereas the larger lesions have been shown to lead to significant orthopaedic deformities and lameness in meat-type chickens. Abnormal, non-uniform bone growth within the area of the lesion leads to increased tibial plateau angle and tibial bowing [9]. This can cause angular and rotational deformities within the bone and may through altered biomechanical forces lead to other bone deformities and gait problems within the hock joint.

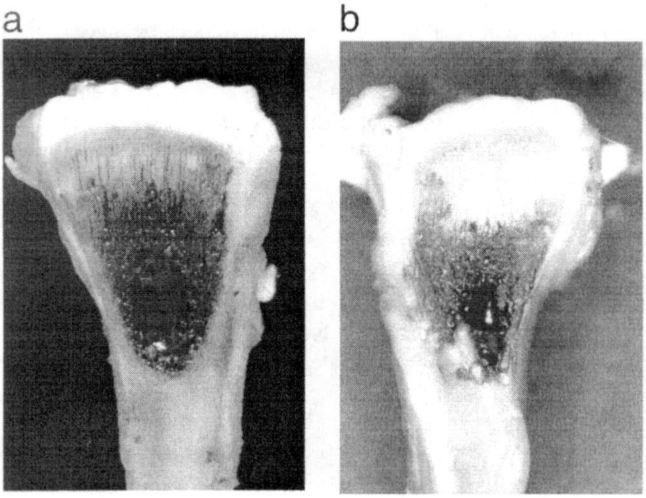

Figure 1. Comparison of normal (a) and dyschondroplastic (b) growth plates. The dyschondroplastic lesion is opaque and avascular and occupies an area of the metaphysis under the growth plate.

Fundamental Mechanisms

Chondrocyte Proliferation

Evidence that the TD lesion emanates from the growth plate and is due to the accumulation of growth plate chondrocytes has been shown clearly in tritiated thymidine incorporation experiments [10]. Four days after radioisotope injection labelled cells that originated within the proliferating zone were found deep in the lesion. In similar and extended studies [11] marked differences in chondrocyte proliferation were not found between normal growth plates and those with moderate lesions. Glucose 6-phosphate dehydrogenase activity has been shown to be closely correlated to chondrocyte proliferation within the epiphyseal growth plate [12] presumably through its role in ribose sugar synthesis via the pentose phosphate pathway. Enzyme activity was not found to be raised in proliferating chondrocytes of severely affected chicks [11]. These results rule out an increased rate of proliferation as a possible mechanism for the accumulation of cartilage observed in TD.

Chondrocyte Differentiation

Alkaline phosphatase (ALP) activity has long been recognised as a marker of chondrocyte differentiation and is likely to be involved in cartilage mineralisation. Decreased ALP activity in the chondrocytes of the lesion [11,13] is possibly secondary to chondrocyte necrosis, which is known to occur [11]. Enzyme activity at the distal edge of the lesion may represent the presence of a repair mechanism (as seen with matrix proteins) and the initiation of mineralisation. This is in accord with ultra structural studies and elemental analysis where mineralisation has been observed in this area of the lesion [14, 15]. Quantification of ALP activity by *in situ*-microdensitometry has indicated that enzyme activity of the prehypertrophic chondrocytes situated proximal to the lesion is higher with increasing severity of the lesion [11]. Given the role of ALP activity in mineralisation, it might have been predicted that ALP activity would have been decreased throughout the TD growth plate. The activity of ALP may be related to the process of differentiation of the chondrocyte phenotype, which in turn is regulated by a variety of autocrine and paracrine growth factors. Taken together these and other results [13] suggest that the impaired mineralisation noted in TD is not a consequence of a lack of ALP activity.

Cartilage Resorption

A possible contributory cause of TD is a failure of growth plate cartilage resorption. Increased osteoclastic resorption can occur either through higher individual osteoclastic activity or an increase in their number. Individual osteoclastic tartrate-resistant acid phosphatase (TRAP) activity was not correlated with lesion severity in 3 week-old chicks fed on a diet imbalanced in calcium and phosphorus [11] This is in broad agreement with the semi-quantitative study of Lawler *et al* [5], which found inconsistent reductions in TRAP activity in osteoclasts of TD. Osteoclast number under the lesion was found to be decreased in bones with severe lesions [11, 16] but this was only found to be significant in the study of Walser [16].

Regulators of Chondrocyte Differentiation

Growth Factors

Peptide growth factors are known to be central to chondrocyte differentiation and recent studies have indicated that the pathogenesis of dyschondroplasia may be linked to abnormal expression of a number of these autocrine/paracrine factors. Transforming growth factor - β (TGF-β is known to regulate chondrocyte differentiat ion and there was a marked decrease in TGF- β3 within the prehypertrophic chondrocytes above the lesion [17]. In areas of repair at the periphery of the TD lesion, chondrocytes adjacent to the invading metaphyseal vessels stain positively for TGF- β3 supporting the concept that TGF- β is important for the induction of angiogenesis. In TD, the intensity of immunocytochemical staining for basic fibroblast growth factor, insulin-like growth factor-I, epidermal growth factor and TGF- β was greatly reduced [18, 19, 20] as was the number of positively IGF and FGF stained chondrocytes [19, 20]. Increased staining of growth factors was noted in areas of repair at the periphery of the lesion and overall the changes in the staining pattern noted in TD appeared very similar to those noted with TGF- β [17].

Parathyroid hormone-related peptide (PTHrP) acts as a negative regulator of growth plate chondrocyte differentiation, thereby mediating the effects of indian hedgehog on bone growth. PTHrP and its receptor (PTHR) have been found to be expressed at discrete locations in the growth plate [21]. PTHrP is essentially absent from the dyschondroplastic lesion and does not appear to be overexpressed in the prehypertrophic chondrocytes proximal to the lesion possibly indicating that PTHrP does not have a role in the inhibition of chondrocyte differentiation noted in TD [22] (Fig. 2). Recent studies on the TD growth plate has also indicated that normal expression of the PTHR gene was observed in the lower proliferating chondrocytes situated proximal to the lesion. No expression was observed within the lesion. Interestingly, PTHR expression was completely absent within rachitic growth plates of vitamin D deficient chicks. This was possibly due to elevated PTH levels and suggests that PTH is not involved in the aetiology of TD [23].

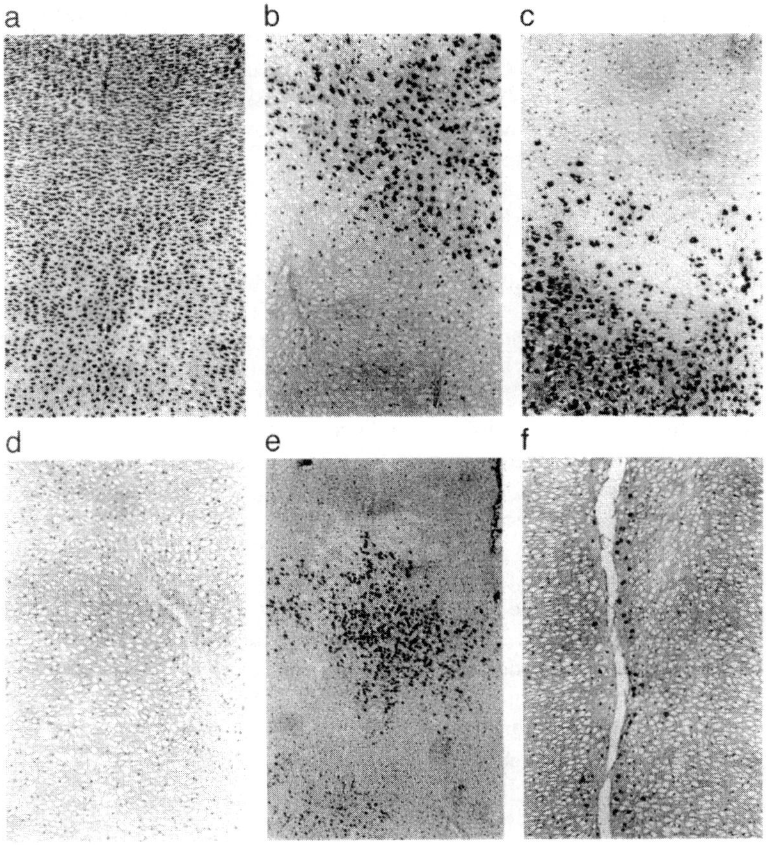

Figure 2. PTHrP immunostaining of TD growth plate. **(a)** Positive staining in proliferating (P) and prehypertrophic (PH) chondrocytes situated above, and proximal to, the lesion **(b & c)** Chondrocyte staining in the proximal (b) and distal (c) part of the lesion. **(d)** Middle of lesion where most chondrocytes failed to stain. **(e)** Small positively stained clumps of cells within the middle of the lesion. **(f)** Chondrocyte staining associated with blood vessel within lesion.

C-myc Protein

Differentiating chick chondrocytes contain significantly more c-myc protein than proliferating chondrocytes, suggesting a role in the differentiating process [24]. In TD, the c-myc content of the prehypertrophic chondrocytes is significantly reduced in comparison to similar cells within normal growth plates. In contrast, there is no significant difference in the c-myc protein content of proliferative chondrocytes of normal and TD birds [17]. This further suggests that the initial steps in chondrocyte differentiation are characterised by increased c-myc protein and a possible link between c-myc, TGF-β and 1,25 -dihydroxyvitamin D_3 (1,25-) in the chondrocyte differentiation process has been proposed ([17, 25]. 1,25-D is known to regulate chondrocyte differentiation and modulates both TGF- β and c-myc, while c-myc alters the expression of the VDR. This suggests that c-myc and TGF- β are essential components of the signal cascade regulated by 1,25 -D in growth plate chondrocytes. The lack of mineralisation in TD may also be related to lower levels of c-myc protein. Mineralisation of the growth plate matrix is mediated by the actions of matrix vesicles, which are formed by a number of processes, which may include apoptosis [26]. C-myc protein is a potent inducer of apoptosis and a decrease in c-myc protein concentration may lead to decreased matrix vesicle synthesis and a failure of mineralisation. Supporting evidence for this hypothesis has been provided by Nie *et al*, [27] who demonstrated there was a defect in the formation of matrix vesicles in TD.

Vitamin D Metabolites

Vitamin D metabolites play a major role in chondrocyte metabolism and appear to be essential for normal bone formation. The actions of this hormone are mediated through its interaction with its nuclear receptor (VDR) and subsequent regulation of gene expression. Receptors for 1,25-D have been located on growth plate chondrocytes suggesting a direct role for this metabolite in chondrocyte metabolism [28]. Although results from *in vitro* chondrocyte culture experiments have been inconsistent, a stimulation of chondrocyte differentiation by 1,25-D has been shown *in vivo* [29]. This capacity for 1,25-D to promote chondrocyte differentiation may be responsible for its ability to normalise growth plate development and prevent TD when given as a dietary supplement [30]. If this were indeed the case it would suggest that the rate of endogenous synthesis of 1,25-D from its precursor 25-D in the kidney might be deficient in the fast growing young broiler. Strains with a greater predisposition to TD have been found to have lower concentrations of plasma 1,25-D [31]. Chondrocytes within the lesion have a reduced number and affinity of VDR [28] and therefore their ability to respond to 1,25-D may be reduced and higher concentrations of 1,25-D may be required to achieve normal effects. The finding that dietary 1,25-D increases chondrocyte VDR number lends support to the hypothesis that 1,25-D may prevent TD by an effect on VDR synthesis.

Extracellular Matrix

As chondrocytes differentiate and hypertrophy concomitant changes occur within the extracellular matrix and evidence suggests that chondrocytes must provide the correct extracellular network and establish cell-matrix interactions to allow progressive differentiation [32]. Although the structure-function relationship is not fully understood it is possible that abnormal matrix synthesis and interactions may contribute to the development of TD.

Collagen type X, due to its synthesis by exclusively hypertrophic chondrocytes in the growth plate, has been extensively studied. Immunohistochemistry has indicated sparse matrix staining within the TD lesion and, where present, was restricted to the proximal part of the lesion and associated, in some cases, with sites of vascular invasion [33]. Collagen type X intracellular staining was also noted within TD chondrocytes, but only in those without positive extracellular matrix staining, suggesting a defect in its secretion or incorporation into the matrix [33, 34]. Bashey *et al* [35] reported collagen type X production within the TD lesion to be less than 50% of the amount found in pure hypertrophic cartilage.

The presence of collagen type X mRNA within the proximal lesion bordering the junction with the normal growth plate [36, 37] is similar to the pattern of distribution achieved by collagen type X immunohistochemistry [33]. This strongly suggests that prior to chondrocyte developmental arrest and formation of the lesion the chondrocytes had initiated differentiation at least in part, to the hypertrophic stage. Less controversy exists about the distribution of collagen type II that is present throughout the normal and TD growth plate with the matrix of the TD lesion staining strongly [33, 36]. Immunostaining of the non-collagenous proteins osteopontin and osteonectin in TD tissue indicates their presence in the matrix distal to the lesion but absence within the lesion itself [37, 38]. This distribution is similar to that of collagen type X and the presence of matrix gene expression distal to the lesion suggests a process of repair and the resumption of normal bone development.

In general, dyschondroplastic cartilage has a similar concentration of proteoglycans to the normal growth plate [39]. Aggrecan has been detected by using antibodies to potential glycosaminoglycan epitopes of aggrecan, namely chondroitin 4 and 6 sulphate (C4S and C6S) and keratan sulphate (KS). Staining for C4 and C6S within normal and TD tissue was of similar intensity and distribution and was localised to the matrix throughout the growth plate [40]. Little KS staining was observed in the hypertrophic zone of the normal growth plate whereas in the dyschondroplastic lesion intense matrix staining was present [37]. This observation is consistent with a change from aggrecan to decorin and biglycan synthesis during normal chondrocyte maturation and the lack of attainment of full chondrocyte hypertrophy in TD.

Collagen undergoes a number of post-translational modifications, one of which is the action of lysyl oxidase to initiate the formation of the mature non-reducible intermolecular crosslinks (pyridinoline and deoxypyridinoline) that confer strength and stability to the growth plate cartilage. The integrity of the collagen network influences biomechanical properties of the growth plate cartilage and may alter its resistance to osteoclastic resorption. Collagen cross-linking in dyschondroplasia has recently been investigated and quantitative studies [41] indicate increased pyridinium crosslinks within the lesion. In normal and TD tissue the pyridinoline concentration was found to be 10-fold higher in the proliferating zone than the mature, collagen type X, containing tissue [41]. This may be an essential adaptation (via increased collagenase activity and collagen turnover) of the matrix for vascular invasion and cartilage resorption to occur. Deoxpyridinoline was only observed in the more differentiated zones of the growth plate and a progressive increase in the concentration of both cross-links was noted from the proximal to the distal parts of the lesion [41]. Normal amounts of pyridinium crosslinks above the lesion suggests that the high concentrations noted in the distal lesion may reflect a tissue that is failing to turn over due to reduced matrix metalloproteinase activity. This may exacerbate the condition but is unlikely to be the primary cause of TD [41].

Programmed Cell Death

The fate of the terminally differentiated hypertrophic chondrocyte is unclear. It is accepted however, that the differentiated chondrocyte must be removed to maintain the steady state thickness of the growth plate. A number of recent studies have investigated chondrocyte apoptosis in TD in order to determine if disruption to this process can explain the accumulation of prehypertrophic chondrocytes within the growth plate. Small TD lesions have been found to contain few, if any, apoptotic chondrocytes whereas in severe lesions numerous apoptotic cells were present [41,43]. Due to the poor accessibility of cartilage to phagocytes, which are responsible for the removal of apoptotic cells, the inactive and avascular TD cartilage may persist and contribute to the pathology of the disorder. In contrast, a further study has shown that the TD lesion is associated with an impairment of apoptosis, which suggests that the accumulated prehypertrophic chondrocytes of the lesion have outlived their normal life span [44]. This lack of apoptosis may be linked to the reduced expression of c-myc noted in the TD growth plate [17] and, in part, be responsible for the retention of the chondrocytes and extracellular matrix observed in TD.

Cell Culture Studies

Chondrocytes isolated from large TD lesions were found to have abnormal morphology and reduced mitogenic activity and proteoglycan synthesis [45]. This diminished metabolic activity may reflect the reduced cell viability within the lesion [14, 22]. In addition, type X collagen synthesis by TD chondrocytes was reduced but mRNA levels were substantially higher than in normal chondrocytes [34]. This discrepancy may be a result of a block in translation or post-translational modification of the protein or alternatively may reflect a defect in the secretion or incorporation of this collagen type into the extracellular matrix. Collagen type X, high levels of ALP activity and matrix mineralisation were all reported in lesion chondrocytes grown in a high cell density culture system. This suggests that, in culture, lesion chondrocytes have the ability to terminally differentiate and mineralise and that the primary abnormality in TD is related to a developmental fault that is only operative *in vivo* [33].

Identification of Candidate Genes

Dyschondroplasia has a strong genetic component that has led to the generation of experimental lines of broilers selected for a high and low incidence of TD [45]. These lines serve as valuable models with which to identify genes involved in the development of dyschondroplasia and to map candidate genes, once their role in initiating the disorder has been established. The characterisation of candidate genes for TD has been concentrated on genes known to be associated with chondrocyte differentiation and hypertrophy. A number of these are markers of hypertrophy, which are down regulated in the TD lesion, such as collagen type X, aggrecan, osteopontin and osteonectin.

Although many studies have identified differential expression of genes or gene products in TD, which have given valuable insights into the nature of the condition, many have been associated with the gross abnormalities of the lesion. As TD involves a failure of chondrocyte differentiation it can be misleading to compare patterns of gene expression in cells isolated from TD lesion and non-lesion tissues directly. Some chondrocyte cell phenotypes will be absent from the lesion, whereas others will be altered. This will give rise to predominantly secondary changes associated with the pathology rather than the aetiology of this disorder. In order to circumvent this problem, we have made use of a strategy that utilises Percoll density gradient centrifugation to separate normal growth plates into cell fractions of differing maturational development [21, 22, 47, 48]. This allows the comparison of TD and non-TD cell fractions containing cells proximal to, and out with, the lesion. These cells would normally progress to the fully hypertrophic phenotype in the normal growth plate, but in TD fail to hypertrophy and give rise to the lesion. Any misexpression of a gene would strongly implicate that gene in the pathogenesis of dyschondroplasia. Using this approach, we have recently screened a number of candidate genes for TD, using semi-quantitative RT-PCR (Table 1) [6]. None of these genes showed any difference in levels of expression between TD and normal growth plate cell fractions. This may suggest that at least quantitatively, at the mRNA level, none of these genes appear to be implicated in initiating dyschondroplasia. These results highlight the importance of studying gene expression in cells proximal to the lesion, as it is in these cells that the perturbation of differentiation that gives rise to TD must first occur.

Table 1. Candidate genes for TD screened for expression levels in Percoll fractionated chondrocytes

1.	Collagen types II and X
2.	B-cadherin
3.	BMPs 2, 4, 5, 6 and 7.
4.	BMP receptor kinase 1, 2 and 3
5.	Ex-FABP
6.	HT7
7.	Parathyroid hormone-related peptide (PTHrP)
8.	Transforming growth factor β1, β2 and β3
9.	PTH/PTHrP receptor
10.	Vitamin D receptor

Conclusions

Despite extensive research the aetiology of dyschondroplasia remains unclear. This is in part due to the multitude of apparently disparate factors that can cause TD, possibly by several distinct mechanisms, but each results in the occurrence of a histologically similar lesion. However, the observation that this disorder develops in the maturing prehypertrophic chondrocytes, which are then unable to differentiate into fully hypertrophic chondrocytes, may be fundamental to its progression. Such chondrocytes would fail to elicit the biochemical changes associated with hypertrophy that are a prerequisite for cartilage vascularisation, mineralisation and resorption by osteoclasts. Many morphological, biochemical and molecular changes occurring in TD have been reported but interpretation of these findings is often difficult as it is unknown if any observed change(s) inhibit the differentiation process or are a result of, and secondary to, the impaired differentiation. If the primary cellular developmental fault is to be identified it is imperative that significance should be attached to changes that occur in the prehypertrophic cells proximal to the lesion and not within the lesion itself.

The identification of the genes involved in the control of growth plate chondrocyte development together with a knowledge of how they influence signalling processes, matrix synthesis and cellular events such as hypertrophy and cell death are essential for a more complete understanding of endochondral bone growth. This information is now being obtained by various approaches and careful interpretation of the accumulating data will enable us to obtain a more comprehensive understanding of chondrocyte development within the growth plate. Such fundamental information will be critical to our understanding of the primary developmental fault that occurs in dyschondroplasia. Identification of the gene(s) involved in the perturbation of chondrocyte differentiation in TD may open the way to elimination of the disorder by genetic selection.

Acknowledgements

This work was supported by the Ministry of Agriculture Fisheries and Food (MAFF) and Biotechnology and BioSciences Research Council (BBSRC).

References

[1] Olsson SE 1978 Osteochondrosis in domestic mammals. Acta Radiol. Suppl. **358:**9-14.

[2] Jeffcott LB, Henson FMD 1998 Studies on growth cartilage in the horse and their application to aetiopathogenesis of dyschondroplasia (osteochondrosis). Vet. Journal **156:**177-192.

[3] Riddell C 1975 Studies on the pathogenesis of tibial dyschondroplasia in chickens. I. Production of similar effects by surgical interference. Avian Diseases. **19:**483-489.

[4] Poulos PW 1978 Tibial dyschondroplasia (osteochondrosis) in the turkey. Acta Radiol. Suppl. **358:**197-227.

[5] Lawler M, Shivers JL, Walser MM 1988 Acid phosphatase activity of chondroclasts from fusarium-induced tibial dyschondroplastic cartilage. Avian Diseases **32:**240-245.

[6] Jefferies D, Houston B, Lester D, Whitehead CC, Thorp BH, Botman M, Farquharson C 2000 Expression patterns of chondrocyte genes cloned by differential display in tibial dyschondroplasia. Biochem. Biophys. Acta **1501:**180-188.

[7] Cermelli S, Zerega B, Carlevaro M, Gentili C, Thorp B, Farquharson C, Cancedda R, Descalzi-Cancedda F 2000 Extracellular fatty acid binding protein (Ex-FABP) modulation by inflammatory agents : "physiological" acute phase response in endochondral bone formation. Uer. J. Cell Biol. **79:**155-164.

[8] Leach RM, Nesheim MC 1965 Nutritional, genetic and morphological studies of an abnormal cartialge formation in young chicks. J. Nutr. **86:**236-244.

[9] Lynch M, Thorp BH, Whitehead CC 1992 Avian tibial dyschondroplasia as a cause of bone deformity. Avian Pathol. **21:**275-285.

[10] Gay CV, Leach RM 1985 Tritiated thymidine uptake in chondrocytes afflicted with tibial dyschondroplasia. Avian Diseases **29:**1224-1229.

[11] Farquharson C, Whitehead C, Rennie JS, Thorp BH, Loveridge N 1992 Cell proliferation and enzyme activities associated with the development of avian tibial dyschondroplasia: An in situ biochemical study. Bone **13:**59-67.

[12] Farquharson C, Loveridge 1990 Cell proliferation within the growth plate of long bones as assessed by bromodeoxyuridine uptake and its relationship to glucose 6-phosphate dehydrogenase activity. Bone Min. **10:**121-129.

[13] Gay CV Anderson RE, Leach RM 1985 Activities and distribution of alkaline phosphatase and carbonic anhydrase in the tibial dyschondroplastic lesion and associated growth plate of chicks. Avian Diseases **29:**812-821.

[14] Hargest TE, Gay CV, Leach RM 1985 Avian tibial dyschondroplasia: 1. Ultrastructure. Amer. J. Pathol. **119:**175-190.

[15] Hargest TE, Leach RM, Gay C.V. 1985 Avian tibial dyschondroplasia: 111. Electron probe analysis. Amer. J. Pathol. **119:**199-209.

[16] Walser MM, Allen NK, Mirocha CJ, Hanlon GF, Newman JA, 1982 A fusarium-induced osteochondrosis (tibial dyschondroplasia) in chickens. Vet. Pathol. **19:**544-550.

[17] Loveridge N, Farquharson C, Hesketh JE, Jakowlew S, Whitehead CC, Thorp BH 1993 The control of chondrocyte differentiation during endochondral bone growth in vivo: changes in TGF-β and the proto-oncogene c-myc. J. Cell Sci. **105:**949-956.

[18] Ren P, Rowland GN, Halper J, 1997 Expression of growth factors in chicken growth plate with special reference to tibial dyschondroplasia. J. Comp. Pathol. **116:**303-320.

[19] Thorp BH, Jakowlew SB., Goddard C, 1995 Avian tibial dyschondroplasia: Local deficiences in growth factors are integral to the aetiopathogenesis. Avian Pathol. **24:**135-148.

[20] Twal WO, Wu J, Gay CV, Leach RM 1996 Immunolocalisation of basic fibroblast growth factor in avian tibial dyschondroplastic cartilage. Poultry Sci. **75:**130-134.

[21] Farquharson C, Jefferies D, Seawright E, Houston B 2001 Regulation of chondrocyte terminal differentiation in the post-embryonic growth plate: the role of the parathyroid hormone-related peptide-Indian hedgehog axis. Endocrinology (in press)

[22] Farquharson C, Seawright E, Jefferies D 2001 Parathyroid hormone-related peptide expression in tibial dyschondroplasia. Avian Pathol (in press)

[23] Bassat SB, Genina O, Lavelin I, Leach RM, Pines M 1999 Parathyroid receptor gene expression by epiphyseal growth plates in rickets and tibial dyschondroplasia. Mol. Cell. Endocrinol. **149:**185-195.

[24] Farquharson C, Hesketh JE, Loveridge N 1992 The proto-oncogene c-myc is involved in cell differentiation as well as cell proliferation: Studies on growth plate chondrocytes in situ. J. Cell. Physiol. **152:**135-144.

[25] Farquharson C, Law AS, Seawright E, Burt DW, Whitehead CC 1996 The expression of transforming growth factor – β by cultured chick growth plate chondrocy tes: differential regulation by 1,25-dihydroxyvitamin D_3. J. Endocrinol. **149:**277-285.

[26] Akisaka T, Gay CV, 1985 The plasma membrane and matrix vesicles of mouse growth plate chondrocytes during differentiation as revealed in freeze fracture replicas. Am. J. Anat. **173:**269-286.

[27] Nie D, Genge BR, Wu LNY, Wuthier RE 1995 Defect in formation of functional matrix vesicles by growth plate chondrocytes in avian tibial dyschondroplasia: Evidence of defective tissue vascularisation. J. Bone Min. Res. **10:**1625-1634.

[28] Berry JL, Farquharson C, Whitehead CC, Mawer EB 1996 Growth plate chondrocyte vitamin D receptor number and affinity are reduced in avian tibial dyschondroplastic lesions. Bone **19:**197-203.

[29] Farquharson C., Whitehead CC, Rennie JS, Loveridge N 1993. In vivo effect of 1,25-dihydroxycholecalciferol on the proliferation and differentiation of avian chondrocytes. J. Bone Min. Res. **8:**1081-1088.

[30] Rennie JS, Whitehead CC, Thorp B 1993 The effect of dietary 1,25-dihydroxycholecalciferol in preventing tibial dyschondroplasia in broilers fed on diets imbalanced in calcium and phosphorus. Brit. J. Nutr. **69:** 809-816.

[31] Parkinson G, Thorp BH, Azoulas J, Vaiano 1996 Sequential studies of endochondral ossification and serum 1,25-dihydroxycholecalciferol in broiler chickens between one and 21 days of age. Res. Vet. Sci. **60:**173-178.

[32] Farquharson C, Berry JL, Mawer EB, Seawright E, Whitehead CC 1998 Ascorbic acid induced chondrocyte terminal differentiation: The role of the extracelluklar matrix and 1,25-dihydroxyvitamin D. Eur. J. Cell Biol. **76:**110-118.

[33] Farquharson C, Berry JL, Mawer EB, Seawright E, Whitehead CC 1995 Regulators of chondrocyte differentiation in tibial dyschondroplasia: An in vivo and in vitro study. Bone **17:**279-286.

[34] Reginato AM, Bashey RI, Rosselot G, Leach RM, Gay CV, Jimenez SA 1998 Type X collagen biosynthesis and expression in avian tibial dyschondroplasia. Osteoarthritis and Cartilage **6:**125-136.

[35] Bashey RI, Leach RM, Gay CV, Jimenez S 1989 Type-X collagen in avain tibial dyschondroplasia. Lab. Invest. **60:**106-112.

[36] Chen Q, Gibney E, Leach RM, Linsenmayer T 1993. Chicken tibial dyschondroplasia: A limb mutant with two growth plates and possible defects of collagen crosslinking. Dev. Dynam. **196:**54-61.

[37] Pines M, Knopov V, Genina O, Hurwitz S, Gerstenfield LC, Leach RM 1998 Development of avian tibial dyschondroplasia: Gene expression and protein synthesis. Calcif. Tissue Int. **63:**521-527.

[38] Wu J, Pines M, Gay CV, Hurwitz S, Leach R, 1996 Immunolocalisation of osteonectin in avian tibial dyschondroplasia. Dev. Dynam. **207:**69-74.

[39] Freedman BD, Gay CV, Leach RM 1985 Avian tibial dyschondroplasia 11. Biochemical changes. Amer. J. Pathol. **119:**191-198.

[40] Farquharson C, Whitehead CC, Loveridge N 1994. Alterations in glycosaminoglycan concentration and sulfation during chondrocyte maturation. Calcif. Tissue Int. **54:**296-303.

[41] Farquharson C, Duncan A, Seawright E, Whitehead CC, Robins SP 1996 Distribution and quantification of pyridinium cross-links of collagen within the different maturational zones of the chick growth plate. Biochim. Biophys. Acta **1290:**250-256.

[42] Praul C, Gay CV, Leach RM 1997 Chondrocytes of the tibial dyschondroplastic lesion are apoptotic. Int. J. Dev. Biol. **41:**621-626.

[43] Rath NC, Huff W, Balog JM, Bayyari GR, Reddy RP 1996 Matrix metalloproteinase activities in avian tibial dyschondroplasia. Poultry Sci. **76:**501-505.

[44] Ohyama K, Farquharson C, Whitehead CC, Shapiro IM 1997 Further observations on programmed cell death in the epiphyseal growth plate: Comparison of normal and dyschondroplastic epiphyses. J. Bone Min. Res. **12:**1647-1656.

[45] Rosselot G, Sokol C, Leach R 1994 Effect of lesion size on the metabolic activity of tibial dyschondroplastic chondrocytes. Poultry Sci. **73:**452-456.

[46] Thorp BH, Ducro B, Whitehead CC, Farquharson C, Sorensen P 1993 Avian tibial dyschondroplasia: the interaction of genetic selection and dietary 1,25-dihydroxycholecalciferol. Avian Pathol. **22:**311-324.

[47] Houston B, Seawright E, Jefferies D, Hoogland E, Lester D, Whitehead CC, Farquharson C 1999 Identification and cloning of a novel phosphatase expressed at high levels in differentiating growth plate chondrocytes. Biochim. Biophys. Acta **1448:**500-506.

[48] Farquharson C, Lester D, Seawright E, Jefferies D, Houston B 1999 Microtubules are potential regulators of growth-plate chondrocyte differentiation and hypertrophy. Bone **25:**405-412.

RUNX2/CBFA1 Mutations in Cleidocranial Dysplasia: Phenotypic and Structure/Function Correlations

Kim McBride, Dobrawa Napierala, Yuqing Chen, Qiping Zheng, Guang Zhou, and Brendan Lee
Department of Molecular and Human Genetics, Baylor College of Medicine
Houston, TX 77030

Abstract. *RUNX2/CBFA1*is a required transcriptional determinant for osteoblast differentiation. Heterozygous mutations in *RUNX2* cause cleidocranial dysplasia (CCD), a human skeletal dysplasia characterized by defective intramembranous ossification, i.e., hypoplastic clavicles and delayed closure of fontanel. Our natural history study of over 90 CCD patients and previous clinical studies together point to significant variable expressivity and an underlying general dysplasia involving endochondral ossification. The clinical spectrum spans from severe CCD with osteopenia, to classic and mild CCD, to patients with only isolated dental anomalies. Moreover, additional genetic loci contributing to the CCD spectrum may exist. Mutation analysis of CCD patients of different ethnic backgrounds have revealed over 50 missense, nonsense, deletion, splicing, and insertion mutations. Most of these mutations disrupt DNA binding *in vitro* and hence result in haploinsufficiency. Some mutations do not affect DNA binding, but instead localize to amino acid residues implicated in mediating protein-protein interaction by X-ray crystallographic studies. Few patients have been reported to have expansion of a unique polyalanine stretch which has been described to harbor a transactivation domain. We have found that another unique domain, the polyglutamine stretch, also adds to the transactivation by *RUNX2*. However, no expansion or contractions of this region have yet been found to date in humans. Together these data show that the clinical spectrum of CCD can be caused by loss of the RUNX2 protein, loss of its DNA binding activity, loss of its transactivation, and perhaps also loss of specific protein-protein interactions.

Cleidocranial Dysostoses or Dysplasia?

The French physicians Pierre Marie and Paul Sainton were credited with the first clinical description of CCD in 1898 [1]. They described patients with the pathognomonic features of delayed ossification of the fontanel, aplasia of the clavicles, and hereditary transmission. Based on these findings, they concluded that the principal characteristics were consistent with a dysostoses of the skeleton, i.e., a developmental malformation of specific skeletal elements which often results from a developmental alteration at a specific time point during embryogenesis. It often connotes the absence of a more generalized dysregulated process in postnatal life. Since that time, multiple additional case reports have been published and we recently completed a large natural history study of 90 CCD patients [2-4]. Interestingly, CCD is a disease recognized since antiquity with reference to many features of CCD in a central character, Thersites, from the Homeric epics, the Illiad and Odyssey [5]. In recent years, the further clinical and radiographic delineation of this phenotype has clearly shown that CCD is a generalized skeletal dysplasia with continued dysregulated skeletal development in the postnatal period. While the defect in craniofacial and clavicular development reflect the requirement for the CCD gene during osteoblast differentiation, the findings of long bone and growth plate alterations point to a concurrent defect in chondrocyte differentiation. Hence, the human phenotype suggests that the CCD gene is required for both endochondral and intramembranous ossification.

CCD is a dominantly inherited disorder characterized by high penetrance and significant intrafamilial variability. Craniofacial features include brachycephalic skull, large head with frontal bossing, depressed nasal ridge, hypertelorism, underdeveloped zygomatic and lacrimal bones, and relative prognathism [6]. There are delayed closure of the fontanel, delayed closure of the sagittal and metopic sutures, multiple wormian bones, and underossification of the cranial base and calvaria. The clavicles are hypoplastic or aplastic and pseudoarthroses may develop in adulthood. Based on the appearance of the clavicles, prenatal diagnosis by high resolution ultrasound may be performed [7]. In addition to the characteristic clavicular and cranial abnormalities, patients have characteristic dental abnormalities which contribute significantly to overall morbidity [8]. CCD patients have delayed loss of primary dentition, supernumerary dentition, delayed eruption of permanent teeth, dentigerous cysts, and malocclusion. This often requires multiple orthodontic surgeries for removal of primary dentition and exposure and/or extraction of permanent teeth. An important and prevalent finding confirmed by Cooper et. al. in their natural history study of CCD includes significant short stature consistent with an underlying dysplastic process. This clinical observation in conjunction with the previously reported radiographic findings involving the epiphyses of long bones, i.e., large capital femoral epiphyses, cone-shaped and pseudoephiphyses of the phalanges, support a basic alteration of growth plate physiology. In fact, the basic embryology of the clavicles directly reflects disorder in both endochondral and intramembranous ossification [9]. The clavicle is the first bone to ossify in the embryonic period. It completes intramembranous ossification at two centers by 7-8 weeks gestation and then secondarily develops growth cartilage at each end. Hence, it is a primarily affected bony element in CCD. The study of the human condition clearly supports that CCD is caused by a generalized dysplasia of the skeleton with both prenatal and postnatal consequences on intramembranous and endochondral ossification.

RUNX2/CBFA1 during Skeletal Development

In 1997, a series of convergent studies identified the runt domain transcription factor *Runx2/Cbfa1* as an important determinant of osteoblast differentiation and as the gene mutated in CCD patients [10-13]. *RUNX2* or *CBFA1*, core binding factor-alpha 1, is one of three mammalian orthologs of the Drosophila *runt* gene. It encodes a non-redundant transcription factor that is required for osteoblast cell fate commitment. *RUNX2* contains an amino-terminal stretch of consecutive polyglutamine and polyalanine repeats (Q/A domain), a RUNT domain, and a carboxyl-terminal proline/serine/threonine-rich (PST) activation domain (Fig. 1) [14]. The RUNT domain is a 128 amino acid polypeptide motif originally described in the Drosophila *runt* gene and it has the unique ability of independently mediating DNA binding and protein dimerization [15]. The first vertebrate ortholog, *CBFA2/RUNX1/AML1*, was found to be translocated in several forms of acute myelogenous leukemia [16,17]. In contrast, *Runx2/Cbfa1* does not play a role during hematopoietic differentiation. Instead it is specifically expressed in pre-osteoblastic and osteoblastic cells. In transfection studies, *Runx2* can directly activate transcription of several bone-specific genes including osteocalcin, *Col1a1*, and osteopontin [18]. *RUNX2* was mapped to human chromosome 6p21, syntenic with the genetic location of cleidocranial dysplasia (CCD) [11]. Interestingly heterozygous disruption of the *Runx2* allele by gene targeting in mice reproduced the CCD phenotype [10,13]. Homozygous mice die at birth and have complete absence of osteoblasts and bone. Recent studies in *Runx2* null mice clearly show additional alterations of the cartilagenous skeleton [19,20]. Mice have defective chondrocyte maturation especially of proximal bones of the appendicular skeleton correlating with the human observations which suggested a central role for *Runx2* during chondrogenesis. Further transgenic mouse studies have confirmed that *Runx2* positively regulates chondrocyte maturation [21,22].

A series of studies over the past three years have dissected the functional domains of *Runx2* both in cell culture and cell-free systems. The PST and the Q/A domains have been shown to harbor transactivation motifs [23]. The carboxy-terminus of the RUNT domain contains a nuclear localization signal important in accumulation of RUNX2 in cell nuclei. The RUNT domain of the homologous RUNX1 protein has been shown to interact with the CBFβ protein at least during hemato poietic development. Both *Runx1* null mice and *Cbfβ*null mice exhibit similar phenotypes, i.e., defective hematopoiesis and embryonic lethality [24,25]. Unfortunately, the early lethality of the *Cbfβ* null mice prior to skeletogenesis has prevented the study of *Cbfβ* during osteoblast differentiation. Whether the runt domain in RUNX2/CBFA1 similarly mediates interaction with CBFβ during skeletogenesis has been a more controversial question. *In vitro* transfection studies in tissue culture have both suggested and argued against a direct interaction *in vivo* [23,26,27] It is likely that *Runx2* functions in a context-dependent manner specified in part by protein-protein interactions within the specific cell type [28]. For example, the TLE/Grg family of proteins including the Drosophila *Groucho* ortholog has been shown to interact with Runx2 specifically via the VWRPY pentapeptide motif in the carboxy terminus of the PST domain [29,30]. These proteins act in a repressor fashion on *Runx2* transactivation. In contrast, SMAD proteins have been demonstrated to augment *Runx2* activity by interaction with the carboxy-terminal half of the PST domain [31,32]. The consequences of these interactions *in vivo* have not yet been clearly defined at least in vertebrate development. Because of the diversity of the allelic series available via human molecular studies of CCD patients, specific mutations with clear *in vivo* phenotypic consequences can be correlated with potential biochemical functions identified in cell culture and cell free systems.

RUNX2 Mutations and Phenotypic Consequences

Including the present report, there have been over 50 different CCD alleles described in the literature [11,12,33-39] (Table 1). Several mutations have recurred in unrelated CCD families. The majority of mutations are missense mutations which affect the highly conserved RUNT domain. In our series, they often cluster in the carboxy-terminal half of the RUNT domain. They are expected to disrupt DNA binding and hence transactivation of target promoters. Both electrophoretic mobility shift assays (EMSA) and transfection studies have confirmed this observation [11,38]. More recently, NMR and X-ray crystallographic studies have identified the RUNT amino acid residues which mediate DNA binding and further showed that the majority of CCD missense mutations affect DNA binding [40-44]. A second large group of mutations include insertion and deletion mutations which cause frame-shift and premature termination. The majority of these mutations likely produce an unstable mRNA secondary to nonsense-mediated RNA decay (NMD) [45]. Similarly nonsense mutations which introduce stop codons prematurely are expected to also activate NMD. All of these mutations cause haploinsufficiency of the mutant allele and hence the classic CCD phenotype. The findings of microdeletions in the human condition as well as a deletion of the entire gene in the *Ccd* radiation-induced mouse mutant demonstrate that haploinsufficiency is the primary mechanism for generating the CCD phenotype [46,47]. However, detailed molecular studies in CCD families have broadened the phenotypic spectrum associated with *RUNX2* mutations to include mild CCD with little to no clavicular involvement and to individuals with isolated dental anomalies, but without the clear craniofacial and clavicular features of classic CCD [36,38,39]. On the other end of the spectrum, severe CCD with significant skeletal deformity including progressive scoliosis as well as osteopenia and recurrent fractures has been reported [2,39]. While CCD patients do not in general have increased fracture rate, a small percentage have a significantly increased susceptibility to fracture and osteopenia. This raises the question of whether this is simply due to stochastic variables, to true genetic modifiers such as the level of expression of the other wild type allele, or to mutations with functional consequence other than haploinsufficiency.

Table 1. RUNX2 mutations in CCD. Mutations are listed according to functional domains. Mutation types are listed: M: missense; F: Frameshift; N: Nonsense; D: Deletion; I: Insertion; S: Splicing. References are as listed Giannott et. al. (37); Golan et. al. (36); Lee et. al. (11); Mundlos et. al. (12); Quack et. al. (39); Tsai et. al. (35); Yokozeki et. al. (34); Zhang et. al. (33); Zhou et. al. (38).

Location	Mutation	Type	Comment	Ref.
N-terminal	90insC (stop in runt)	F	Mild CCD	Zhou
Poly Q/A domain	186ins16(stop in runt)	F		Mundlos
	220del173 (stop in runt)	F		Zhou
	222ins30	I	In-frame duplication of 10 alanines which abolishes transactivation	Mundlos
Runt domain	L113R	M		Quack
	S118R	M		Quack
	F121C	M	Mild CCD but classic CCD when found as compound heterozygote with G511S	Quack
	C123R	M		Quack
	S128F	M		Yokozeki
	397delAAC	D	Abolished DNA binding	Zhou
	R148G	M	4 family members with classic CCD, 1 with dental anomalies only	Golan
	D161X	N		Quack
	G166X	N		Zhou
	R169Q	M		Zhou
	M175R	M	Abolishes DNA binding and transactivation	Lee
	532delC	F		Zhang
	539del10	F		Mundlos

	542delG(stop in PST)	F	CCD plus fractures	Present
	553delCT(stop in runt)	F		Zhou
	R190P	M		Present
	R190Q	M	Abolished DNA binding	Zhou
	R190W	M		Giannoti
	S191N	M	Abolished DNA binding and transactivation	Lee
	R193L	M		Present
	R193C	M		Zhou
	R193X	N		Zhang, Quack
	F197S	M		Zhang
	L199F	M	Abolishes DNA binding and transactivation	Zhou
	T200A	M	Father with dental anomalies, 2 children with classic CCD – no effect on DNA binding – decreases potential Cbfβ interaction	Zhou
	T205R	M		Quack
	Q209R	M		Zhou
	Q209H	M		Present
	636delC(stop in runt)	F		Zhou
	R225Q	M		Zhou, Quack
	R225W	M	Interferes with nuclear accumulation	Quack
	866insC(stop in PST)	F		Present
	IVS5+lG->T in-frame exon skipping	S	In-frame deletion of last 35 amino acids of runt domain	Zhou, Zhang
PST - domain	824delG (stop in PST)	F		Quack
	Q284X	N		Present
	Q292X	N		Present
	884delC(stop in PST)	F		Zhou
	887delC (stop in PST)	F		Quack
	W297X	N		Mundlos
	915delC (stop in PST)	M		Quack
	960delG(stop in PST)	F		Present
	1127insT(stop in PST)	F		Quack

	1157delG(stop in PST)	F		Quack
	R391X	N	Classic CCD plus brachydactyly (Tsai); abolishes SMAD interaction; residual transactivation potential	Zhou, Zhang, Tsai
	1205insC(stop in PST)	F	Severe CCD with fractures	Quack
	1379insC(stop in PST)	F	Mild CCD	Quack
	G511S	M	No phenotype alone (may be neutral polymorphism)	Quack

RUNX2 Mutations and Structure/Function Correlates

Several mutations described by us and others point to potential alternative pathogenetic mechanisms and lend insight into the *in vivo* significance of some of the biochemical studies performed to date on *Runx2* (Fig. 1). We previously reported a 90insC mutation in a family with mild CCD [38]. In general, a frameshift mutation early in the transcript should cause premature termination and haploinsufficiency secondary to NMD. However, in this case, an alternative in-frame translation start site exists downstream of the insertion. Hence, a shorter RUNX2 peptide may be generated *in vivo* which retains much of the biological activity of RUNX2. In fact, at least in *in vitro* transcription/translation studies, this alternative ATG can be used, albeit much less efficiently [23]. This may account for a hypomorphic effect of this mutation which otherwise would be expected to produce classic CCD due to haploinsufficiency. In contrast a late insertion mutation described by Quack et. al., 1205insC, has been associated with a severe perinatal phenotype with fractures and osteopenia. Interestingly, this mutation causes a frame-shift and premature termination in the final exon. It is likely that NMD is not active in this case and that a protein lacking the carboxyterminal SMAD interaction domain and the VWRPY protein interaction motif may be made. This product may act in a dominant negative fashion further decreasing *in vivo RUNX2* transactivation of target genes. In a transgenic mouse mutant generated by Ducy et. al. a dominant negative truncated form of *Runx2* harboring only the RUNT domain produced an osteopenia phenotype [48]. We have a similar mutation, 1224insC downstream of this in a fetal case of CCD associated with significant alteration of the growth plate including hypoplasia of the zone of hypertrophy (Zheng et. al. submitted). Zhang et. al in fact reported a recurrent nonsense mutation, R391X, in the PST domain upstream of these mutations and which they demonstrate to disrupt SMAD interaction and transactivation [49]. These patient mutations together underscore the importance of the carboxy-terminal PST domain for protein-protein interactions and the potential for dominant negative acting mutants if the mutant mRNA is stable. Several mutations in CCD patients affecting the carboxy-terminal of the RUNT domain have highlighted the importance of these amino acid residues for DNA binding and for nuclear localization. The recurrent missense mutations involving R225 have been reported to both disrupt DNA binding (R225Q) and the putative nuclear localization signal of RUNX2 [38,39]. In fact, immunolocalization studies showed that the R225W mutation prevents nuclear localization [39]. A splice site mutation which deletes the carboxy-terminal portion of the RUNT domain and 35 amino acids, but which leaves the rest of the molecule in-frame also produced CCD [33,38].

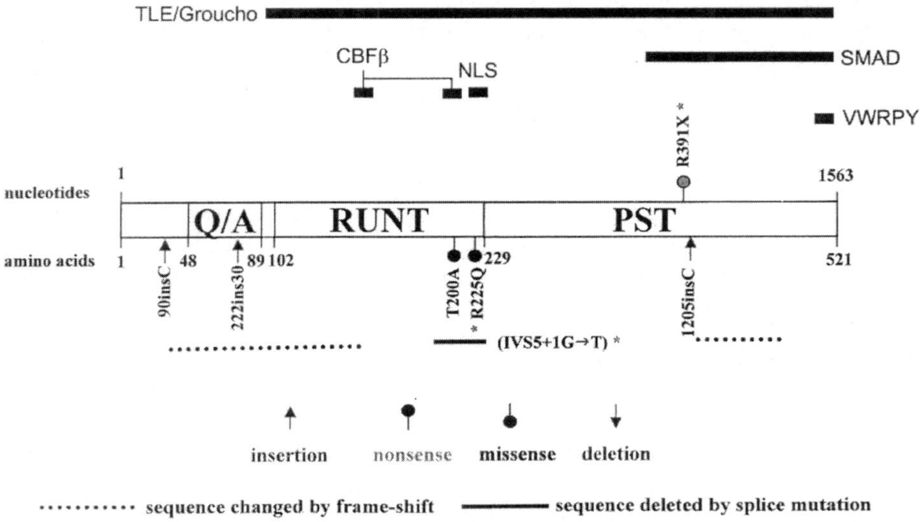

Figure 1. Schematic of RUNX2 functional domains, protein interactions, and representative human mutations. VWRPY interaction motif, putative SMAD, TLE/Groucho, and potential CBFβ interactions are highlighted. Nuclear localization signal (NLS) is also shown. The respective mutation types are listed as shown in the legend. Asterisk denotes mutations which have been reported in multiple patients.

Interestingly, no missense mutations affecting the stretch of RUNT amino acids from 156-165 have been reported. NMR, X-ray crystallography, and *in vitro* mutagenesis studies have identified these residues to be required in potential RUNT-CBFβ interaction [40 -44]. While this may suggest that CCD is not caused by a potential disruption of RUNX2/CBFβ interaction, we have identified a unique T200A missense substitution which may disrupt a second domain of RUNX2/CBFβ interaction as identified by X -ray crystallography [44]. In biochemical studies, this mutation did not affect either DNA binding or transactivation [38]. Moreover, *in vitro* mutagenesis of this position in the RUNT domain decreased interaction with CBFβ *in vitro* [43]. Interestingly, this mutation was associated with mild CCD and isolated dental anomalies suggesting that the mutation may in fact cause a hypomorphic effect on *Runx2* transactivation and hence a mild phenotype. Since CBFβ interaction with RUNT increases its affinity to the DNA target, disruption of this interaction may decrease transactivation of target genes in a hypomorphic manner and hence produce a mild CCD phenotype or only dental anomalies. Another human mutation has underscored the importance of the polyalanine stretch in the transactivation potential of *RUNX2*. While contractions of the 17 consecutive alanines in the Q/A domain to 11 consecutive alanines is a neutral polymorphism in the population, expansion of this region by 10 alanines (222ins30) will cause loss of transactivation and hence CCD [12,23]. In contrast, we have not observed polymorphism in the 23 consecutive glutamines immediately amino terminal to the polyalanine stretch in over 100 individuals. This may be due in part to the finding of several non CAG glutamine codons which interrupt the CAG stretch. We have, in fact, expanded this polyglutamine *in vitro* to 34 and 72 consecutive glutamines without deleterious effects on RUNX2-mediated transactivation *in vitro*. However, deletion of 21 consecutive glutamines decreased transactivation of target genes by more than 50% (unpublished data, Zhou G. et. al.). It is unclear what the phenotypic consequences of such mutations might be in the human skeleton: whether a gain of function mutation might occur as is observed in the polyglutamine expansion neurodegenerative diseases [50].

The study of the human phenotype has also pointed to the potential of a second genetic locus associated with CCD. In two unrelated families, a CCD-like phenotype was reported to involve cytogenetic rearrangement of 8q22 [51]. In the first family, there is a chromosome 8q22.1 and 10p12.3 balanced translocation in both a mother and daughter with CCD. In the second case, a CCD patient was found to carry a duplication of 8q13.3-8q22.1. An attractive hypothesis would be that either gain of function or loss of function of a gene on 8q22.1 affects *Runx2* transactivation. Since no locus heterogeneity has been reported in the CCD families studied to date, it may be that simple loss of function is not the cause. Instead, the cytogenetic rearrangement may lead to activation of the gene in a gain of function mutation. Proteins which interact with RUNX2 to specify its context-dependent activity would be superb candidates for this second locus.

The correlation of CCD with *RUNX2* mutations was one of several key studies which highlighted the importance of *Runx2* action in osteoblast differentiation. Further study of the mouse and human phenotype have now underscored the emerging role of *Runx2* during chondrocyte differentiation and hypertrophy. A survey of the human mutations to date now has further correlated elegant biochemical studies on RUNX2 protein interaction and transactivation domains, as well as the RUNT protein structure. Ultimately, the study of CCD-like phenotypes may identify additional genes which modify RUNX2 action and account for its context-dependent function.

Acknowledgements

We thank Olivia Hernandez for administrative assistance. This work was supported by the National Institutes of Health AR 44738 and the March of Dimes Birth Defects Foundation. Q.Z. is a Arthritis Foundation postdoctoral fellow.

References

[1] Anonymous 1968 The classic: Marie, P., and Sainton P.: Sur la dysostose cleido-cranienne herediataire, Rev. neurol. 6:835, 1898. On hereditary cleido-cranial dysostosis. Clinical Orthopaedics & Related Research **58:**5-7.

[2] Cooper SC, Lee B, Zhou G, Flaitz C, Hecht J 2001 A naturaly history of cleidocranial dysplasia. American Journal of Medical Genetics **in press**.

[3] Chitayat D, Hodgkinson KA, Azouz EM 1992 Intrafamilial variability in cleidocranial dysplasia: a three generation family. American Journal of Medical Genetics **42**(3)**:**298-303.

[4] Mundlos S 1999 Cleidocranial dysplasia: clinical and molecular genetics. J Med Genet **36**(3)**:**177-82.

[5] Altschuler EL 2001 Cleidocranial dysostosis and the unity of the Homeric epics: an essay. Clin Orthop (383)**:**286-9.

[6] Jones KL 1997 Smith's Recognizable Patterns of Human Malformation, 5th ed. W.B. Saunders Company, Philadelphia, pp 408-409.

[7] Hamner LH, 3rd, Fabbri EL, Browne PC 1994 Prenatal diagnosis of cleidocranial dysostosis. Obstetrics & Gynecology **83**(5 Pt 2)**:**856-7.

[8] Becker A, Lustmann J, Shteyer A 1997 Cleidocranial Dysplasia .1. General Principles of the Orthodontic and Surgical Treatment Modality. American Journal of Orthodontics & Dentofacial Orthopedics **111**(1)**:**28-33.

[9] Gardner E 1968 The embryology of the clavicle. Clinical Orthopaedics & Related Research **58:**9-16.

[10] Komori T, Yagi H, Nomura S, Yamaguchi A, Sasaki K, Deguchi K, Shimizu Y, Bronson RT, Gao YH, Inada M, Sato M, Okamato R, Kitamura Y, Yoshiki S, Kishimoto T 1997 Targeted disruption of Cbfa1 results in a complete lack of bone formation owing to maturational arrest of osteoblasts. Cell **89:**755-764.

[11] Lee B, Thirunavukkarasu K, Zhou L, Pastore L, Baldini A, Hecht J, Geoffroy V, Ducy P, Karsenty G 1997 Missense mutations abolishing DNA binding of the osteoblast-specific transcription factor OSF2/CBFA1 in cleidocranial dysplasia. Nature Genetics **16**(3):307-310.

[12] Mundlos S, Otto F, Mundlos C, Mulliken JB, Aylsworth AS, Albright S, Lindhout D, Cole WG, Henn W, Knoll JHM, Owen MJ, Mertelsmann R, Zabel BU, Olsen BR 1997 Mutations involving the transcription factor CBFA1 cause cleidocranial dysplasia. Cell **89**(5):773-779.

[13] Otto F, Thornell AP, Crompton T, Denzel A, Gilmour KC, Rosewell IR, Stamp GWH, Beddington RSP, Mundlos S, Olsen BR, Selby PB, Owen MJ 1997 Cbfa1, a candidate gene for cleidocranial dysplasia syndrome, is essential for osteoblast differentiation and bone development. Cell **89**(5):765-771.

[14] Geoffroy V, Corral DA, Zhou L, Lee B, Karsenty G 1998 Genomic organization, expression of the human CBFA1 gene, and evidence for an alternative splicing event affecting protein function. Mammalian Genome **9**(1):54-7.

[15] Wheeler JC, Shigesada K, Gergen JP, Ito Y 2000 Mechanisms of transcriptional regulation by Runt domain proteins. Semin Cell Dev Biol **11**(5):369-75.

[16] Downing JR, Higuchi M, Lenny N, Yeoh AE 2000 Alterations of the AML1 transcription factor in human leukemia. Semin Cell Dev Biol **11**(5):347-60.

[17] Downing JR 2001 AML1/CBFbeta transcription complex: its role in normal hematopoiesis and leukemia. Leukemia **15**(4):664-5.

[18] Ducy P, Zhang R, Geoffroy V, Ridall AL, Karsenty G 1997 Osf2/Cbfa1: A transcriptional activator of osteoblast differentiation. Cell **89**:747-754.

[19] Enomoto H, Enomoto-Iwamoto M, Iwamoto M, Nomura S, Himeno M, Kitamura Y, Kishimoto T, Komori T 2000 Cbfa1 is a positive regulatory factor in chondrocyte maturation. J Biol Chem **275**(12):8695-702.

[20] Kim IS, Otto F, Zabel B, Mundlos S 1999 Regulation of chondrocyte differentiation by Cbfa1. Mech Dev **80**(2):159-70.

[21] Ueta C, Iwamoto M, Kanatani N, Yoshida C, Liu Y, Enomoto-Iwamoto M, Ohmori T, Enomoto H, Nakata K, Takada K, Kurisu K, Komori T 2001 Skeletal malformations caused by overexpression of Cbfa1 or its dominant negative form in chondrocytes. J Cell Biol **153**(1):87-100.

[22] Takeda S, Bonnamy JP, Owen MJ, Ducy P, Karsenty G 2001 Continuous expression of Cbfa1 in nonhypertrophic chondrocytes uncovers its ability to induce hypertrophic chondrocyte differentiation and partially rescues Cbfa1-deficient mice. Genes Dev **15**(4):467-81.

[23] Thirunavukkarasu K, Mahajan M, McLarren KW, Stifani S, Karsenty G 1998 Two domains unique to osteoblast-specific transcription factor Osf2/Cbfa1 contribute to its transactivation function and its inability to heterodimerize with Cbfbeta. Mol Cell Biol **18**(7):4197-208.

[24] Niki M, Okada H, Takano H, Kuno J, Tani K, Hibino H, Asano S, Ito Y, Satake M, Noda T 1997 Hematopoiesis in the Fetal Liver Is Impaired By Targeted Mutagenesis of a Gene Encoding a Non-Dna Binding Subunit of the Transcription Factor, Polyomavirus Enhancer Binding Protein 2/Core Binding Factor. Proceedings of the National Academy of Sciences of the United States of America **94**(11):5697-5702.

[25] Wang Q, Stacy T, Binder M, Marin-Padilla M, Sharpe AH, Speck NA 1996 Disruption of the Cbfa2 gene causes necrosis and hemorrhaging in the central nervous system and blocks definitive hematopoiesis. Proceedings of the National Academy of Sciences of the United States of America **93**(8):3444-9.

[26] Harada H, Tagashira S, Fujiwara M, Ogawa S, Katsumata T, Yamaguchi A, Komori T, Nakatsuka M 1999 Cbfa1 isoforms exert functional differences in osteoblast differentiation. J Biol Chem **274**(11):6972-8.

[27] Li LH, Gergen JP 1999 Differential interactions between Brother proteins and Runt domain proteins in the Drosophila embryo and eye. Development **126**(15):3313-22.

[28] Javed A, Barnes GL, Jasanya BO, Stein JL, Gerstenfeld L, Lian JB, Stein GS 2001 runt homology domain transcription factors (Runx, Cbfa, and AML) mediate repression of the bone sialoprotein promoter: evidence for promoter context-dependent activity of Cbfa proteins. Mol Cell Biol **21**(8):2891-905.

[29] McLarren KW, Lo R, Grbavec D, Thirunavukkarasu K, Karsenty G, Stifani S 2000 The mammalian basic helix loop helix protein HES-1 binds to and modulates the transactivating function of the runt-related factor Cbfa1. J Biol Chem **275**(1):530-8.

[30] Aronson BD, Fisher AL, Blechman K, Caudy M, Gergen JP 1997 Groucho-Dependent and - Independent Repression Activities of Runt Domain Proteins. Molecular & Cellular Biology **17**(9):5581-5587.

[31] Bae SC, Lee KS, Zhang YW, Ito Y 2001 Intimate relationship between TGF-beta/BMP signaling and runt domain transcription factor, PEBP2/CBF. J Bone Joint Surg Am **83-A**(Suppl 1(Pt 1))**:**S48-55.

[32] Lee KS, Kim HJ, Li QL, Chi XZ, Ueta C, Komori T, Wozney JM, Kim EG, Choi JY, Ryoo HM, Bae SC 2000 Runx2 is a common target of transforming growth factor beta1 and bone morphogenetic protein 2, and cooperation between Runx2 and Smad5 induces osteoblast-specific gene expression in the pluripotent mesenchymal precursor cell line C2C12. Mol Cell Biol **20**(23)**:**8783-92.

[33] Zhang YW, Yasui N, Kakazu N, Abe T, Takada K, Imai S, Sato M, Nomura S, Ochi T, Okuzumi S, Nogami H, Nagai T, Ohashi H, Ito Y 2000 PEBP2alphaA/CBFA1 mutations in Japanese cleidocranial dysplasia patients. Gene **244**(1-2)**:**21-8.

[34] Yokozeki M, Ohyama K, Tsuji M, Goseki-Sone M, Oida S, Orimo H, Moriyama K, Kuroda T 2000 A case of Japanese cleidocranial dysplasia with a CBFA1 frameshift mutation. J Craniofac Genet Dev Biol **20**(3)**:**121-6.

[35] Tsai FJ, Wu JY, Lin WD, Tsai CH 2000 A stop codon mutation in the CBFA 1 gene causes cleidocranial dysplasia. Acta Paediatr **89**(10)**:**1262-5.

[36] Golan I, Preising M, Wagener H, Baumert U, Niederdellmann H, Lorenz B, Mussig D 2000 A novel missense mutation of the CBFA1 gene in a family with cleidocranial dysplasia (CCD) and variable expressivity. J Craniofac Genet Dev Biol **20**(3)**:**113-20.

[37] Giannotti A, Tessa A, Patrono C, Florio LD, Velardo M, Dionisi-Vici C, Bertini E, Santorelli FM 2000 A novel CBFA1 mutation (R190W) in an Italian family with cleidocranial dysplasia. Hum Mutat **16**(3)**:**277.

[38] Zhou G, Chen Y, Zhou L, Thirunavukkarasu K, Hecht J, Chitayat D, Gelb BD, Pirinen S, Berry SA, Greenberg CR, Karsenty G, Lee B 1999 CBFA1 mutation analysis and functional correlation with phenotypic variability in cleidocranial dysplasia. Hum Mol Genet **8**(12)**:**2311-6.

[39] Quack I, Vonderstrass B, Stock M, Aylsworth AS, Becker A, Brueton L, Lee PJ, Majewski F, Mulliken JB, Suri M, Zenker M, Mundlos S, Otto F 1999 Mutation analysis of core binding factor A1 in patients with cleidocranial dysplasia. Am J Hum Genet **65**(5)**:**1268-78.

[40] Nagata T, Gupta V, Sorce D, Kim WY, Sali A, Chait BT, Shigesada K, Ito Y, Werner MH 1999 Immunoglobulin motif DNA recognition and heterodimerization of the PEBP2/CBF Runt domain. Nat Struct Biol **6**(7)**:**615-9.

[41] Warren AJ, Bravo J, Williams RL, Rabbitts TH 2000 Structural basis for the heterodimeric interaction between the acute leukaemia-associated transcription factors AML1 and CBFbeta. Embo J **19**(12)**:**3004-15.

[42] Bravo J, Li Z, Speck NA, Warren AJ 2001 The leukemia-associated AML1 (Runx1)--CBF beta complex functions as a DNA-induced molecular clamp. Nat Struct Biol **8**(4)**:**371-8.

[43] Nagata T, Werner MH 2001 Functional mutagenesis of AML1/RUNX1 and PEBP2 beta/CBF beta define distinct, non-overlapping sites for DNA recognition and heterodimerization by the Runt domain. J Mol Biol **308**(2)**:**191-203.

[44] Tahirov TH, Inoue-Bungo T, Morii H, Fujikawa A, Sasaki M, Kimura K, Shiina M, Sato K, Kumasaka T, Yamamoto M, Ishii S, Ogata K 2001 Structural analyses of DNA recognition by the AML1/Runx-1 Runt domain and its allosteric control by CBFbeta. Cell **104**(5)**:**755-67.

[45] Frischmeyer PA, Dietz HC 1999 Nonsense-mediated mRNA decay in health and disease. Hum Mol Genet **8**(10)**:**1893-900.

[46] Mundlos S, Mulliken JB, Abramson DL, Warman ML, Knoll JH, Olsen BR 1995 Genetic mapping of cleidocranial dysplasia and evidence of a microdeletion in one family. Human Molecular Genetics **4**(1)**:**71-5.

[47] Mundlos S, Huang LF, Selby P, Olsen BR 1996 Cleidocranial dysplasia in mice. Annals of the New York Academy of Sciences **785:**301-2.

[48] Ducy P, Starbuck M, Priemel M, Shen J, Pinero G, Geoffroy V, Amling M, Karsenty G 1999 A Cbfa1-dependent genetic pathway controls bone formation beyond embryonic development. Genes Dev **13**(8)**:**1025-36.

[49] Zhang YW, Yasui N, Ito K, Huang G, Fujii M, Hanai J, Nogami H, Ochi T, Miyazono K, Ito Y 2000 A RUNX2/PEBP2alpha A/CBFA1 mutation displaying impaired transactivation and Smad interaction in cleidocranial dysplasia. Proc Natl Acad Sci U S A **97**(19)**:**10549-54.

[50] Orr HT 2001 Beyond the Qs in the polyglutamine diseases. Genes Dev **15**(8)**:**925-32.

[51] Brueton LA, Reeve A, Ellis R, Husband P, Thompson EM, Kingston HM 1992 Apparent cleidocranial dysplasia associated with abnormalities of 8q22 in three individuals. American Journal of Medical Genetics **43**(3)**:**612-8

The Growth Plate
I.M. Shapiro et al. (Eds.)
IOS Press, 2002

BMP-Regulated Chondrocyte Hypertrophy

Phoebe S. Leboy, Giovi Grasso-Knight, Marina D'Angelo and Sherrill Adams
Department of Biochemistry, School of Dental Medicine, University of Pennsylvania, Philadelphia, PA 19104-6003

Abstract. Bone morphogenetic proteins (BMPs) not only promote the expression of osteoblast-specific genes, but also induce the maturation of pre-hypertrophic chondrocytes. They act by binding to heterodimeric BMP-specific receptors which, in turn, phosphorylate intracellular Smad proteins capable of functioning as transcription factors. We have characterized a BMP-responsive region of the type X collagen gene which contains several potential Smad response elements. When type X promoter constructs are transfected into pre-hypertrophic chondrocytes from the upper sternum of day 15 chick embryos, maximal transcription requires both activated Smad 1 or 5 and the transcription factor Runx2. These studies suggest that Runx2 serves as a co-modulator with BMP-activated Smads for transcriptional activation of genes induced during chondrocyte hypertrophy. The ability of BMPs to induce hypertrophy is normally restricted to pre-hypertrophic chondrocytes, and addition of ascorbate will further increase the rate of hypertrophy. In contrast, chondrocytes from the lower region of embryonic sternum (LSC), which does not undergo endochondral bone formation during development, do not respond to BMPs or ascorbate. However, ascorbate-treated LSC expressing constitutively active forms of the BMP receptors ALK3 (BMPR-IA) or ALK6 (BMPR-IB) showed elevated expression of alkaline phosphatase and type X collagen by day 7 and nodules containing hydroxyapatite at day 14. These cultures also activated reporter constructs controlled by the BMP-responsive region from the type X collagen promoter. Inability of exogenous BMPs to induce maturation in LSC was correlated with high levels of mRNA for the secreted BMP-binding protein, noggin. The ascorbate effect is correlated with increased levels of Runx2. These studies imply that suppression of hypertrophy in lower sternal chondrocytes is mediated by secretion of BMP-binding proteins, as well as low levels of the co-modulator Runx2. We also provide evidence that retinoic acid stimulation of hypertrophy is via the BMP signaling pathway.

Introduction

It has been over 30 years since Urist [1] demonstrated that demineralized bone powder contained components which could induce ectopic bone formation. With the cloning of bone morphogenetic proteins (BMPs), it became clear that the observed bone induction was mediated by proteins which were part a large family of growth and differentiation growth factors, the TGF-β superfamily. [2] The BMPs are a group of 100-140 amino acid secreted polypeptides which are active as homodimers. Unlike classical growth factors which influence cell proliferation, BMPs function primarily as differentiation factors. Given the importance of BMPs in early embryonic patterning, their appearance at areas of epithelial-mesenchymal interactions, and the fact that they are involved in the development of nearly all organs and tissues including nervous system, somites, lung, kidney and gonads,[3] a more appropriate term would probably be "body morphogenetic proteins". Nonetheless, several BMPs are clearly implicated in the differentiation of skeletal tissue. High levels of BMP-2, -4, -6, and -7 are found in both osteoblasts and maturing chondrocytes.[4] These

BMPs can promote both osteogenesis [5] and chondrogenesis, [6;7] and will induce the maturation of growth plate chondrocytes.[8-10]

Consistent with their distinctive role as differentiation factors, members of the TGF-β family share a unique intracellular signaling mechanism; ligand binding and activation of receptors induces phosphorylation of a group of intracellular transcription factors known as Smads (Fig. 1). These receptor-activated Smads (R-Smads) associate with a "co-activator" Smad 4 in the cytoplasm,[11;12] allowing the activated Smad complex to translocate to the nucleus where it participates in transcriptional regulation. [11-13] The activity of the Smad signal transduction pathway is modulated by several inhibitory factors: Smad 6 competes for Smad4 co-activator, Smad7 competes for activated receptor, and BAMBI is a pseudoreceptor which dimerizes with type I subunit of the BMP receptor.[12;14]

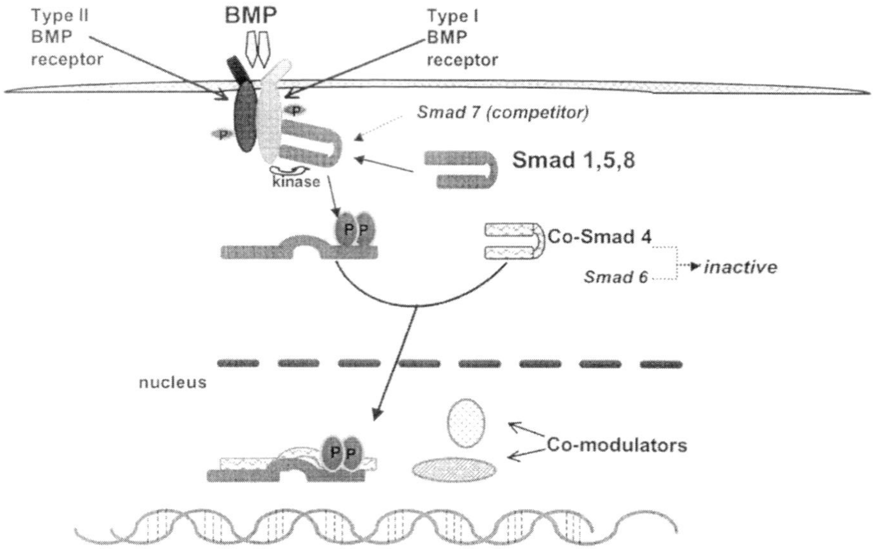

Figure 1. Mechanism of Smad-mediated BMP signaling.

Receptors for the TGF-β superfamily are trans -membrane cell surface heterodimers, containing both type I and type II components. There are 3 general classes of receptors for members of the TGF-β superfamily: one set for TGF-βs, another set for activins and a third set for BMPs.[15] However, recent evidence suggests considerable overlap in receptor utilization, with BMPs capable of binding to several activin type II receptors as well as BMPR-II.[12] Ligand binding to the receptor dimer permits the type II serine/threonine kinase to activate the type I kinase which, in turn, phosphorylates an R- Smad.[16] The R-Smads downstream of TGF-β signaling are Smads 2 and 3, while activated BMP receptors phosphorylate Smads 1,5,and 8.

Although activated Smads have DNA binding activity, it is increasingly apparent that transcriptional regulation by activated Smads involves interaction with one or more additional transcription factors. Smads bind to DNA with relatively low affinity, and the consensus sequence for binding does not provide high specificity; therefore, they are unlikely to be effective transcriptional regulators by themselves.[11,12] Furthermore, the widespread utilization of BMPs in development suggests that co-modulators are required to provide tissue specificity. The mechanism by which BMPs interact with another transcription factor to regulate expression of the osteopontin promoter has recently been elucidated by Cao and co-workers. In an elegant series of experiments, this group has demonstrated that osteopontin expression is repressed by the homeodomain-containing protein Hoxc-8; when Smad1 is activated by BMP it binds to Hoxc-8, relieving repression and permitting osteopontin gene transcription. [17;18] Since osteopontin is produced both by osteoblasts and by a variety of non-skeletal tissues, we do not know whether this mechanism is generally applicable to the control of skeletal-specific genes. However, accumulating evidence based on TGF-β signaling suggests that there are a large number of stimulus-specific and cell-specific transcription factors which can act as co-modulators for activated Smads.[11;12] Studies of the osteocalcin promoter have shown synergy between TGF-β and vitamin D stimulation resulting from a combination of activated Vitamin D receptor and TGF-β-stimulated Smad 3.\pard cs2[19] Similarly, the collagenase-1 promoter requires both Smad3 binding to DNA and cooperativity with the c-jun transcription factor[20].

Another group of transcription factors which interact with Smads are the Runx family. [21;22] The gene for Runx2 (Cbfa1) is essential for bone development, since mice lacking functional Runx2 developed neither intramembranous bone nor endochondral bone.[23;24] Although originally reported to be osteoblast-specific, Runx2 is also expressed in pre-hypertrophic and hypertrophic chondrocytes, [25] and Runx2-deficient mice which lack bone also show defects in chondrocyte maturation [26] Direct evidence for Smad-Runx interactions emerged in studies examining TGF-β induction of immunoglobulin expression; both Runx1 and Runx3 were shown to complex with activated Smad 3.[21;27] Furthermore, cells expressing a Runx2 mutation which prevents Smad-Runx binding lose the ability to undergo BMP-induced osteogenesis,[28] implying that Runx interaction with Smads is essential for BMP responsiveness.

Since BMPs induce maturation of pre-hypertrophic chondrocytes, and Runx2 expression is elevated in pre-hypertrophic and hypertrophic chondrocytes, we have recently examined the possibility that BMP-activated Smads cooperate with Runx2 to induce hypertrophy-related genes. We have also explored conditions under which activation of the BMP signal transduction system will induce hypertrophy in immature chondrocytes.

Materials and Methods

Cell Culture

Cells were isolated from the lower (caudal) or upper (cephalic) one-third portions of sternae from 15-day chick embryos (B&E Eggs, Stevens, PA) and cultured as described previously. [10] Recombinant human BMP-2 (kindly provided by Genetics Institute, Cambridge, MA) was added to cultures where appropriate at a final concentration of 30 ng/ml. Ascorbate-supplemented cultures contained 75μM ascorbate phosphate (Wako Pure Chemical Industries, Ltd., Japan). from day 2 until day 4, and 150μM thereafter. Trans-retinoic acid was added at a final concentration of 35nM.

Luciferase Assays for Measuring Type X Promoter Activity

The chick type X collagen gene contains a "b2" region at -2649 to -2007 which, along with a 640bp proximal promoter, permits BMP-induced transcription in pre-hypertrophic upper sternal chondrocytes.[29] This b2/640 promoter, placed upstream of a Renilla luciferase luciferase reporter gene (pRL, Promega, Madison WI), was transfected into sternal chondrocytes and promoter activity was assessed by measuring luciferase activity as described previously.[29]

Infection of Chondrocytes with Mutated BMP Receptors Expressed in Retroviral (RCAS) Vectors
Type I BMP receptors were expressed in chondrocytes using RCAS vectors as described previously.[30] The constitutively active ALK3 (BMPR-IA) and ALK6 (BMPR-IB) mutants cloned into RCAS were prepared by Dr. Lee Niswander, Sloan-Kettering Institute.[31;32] For sequential infection and transfection of lower sternal chondrocytes, medium containing unconcentrated RCAS virus was added to the cultures at the time of plating, and 75 μM ascorbate phosphate (Wako Chemicals, Richmond VA) added to appropriate wells at day 1. The infection was allowed to spread throughout the chondrocytes until day 7. Cells were then rinsed once with Hank's Buffered Saline Solution, transfected with plasmids as described above, and cultured with or without 150 μM ascorbate phosphate until harvested at day 9.

Assays for Hypertrophy
Alkaline phosphatase activity, Northern blots for type X collagen and alkaline phosphatase mRNA, and DNA determinations were performed as described previously.[29]

Scanning Electron Microscopy and Mineralization Analysis
SEM analysis was performed on cells plated in 6-well Nunc culture dishes at 2.4×10^4 cells/cm^2. Inorganic phosphate (2.5mM) was added starting at day 7 to promote mineralization. The cell layers were rinsed twice with HBSS, fixed for 30 minutes with 500μl Karnovsky solution then washed with increasing ethanol concentrations followed by two rinses at 100% ethanol. After the addition of 500 μl hexane, the wells were wrapped in Parafilm which was punctured to allow the hexane to evaporate overnight. The plates were then inverted, the wells were cut out, and affixed to stubs with quick dry colloidal silver. Once the silver had dried overnight the samples were carbon-coated using a Desk II Denton Vacuum Carbon Accessory and Denton Vacuum Machine. The samples were visualized using a JEOL scanning electron microscope equipped with a KEVEX detector for determining calcium and phosphate levels.

Results

We have used a construct in which the luciferase reporter gene was regulated by a BMP-responsive region of the type X collagen promoter [29] to examine the role of Runx2 in BMP activation of type X collagen synthesis. These studies, using pre-hypertrophic chondrocytes which show BMP-stimulated type X collagen synthesis, indicate that over-expression of Runx2 in the absence of BMPs has no effect on the activity of the type X promoter. However, when Smads are activated with exogenous BMP (Fig. 2), over-expression of Runx2 markedly increases activity of the type X collagen promoter.[33] These results imply that Runx2 functions in conjunction with BMP-activated Smads to activate transcription of the type X collagen gene.

Since retinoic acid (RA) has been reported to stimulate maturation of pre-hypertrophic chondrocytes,[34;35] we have also examined the ability of this compound to regulate type X collagen promoter activity in these cells. As shown in Fig. 3, 35nM RA stimulates the same b2/640 region which is stimulated by BMPs. In the presence of a constitutively active form of the type I BMP receptor ALK6, which stimulates activity of the b2/640 promoter region, no further RA stimulation is observed. Furthermore, RA stimulation is abolished by a dominant negative mutant form of ALK-6. These observations suggest that retinoids increase type X collagen expression via the BMP signaling pathway.

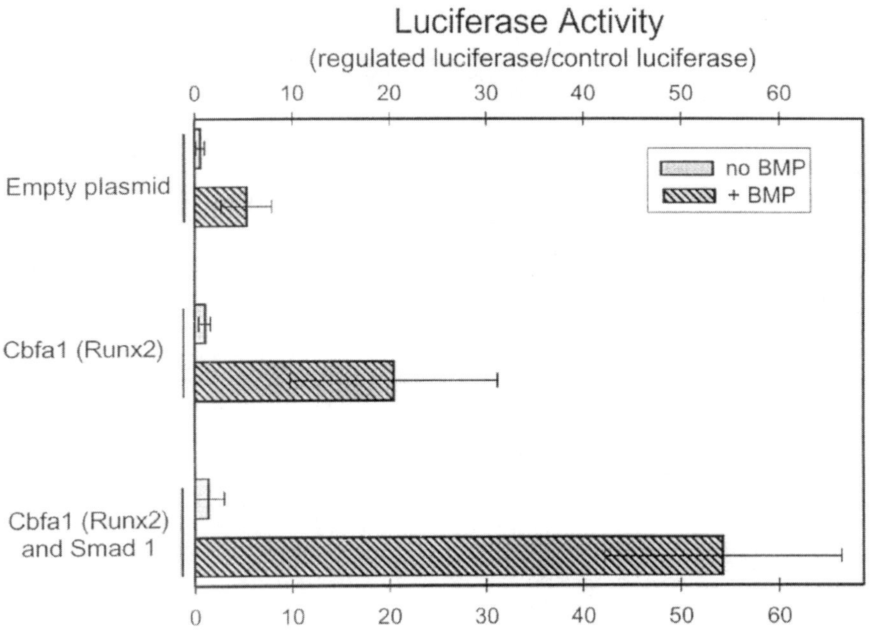

Figure 2. Cbfa1 stimulates activity of the b2/640 region from the type X collagen promoter in the presence of BMP-activated Smads.

While BMPs act to promote hypertrophy of chondrocytes from cartilage regions destined for endochondral bone formation, they are less effective in inducing maturation of chondrocytes from other regions. In pre-hypertrophic chondrocytes from the upper sternal region of day 15 chick embryos, BMPs markedly increase expression of mRNA for type X collagen, alkaline phosphatase, and MMP-13.[10;36] However, parallel cultures of chondrocytes derived from the lower sternum, which does not undergo hypertrophy during development, show no effects of BMP on these markers of hypertrophy.[10;29] The efficacy of BMPs is modulated by several secreted proteins which bind BMPs and prevent receptor activation. One major difference between upper and lower sternal chondrocytes is the expression of the BMP-binding protein noggin: cells from the lower sternum express high levels of noggin mRNA while those from pre-hypertrophic regions of the upper sternum show undetectable levels.[30] If high levels of noggin were solely responsible for the observed differences in BMP responsiveness, providing constitutively active (CA) BMP receptors ALK3 (BMPR-IA) or ALK6 (BMPRI-B) should by-pass the noggin effect. However, over-expressing constitutively active ALK3 or ALK6 in lower sternal chondrocytes was not sufficient to permit hypertrophy.[30] We have therefore examined additional factors which might relieve the suppression of hypertrophy in chick embryo lower sternal chondrocytes.

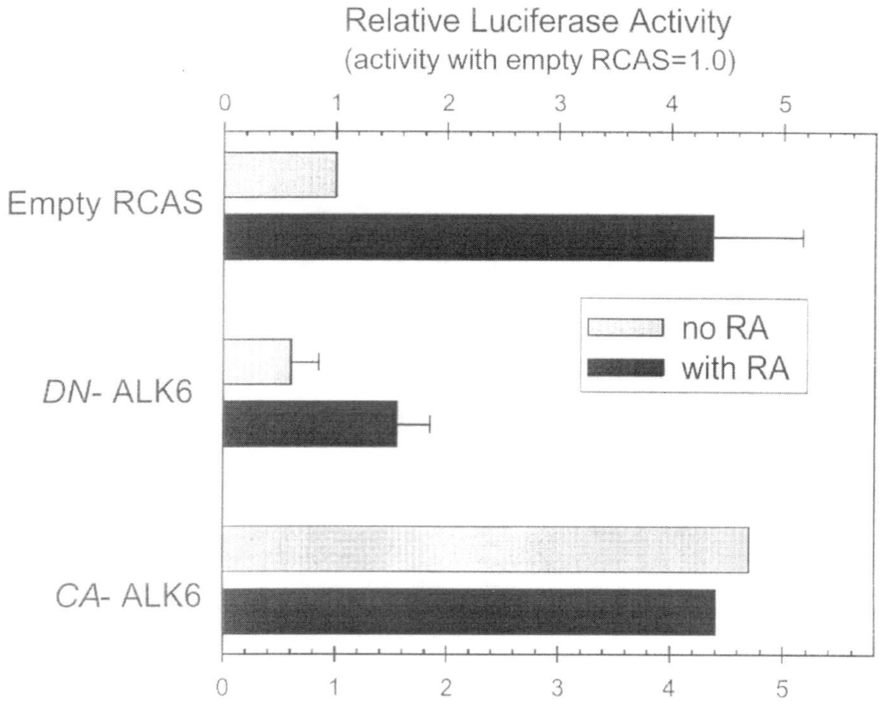

Figure 3. Retinoic acid (RA) effects on the b2/640 type X promoter in upper sternal chondrocytes, with and without expression of mutant BMP receptors. In the absence of BMP signaling (empty RCAS), 35nM RA increases promoter activity. However, a dominant negative BMP receptor (DN-ALK6) blocks this stimulation, while a constitutively active receptor (CA-ALK6) mimics the effect of RA. RA was added 24h before the luciferase assay.

Chondrocytes from the lower one/third of day15 chick embryo sternae were infected with RCAS retrovirus containing constitutively active ALK3 or ALK6 and cultured for 7-9 days in the presence or absence of 0.1mM ascorbate phosphate. Parallel uninfected cultures were maintained in the presence of 35ng/ml BMP-2. In the absence of ascorbate, neither exogenous BMP nor constitutively active BMP receptors promoted high level alkaline phosphatase activity, confirming previous results. However, addition of ascorbate to cultures expressing constitutively active BMP receptor ALK-6 yielded a significant increase in enzyme levels (Fig. 4). In contrast, ascorbate plus exogenous BMP-2 was markedly less effective. Like upper sternal chondrocytes, [10] lower sternal chondrocytes showed a greater response to either BMP or constitutively active ALK-6 when transferred to serum-free conditions than when cultured continuously with serum (Fig 4), and constitutively active ALK-3 was slightly less effective (data not shown).

The ability of induced lower sternal chondrocytes to produce mineralized matrix was analyzed by scanning electron microscopy and electron diffraction analysis (EDAX) of mineral in 14-18 day cultures supplemented with 2.5mM P_i. Cells cultured with ascorbate and P_i which had been infected with control RCAS virus showed no nodule formation and no regions of hydroxyapatite formation after 18 days (Fig. 5A). However, ascorbate-treated cultures in which cells were infected with either ALK3 (Fig. 5B) or ALK6 (Fig. 5C) showed extensive nodule formation by day 14. The mineral associated with these nodules had a Ca/P ratio of 1.74-1.75, indicating the formation of hydroxyapatite.

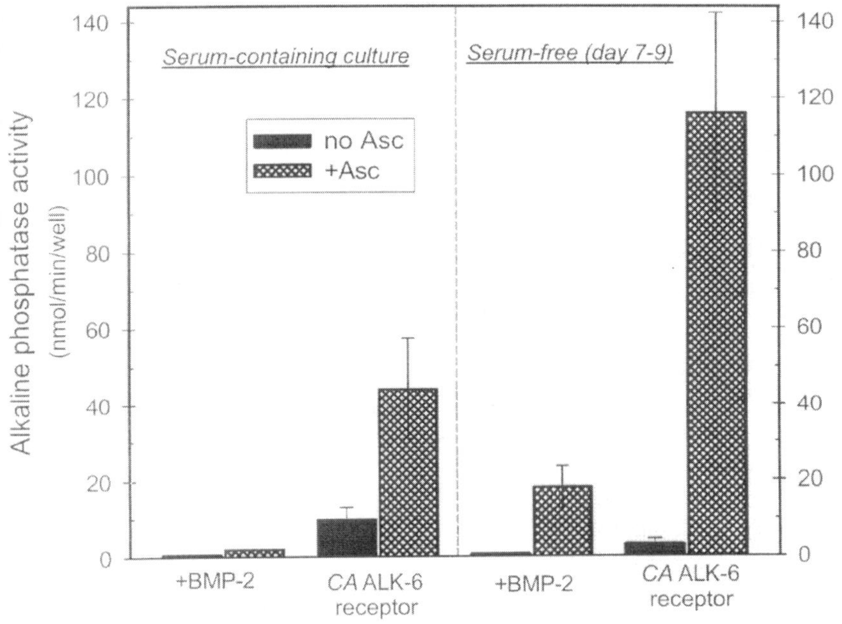

Figure 4. Alkaline phosphatase induction in lower sternal chondrocytes infected with RCAS virus containing constitutively-active (CA) BMP receptors and cultured in the presence of 100μM ascorbate phosphate Transferring lower sternal chondrocytes to serum-free conditions increases alkaline phosphatase activity stimulated by ascorbate plus either exogenous BMP or constitutively active ALK6.

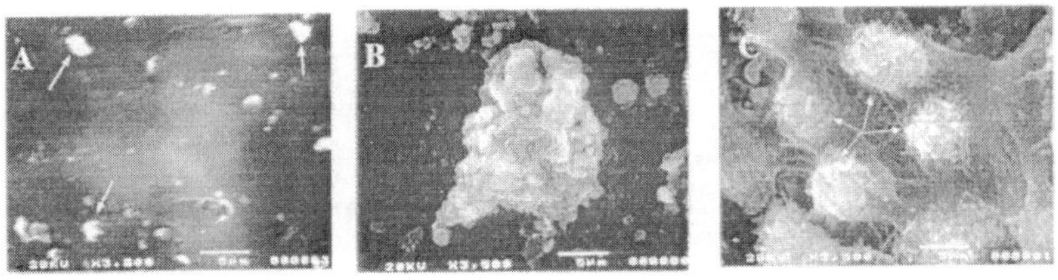

Figure 5. Scanning electron micrographs of lower sternal chondrocytes cultured with ascorbate and 2.5mM P_i for 14-18 days. A. Cells were infected at day 1 with empty RCAS virus and cultured until day 18. No nodules were formed and no regions of hydroxyapatite were detected. Arrows point to high density regions which were not enriched in Ca and P_i. B. Cells infected with virus containing ALK3 and cultured until day 14. Shown is an unusually large, condensed mineralized nodule. C. Cells infected with virus containing ALK6 and cultured until day 14. Arrows indicate more typical smaller nodules which contained hydroxyapatite mineral detected by EDAX analysis.

To confirm that the cells were expressing other markers of hypertrophy, RNA was prepared from day 9 cultures of lower sternal chondrocytes and analyzed by Northern blots. Figure 6 presents data from Northern blots probed with cDNAs for chick alkaline phosphatase and type X collagen. With exogenously added BMP-2 (left panels), ascorbate showed a modest induction of both type X collagen mRNA (upper panel), but there was little detectable alkaline phosphatase mRNA (lower panel). In contrast, constitutively active BMP receptor ALK6, combined with ascorbate, induced both alkaline phosphatase and type X collagen mRNAs.

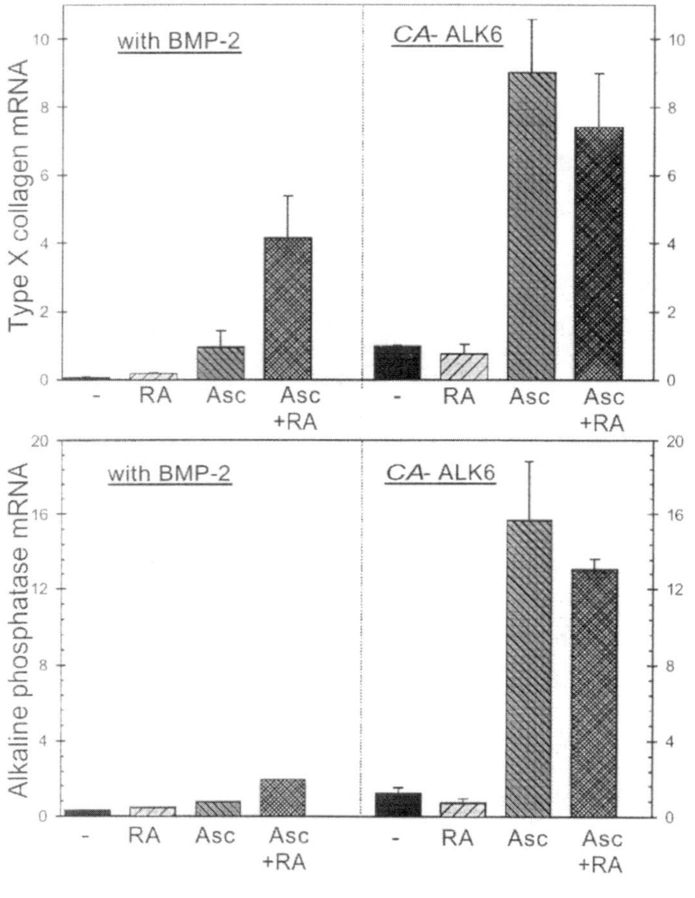

Figure 6. Levels of mRNA for alkaline phosphatase and type X collagen in lower sternal chondrocytes cultured for 9 days. At day 1, cells were infected with RCAS-ALK6 or treated with BMP-2. Ascorbate phosphate(100μM) was added starting at day 1, and 35nM RA was added starting at day 4. Results are the average of densitometry scans of two Northern blots from 2 independent preparations of RNA.

Parallel experiments were carried out with lower sternal chondrocytes exposed to 35nM retinoic acid (RA), with and without ascorbate. RA alone had little effect on either alkaline phosphatase or type X collagen mRNA levels (Fig. 6). Similarly, RA plus ascorbate did not increase mRNA levels in lower sternal chondrocytes expressing constitutively active BMP receptor ALK-6 beyond that observed with CA-ALK6 plus ascorbate. The activity of type X promoter-luciferase constructs in lower sternal chondrocytes infected with RCAS virus containing CA-ALK6 was also examined in the presence of ascorbate with and without retinoic acid (Fig. 7). Cells expressing the constitutively active receptor and cultured with ascorbate for 7 days showed a significant increase in type X promoter activity, but adding retinoic acid produced no further increase. These results are consistent with the Northern blot measurements, and argue that when lower sternal chondrocytes are induced to hypertrophy with ascorbate and CA-BMP receptor, retinoic acid treatment has no further effect.

It is noteworthy that, although RA has no effect on ascorbate-treated LSC which express constitutively active ALK6, the retinoid can modestly increase alkaline phosphatase and type X collagen mRNA in parallel cultures treated with exogenous BMP (Fig. 6, left panels). The ability of RA to augment the effects of exogenous BMP, but not constitutively active BMP receptor, suggests that the retinoid may decrease levels of BMP-binding proteins.

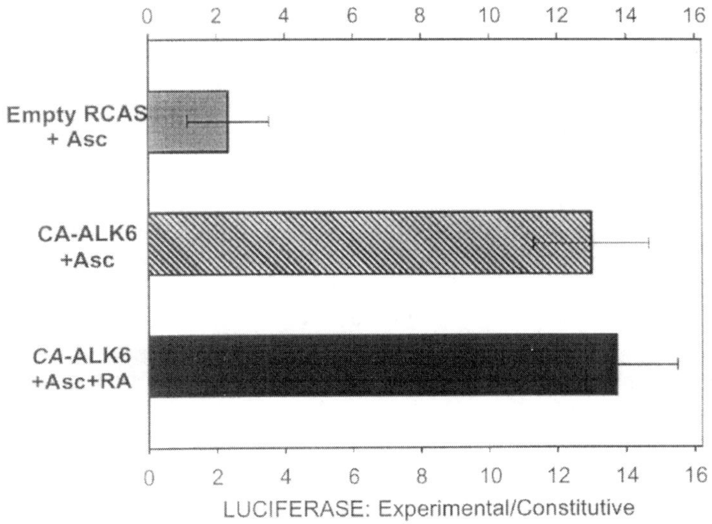

Figure 7. Activity of the b2/640 region of the type X collagen promoter in LSC cultured with ascorbate and expressing a constitutively active BMP receptor. Cultures were infected with virus and treated with ascorbate1 day after plating, transfected with b2/640-luciferase plasmid at day 6, and luciferase assayed 24h later.

Discussion

Our studies examining type X collagen promoter activity in upper sternal chondrocytes demonstrate that Runx2 increases promoter activity in the presence of BMPs. This implies that transcription of type X collagen is regulated by an interaction of BMP-activated Smads with Runx2. An important role for RUNX2 in chondrocyte hypertrophy has been indicated by studies carried out by Komori and co-workers. Enomoto et al [37] demonstrated that antisense oligonucleotides for RUNX2 severely reduced type X expression in a chondrogenic cell line. In addition, studies reported at this meeting show that transgenic mice over-expressing wild-type Runx2 show accelerated hypertrophy while mice over-expressing a truncated dominant negative Runx2 show suppression of hypertrophy and absence of type X collagen.[38] Our studies indicate that over-expression of Runx2 has no effect on promoter activity unless BMP is present, arguing that the Runx2 functions as a co-modulator with BMP-activated Smads. rather than acting independently as a transcription factor for type X collagen gene expression.

Studies with Runx2-deficient mice had demonstrated that Runx2 was required for bone formation,[23;24] and that it acted as a transcription factor to control the expression of several osteoblast-specific genes.[39] Komori et al[23] noted that addition of BMP-2 permitted low level expression of alkaline phosphatase and osteocalcin expression in cells from Runx2 deficient mice, suggesting that BMP-activated Smads might partially relieve the effect of the Runx mutation. More recent studies, with a Runx2 mutation which results in truncated protein, have shown that truncated Runx2 is unable to stimulate the osteocalcin promoter in C2C12 myoblasts, and this is correlated with its inability to bind to activated Smads.[28] These observations suggest that the Runx2 requirement for osteogenesis involves an activated Smad-Runx2 complex. However, Ducy et al [39] reported Runx2 up-regulation of osteocalcin, osteopontin and bone sialoprotein in the absence of exogenous BMP. It is therefore an unresolved question whether osteoblast-specific genes can be activated by Runx2 in the absence of BMP signaling or whether these genes, like type X collagen, require activated Smads to form a complex with Runx2.

While there is ample evidence that BMPs induce hypertrophy, there are several other factors that have also been useful as inducers, including retinoic acid, thyroxine, and ascorbate. We have now provided data indicating that retinoic acid may induce hypertrophy via a mechanism leading to increased signaling by BMP-activated Smads (Figs. 3, 6, and 7). Evidence is accumulating that thyroxine also works by way of the BMP signaling pathway [10] (T. Ballock, personal communication). However, several lines of evidence suggest that BMPs and ascorbate act by different mechanisms. Analyses of pre-hypertrophic upper sternal chondrocytes demonstrated that the effect of ascorbate and BMP was additive in pre-hypertrophic chondrocytes.[10;36] Our studies with lower sternal chondrocytes, derived from cartilage which does not undergo hypertrophy during skeletal development, now provide further evidence that BMPs and ascorbate promote hypertrophy via independent pathways. These cells cannot be induced to mature unless they are provided with both a constitutively active BMP receptor and ascorbate. The fact that exogenous BMP is relatively ineffective with lower sternal chondrocytes compared to either of the constitutively active type I BMP receptors implies either that the cells lack a functional type I receptor or that an inhibitor is blocking the ability of BMP to bind to its receptor. Since previous studies have shown that both ALK3 and ALK6 are expressed in lower sternal chondrocytes, [30] the first possibility is excluded. However, large amounts of mRNA coding for the BMP-sequestering protein noggin are present both in cultured lower sternal chondrocytes [30] and in proliferating chondrocytes of the growth plate.[40] These observations suggest that high levels of a BMP-sequestering protein produced by non-hypertrophying chondrocytes function to prevent the BMP stimulation of these cells, and providing the lower sternal chondrocytes with constitutively active ALK3 or ALK6 circumvents the inhibitory effects. Comparison of ALK3 and ALK6 at day 7 (Fig. 1) indicates that ALK3 is less effective than ALK6; this is also true for upper sternal chondrocytes.[30]

Since lower sternal chondrocytes expressing constitutively active BMP receptors also require ascorbate for induction of hypertrophy, these cells apparently have an additional block preventing hypertrophy which is overcome by ascorbate treatment. Runx2 levels are low in immature chondrocytes and increase in growth plate chondrocytes prior to hypertrophy.[37;41] Furthermore, transgenic mice expressing RUNX2 under the control of the type II collagen promoter show hypertrophy and endochondral ossification in cartilage which does not normally ossify.[41] Preliminary data from our laboratory indicate that cultured lower sternal chondrocytes have very low levels of Runx2 mRNA in the absence of ascorbate, but elevated Runx2 mRNA when cultured with ascorbate. It is therefore plausible that the ascorbate requirement for hypertrophy of lower sternal chondrocytes is associated with its ability to increase expression of Runx2.

Acknowledgements

We gratefully acknowledge gifts of BMP from Genetics Institute, Cambridge, MA, and retroviral BMP receptor constructs from Dr. Lee Niswander, Memorial Sloan-Kettering Institute, New York, NY. This work was supported by NIH grants R01 AR40075 and R01 AR44692.

References

[1] Urist,M.R. 1965 Bone: formation by autoinduction.Science 150: 893-899.
[2] Celeste,A.J., Iannazzi,J.A., Taylor,R.C., Hewick,R.M., Rosen,V., Wang,E.A., and Wozney,J.M. 1990 Identification of transforming growth factor-ß family members present in bone-inductive protein purified from bovine bone. Proc.Natl.Acad.Sci.USA 87: 9843-9847.
[3] Hogan,B.L.M. 1996 Bone morphogenetic proteins: Multifunctional regulators of vertebrate development. Genes Dev. 10: 1580-1594.
[4] Anderson,H.C., Hodges,P.T., Aguilera,X.M., Missana,L., and Moylan,P.E. 2000 Bone morphogenetic protein (BMP) localization in developing human and rat growth plate, metaphysis, epiphysis, and articular cartilage. J.Histochem.Cytochem. 48: 1493-1502.
[5] Wozney,J.M. and Rosen,V. 1998 Bone morphogenetic protein and bone morphogenetic protein gene family in bone formation and repair. Clinical Orthopaedics and Related Research 346: 26-37.
[6] Pizette,S. and Niswander,L. 2000 BMPs are required at two steps of limb chondrogenesis: Formation of prechondrogenic condensations and their differentiation into chondrocytes. Dev.Biol 219: 237-249.
[7] Yi,S.E., Daluiski,A., Pedersen,R.A., Rosen,V., and Lyons,K.M. 2000 The type I BMP receptor BMPR-IB is required for chondrogenesis in the mouse limb. Development 127: 621-630.
[8] Grimsrud,C.D., Romano,P.R., D'Souza,M., Puzas,J.E., Reynolds,P.R., Rosier,R.N., and O'Keefe,R.J. 1999 BMP-6 is an autocrine stimulator of chondrocyte differentiation. J.Bone Miner.Res. 14: 475-482.
[9] Enomoto-Iwamoto,M., Iwamoto,M., Mukudai,Y., Kawakami,Y., Nohno,T., Higuchi,Y., Takemoto,S., Ohuchi,H., Noji,S., and Kurisu,K. 1998 BMP signaling is required for maintenance of differentiated phenotype, control of proliferation, and hypertrophy in chondrocytes. J.Cell Biol. 140: 409-418.
[10] Leboy,P.S., Sullivan,T.A., Nooreyazdan,M., and Venezian,R.A. 1997 Rapid chondrocyte maturation by serum-free culture with BMP-2 and ascorbic acid. J.Cell.Biochem. 66: 394-403.
[11] Attisano,L. and Wrana,J.L. 2000 Smads as transcriptional co-modulators. Curr.Opin.Cell Biol. 12: 235-243.
[12] Massagué,J. and Wotton,D. 2000 Transcriptional control by the TGF-β/Smad signaling system . EMBO J. 19: 1745-1754.
[13] ten Dijke,P., Miyazono,K., and Heldin,C.H. 2000 Signaling inputs converge on nuclear effecters in TGF-β signaling. Trends Biochem.Sci. 25: 64-70.
[14] Wrana,J.L. 2000 Regulation of Smad activity. Cell 100: 189-192.
[15] Cheifetz,S. 1999 BMP receptors in limb and tooth development. Crit.Rev.in Oral Biol.and Med. 10: 182-198.
[16] Ducy,P. and Karsenty,G. 2000 The family of bone morphogenetic proteins. Kidney Int. 57: 2207-2214.
[17] Wan,M., Shi,X.M., Feng,X., and Cao,X. 2001 Transcriptional mechanisms of bone morphogenetic protein-induced osteoprotegrin gene expression. J.Biol.Chem. 276: 10119-10125.
[18] Yang,X.L., Ji,X.H., Shi,X.M., and Cao,X. 2000 Smad1 domains interacting with hoxc-8 induce osteoblast differentiation. J.Biol.Chem. 275: 1065-1072.
[19] Subramanian,N., Leong,G., Cock,T.-A., Flanagan,J.L., Fong,C., Eisman,J., and Kouzmenko,A. 2001 Cross-talk between 1,25-dihydroxyvitamin D3 and TGF-ß signaling requires binding of VDR and Smad3 proteins to their cognate DNA recognition elements. J.Biol.Chem. 276: 15741-15746.
[20] Qing,J., Zhang,Y., and Derynck,R. 2000 Structural and functional characterization of the TGF- β -induced Smad3/c-Jun transcriptional cooperativity. J.Biol.Chem. 275: 38802-38812.

[21] Hanai,J., Chen,L.F., Kanno,T., Ohtani-Fujita,N., Kim,W.Y., Guo,W.H., Imamura,T., Ishidou,Y., Fukuchi,M., Shi,M.J., Stavnezer,J., Kawabata,M., Miyazono,K., and Ito,Y. 1999 Interaction and functional cooperation of PEBP2/CBF with Smads - Synergistic induction of the immunoglobulin germline Cα promoter. J.Biol.Chem. 274: 31577-31582.

[22] Westendorf,J.J. and Hiebert,S.W. 1999 Mammalian runt-domain proteins and their roles in hematopoiesis, osteogenesis, and leukemia. J.Cell.Biochem. S32-33: 51-58.

[23] Komori,T., Yagi,H., Nomura,S., Yamaguchi,A., Sasaki,K., Deguchi,K., Shimizu,Y., Bronson,R.T., Gao,Y.H., Inada,M., Sato,M., Okamoto,R., Kitamura,Y., Yoshiki,S., and Kishimoto,T. 1997 Targeted disruption of Cbfa1 results in a complete lack of bone formation owing to maturational arrest of osteoblasts. Cell 89: 755-764.

[24] Otto,F., Thornell,A.P., Crompton,T., Denzel,A., Gilmour,K.C., Rosewell,I.R., Stamp,G.W.H., Beddington,R.S.P., Mundlos,S., Olsen,B.R., Selby,P.B., and Owen,M.J. 1997 Cbfa1, a candidate gene for cleidocranial dysplasia syndrome, is essential for osteoblast differentiation and bone development. Cell 89: 765-771.

[25] Kim,I.S., Otto,F., Zabel,B.U., and Mundlos,S. 1999 Regulation of chondrocyte differentiation by Cbfa1. Mech.Dev. 80: 159-170.

[26] Inada,M., Yasui,T., Nomura,S., Miyake,S., Deguchi,K., Himeno,M., Sato,M., Yamagiwa,H., Kimura,T., Yasui,N., Ochi,T., Endo,N., Kitamura,Y., Kishimoto,T., and Komori,T. 1999 Maturational disturbance of chondrocytes in Cbfa1-deficient mice. Dev.Dyn. 214: 279-290.

[27] Pardali,E., Xie,X.Q., Tsapogas,P., Itoh,S., Arvanitidis,K., Heldin,C.H., Ten,D., Grundström,T., and Sideras,P. 2000 Smad and AML proteins synergistically confer transforming growth factor β1 responsiveness to human germ-line IgA genes. J.Biol.Chem. 275: 3552-3560.

[28] Zhang,Y.W., Yasui,N., Ito,K., Huang,G., Fujii,M., Hanai,J.I., Nogami,H., Ochi,T., Miyazono,K., and Ito,Y. 2000 A RUNX2/PEBP2α A/CBFA1 mutation displaying impaired transactivation and smad interaction in cleidocranial dysplasia. Proc.Natl.Acad.Sci.USA 97: 10549-10554.

[29] Volk,S.W., LuValle,P., Leask,T., and Leboy,P.S. 1998 A BMP responsive transcriptional region in the chicken type X collagen gene. J.Bone Miner.Res. 13: 1521-1529.

[30] Volk,S.W., D'Angelo,M., Diefenderfer,D., and Leboy,P.S. 2000 Utilization of BMP receptors during chondrocyte hypertrophy. J.Bone Miner.Res. 15: 1630-1639.

[31] Zou,H.Y., Wieser,R., Massagué,J., and Niswander,L. 1997 Distinct roles of type I BMP receptors in the formation and differentiation of cartilage. Genes & Development 11: 2191-2203.

[32] Zou,H.Y. and Niswander,L. 1996 Requirement for BMP signaling in interdigital apoptosis and scale formation. Science 272: 738-741.

[33] Leboy,P.S., Grasso-Knight,G., D'Angelo,M., Volk,S.W., Lian,J.B., Drissi,H., Stein,G.S., and Adams,S.L. 2001 Smad-Runx interactions during chondrocyte maturation. J.Bone Joint Surg.[Am.] 83A: S15-S22.

[34] Koyama,E., Golden,E.B., Kirsch,T., Adams,S.L., Chandraratna,R.A., Michaille,J.J., and Pacifici,M. 1999 Retinoid signaling is required for chondrocyte maturation and endochondral bone formation during limb skeletogenesis. Dev.Biol 208: 375-391.

[35] Iwamoto,M., Yagami,K., Shapiro,I.M., Leboy,P.S., Adams,S.L., and Pacifici,M. 1994 Retinoic acid is a major regulator of chondrocyte maturation and matrix mineralization. Microsc.Res.Tech. 28: 483-491.

[36] D'Angelo,M., Yan,Z., Nooreyazdan,M., Pacifici,M., Sarment,D.P., Billings,P.C., and Leboy,P.S. 2000 MMP-13 is induced during chondrocyte hypertrophy. J.Cell.Biochem. 77: 678-693.

[37] Enomoto,H., Enomoto-Iwamoto,M., Iwamoto,M., Nomura,S., Himeno,M., Kitamura,Y., Kishimoto,T., and Komori,T. 2000 Cbfa1 is a positive regulatory factor in chondrocyte maturation. J.Biol.Chem. 275: 8695-8702.

[38] Ueta,C., Iwamoto,M., Kanatani,N., Yoshida,C., Liu,Y., Enomoto-Iwamoto,M., Ohmori,T., Enomoto,H., Nakata,K., Takada,K., Kurisu,K., and Komori,T. 2001 Skeletal malformations caused by overexpression of Cbfa1 or its dominant negative form in chondrocytes. J.Cell Biol. 153: 87-99.

[39] Ducy,P., Zhang,R., Geoffroy,V., Ridall,A.L., and Karsenty,G. 1997 Osf2/Cbfa1: A transcriptional activator of osteoblast differentiation. Cell 89 : 747-754.

[40] Kameda,T., Koike,C., Saitoh,K., Kuroiwa,A., and Iba,H. (1999) Developmental patterning in chondrocytic cultures by morphogenic gradients: BMP induces expression of Indian hedgehog and Noggin. Genes to Cells 4, 175-184

[41] Takeda,S., Bonnamy,J.P., Owen,M.J., Ducy,P., and Karsenty,G. 2001 Continuous expression of Cbfa1 in nonhypertrophic chondrocytes uncovers its ability to induce hypertrophic chondrocyte differentiation and partially rescues Cbfa1-deficient mice. Genes Dev. 15: 467-481.

Dual Roles of the Wnt Antagonist, Frzb-1 in Cartilage Development

Motomi Enomoto-Iwamoto[1], Jirouta Kitagaki[1], Eiki Koyama[3], Yoshihiro Tamamura[2], Naoko, Kanatani[4], Toshihisa Komori[4], Tsutomu Nohno[5], Maurizio Pacifici[3] and Masahiro Iwamoto[2]

Departments of [1]Molecular, Cell and Tumor Biology and [2]Oral Anatomy & Developmental Biology, Osaka University Faculty of Dentistry, Osaka 565-0871, Japan; [3]Department of Anatomy and Histology, School of Dental Medicine, University of Pennsylvania, PA 19104, USA; [4]Department of Molecular Medicine, Osaka University Medical School, Suita, Osaka 565-0871, Japan; [5]Department of Molecular Biology, Kawasaki Medical School, Kurashiki 701-0192, Japan.

Abstract. Members of the Wnt family, secreted signaling proteins fulfill important functions in various developmental processes. These proteins exert their action by binding to Frizzled receptors. One of the intercellular Wnt signals is transduced by the interaction of β-catenin with TCF/LEF transcription factors. Frzb-1, a secreted form of Frizzled, antagonizes some Wnt actions. Frzb-1 and several members of the Wnt family are expressed in developing cartilage during skeletogenesis, suggesting that Wnt signals play a role in chondrogenesis. However, much remains unclear about the function of Wnt proteins and Frzb-1 in skeletal cells. In this study, we examined Frzb-1 roles in regulating cartilage development. Frzb-1 expression was initially found in cells undergoing chondrogenesis, and was strongly expressed in epiphyseal articular chondrocytes and prehypertrophic chondrocytes. Frzb-1 was sharply downregulated in hypertrophic chondrocytes of the growth plate, whereas RCAS-driven Wnt 8 expression inhibited chondrogenic differentiation in limb bud cultures; in addition inhibition was blocked by co-expression of Frzb-1. In cultured chondrocytes overexpression of Frzb-1, or a dominant-negative form of LEF-1, suppressed cell maturation, while overexpression of Wnt-8 or a constitutively-active LEF-1 strongly promoted maturation. Our findings demonstrate that Frzb-1 is a powerful and direct modulator of the chondrocyte phenotype. The data imply that Wnt signaling is involved in inhibition of chondrogenesis and the progression of endochondral ossification, and that Frzb-1 may negatively regulate both these events.

Introduction

The Wnt gene family, consisting of more than 17 related gene members play important roles in pattern formation and organogenesis during both vertebrate and invertebrate embryonic development (for reviews see [1-3]). The receptors of Wnt proteins, Frizzled(s), encode a seven-spanning transmembrane domain and a cysteine-rich domain (CRD) in the extracellular domain, of the protein. Wnt signaling is transmitted in at least two ways, one of which is dependent on, and the other independent of β-catenin (for reviews see [1,2,4,5]). The β-catenin-dependent pathway requires release of β-catenin from a complex with glycogen synthase kinase 3 β, Axin, and adenomatus polyosis coli (APC) tumor suppressor proteins, and by stabilization of β-catenin inactivation of the ubiquitin-proteasome pathway [1,2,4]. The stabilized β-catenin interacts with the lymphoid enhancer factor-1/T-cell factor (LEF-1/TCF) family of transcription factors and translocates to the nucleus, where it activates specific target genes [1,2,4].

Secreted proteins that are capable of binding to Wnt molecules and compete with Wnt actions are present in mammals [6-9], birds [10,11] and frogs [7,12]. These proteins share sequence similarity with the CRD domain of the Frizzled receptors, but lack the transmembrane domain. It is likely that these proteins are secreted in soluble forms and inhibit the binding of Wnt to Frizzled in a competitive manner. Among these soluble forms of Frizzled proteins, Frzb-1 (also called sFrp3) [7,8,12] is one of the most characterized molecules. Frzb-1 was first identified, and purified from, bovine cartilage extracts [8], and expression of Frzb-1 was found to be associated with chondrogenesis in the developing limb [10,11,13,14]. Accordingly, Frzb-1 may antagonistically regulate Wnt signaling in cartilage development. In addition, several Wnt molecules including Wnt- 4 [15], Wnt-5a [15,16], Wnt-7a [17,18] and Wnt-14 [19] have been shown to be implicated in limb patterning and cartilage development. The objective of this study was to investigate the action of Wnt signals on chondrocytes and the involvement of Frzb-1 in regulation of Wnt actions in chondrogenesis.

Materials and Methods

In Situ Hybridization

Tissue section in situ hybridization was carried out using digoxigenin-conjugated to ^{35}S -labeled riboprobes as described previously [20]. The chick Frzb-1 clone has been described elsewhere [11].

Construction of Frzb-1, Wnt-8 and LEF-1 Mutant Viral Vectors

Full length murine Frzb-1(sfrp3) (kindly provided by Dr. J. Nathans, Johns Hopkins University Medical School, MD) [9] and mouse Wnt-8 cDNA (kindly provided by Dr. P. Chambon, IGBMC, Strasbourg, France) [21] were subcloned into RCASBP(A) and RCASBP(B) retroviral vectors [22] (provided by Dr. C. L. Cepko, Harvard Medical School, MA), respectively. Sequences encoding DN-LEF, which lacks amino acids 7-264 of murine LEF-1, and CA-LEF, which includes amino acids 695-781 of β-catenin fused to the C-terminus of DN-LEF, were subcloned into the RCAS (A) vector. Both LEF constructs were kindly provided by Dr. A. Hecht (Max-Plank-Institute of Immunology, Germany) [23]. Recombinant viral particles were prepared in chick dermal fibroblasts as described previously [24].

Histochemistry and Immunochemistry

Proteoglycan accumulation in cell layers was histochemically visualized as described elswhere [24]. Alkaline phosphatase (ALPase) was determined as described earlier. For immunocytochemistry, chondrocyte cultures were fixed in 3.7% formaldehyde, permeabilized with 0.05% triton-X 100 in PBS, and incubated with a 1:200 dilution of rabbit polyclonal antibodies raised against a synthetic peptide of human β-catenin (amino acids 680-781) (Santa Cruz Biotechnology Inc., Santa Cruz, CA). The bound antibodies were visualized by incubation with a biotinylated secondary antibody (Vector Laboratories, Burlingame CA) followed by Cy3-conjugated Streptavidin (Jackson ImmunoResearch, West Grove, PA).

Cell Cultures and Viral Infection

Cells were isolated from limb buds of 4-day-old chick embryos. 4 x 10^5 cells in 20μl were spotted on 6-well plates and cultured in Ham's F-12 medium containing 10% fetal bovine serum (FBS). Chondrocytes were isolated from sterna of 17-day-old embryos and cultured in high-glucose DMEM containing 10% FBS as described previously [24]. Where indicated, the freshly isolated limb bud cells or chondrocytes were infected with concentrated viral particles.

Results

Expression of Frzb-1 and Wnts in Developing Cartilage

As reported previously, Frzb-1 is initially expressed in the central core of limb buds, where chondrogenesis begins, and then becomes localized to the precartilaginous condensed mesenchyme and cartilage elements [10,11]. Distribution of Frzb-1 transcripts during chondrocyte differentiation was examined in detail using 10-day-old chick embryo metatarsus. Frzb-1 transcripts were intensely localized to the epiphysis of the metatarsus and also found in the prehypertrophic chondrocytes, whereas Frzb-1 expression was sharply downregulated in hypertrophic chondrocytes (Fig. 1).

To investigate which Wnt species interact with Frzb-1, we isolated total RNA from the sternal cartilage of 16-day-old chick embryos and carried out RT-PCR with Wnt specific primers [25]. When the amplified cDNAs were sequenced, it was noted that chick cartilage expresses at least five kinds of Wnt molecules: Wnt-4, -5b, -7a, -8 and -11, of which only Wnt-8 has been reported to bind and compete with Frzb-1 [7,12,26].

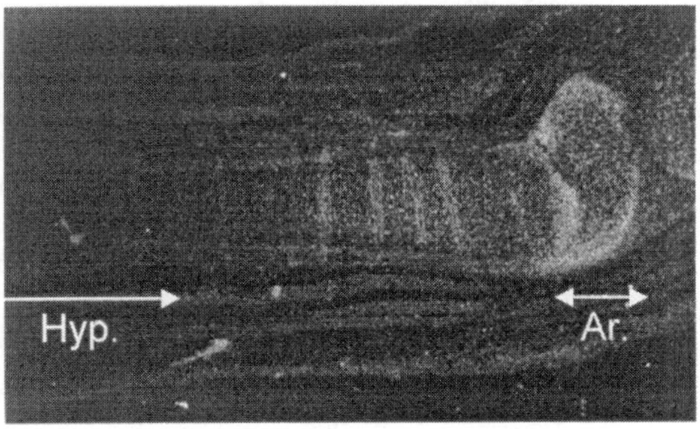

Figure 1. Gene expression of Frzb-1 in the developing limb. The longitudinal section of Day 10 metatarsus was processed for in situ hybridization to detect Frzb-1 gene. The signals were detected from articular cartilage through prehypertrophic zone of the growth plate, but sharply disappeared in the hypertrophic zone.

Forced Expression of Frzb-1 and Wnt 8 in Limb Bud Cell Cultures

The data described above suggested that Frzb-1 may play some part in the initiation of chondrogenesis and chondrocyte maturation during endochondral ossification. To investigate the role of Frzb-1 and Wnt signals in cartilage development, we forced the expression of Frzb-1 and Wnt-8, (a known target of Frzb-1) in limb bud cells and chondrocytes in culture. First, we infected limb bud cells with the virus encoding mouse

Wnt-8 and/or mouse Frzb-1. The Wnt-8 virus-infected cultures (Fig. 2C) showed much less Alcian Blue staining than the control cultures (Fig. 2A), suggesting that Wnt-8 inhibited proteoglycan accumulation. Frzb-1 virus-infected cultures exhibited slightly stronger staining (Fig. 2B). We noted that the inhibition of accumulation of proteoglycan by Wnt-8 virus infection was overcome by co-infection with the Frzb-1 virus (Fig. 2D). These findings indicate that Wnt-8 inhibits chondrogenic differentiation in limb bud cultures and that Frzb-1 counteracts this inhibitory effect of Wnt-8. The inhibition of chondrogenic differentiation by Wnt-8 was confirmed by measurement of ^{35}S-sulfate incorporation into glycosaminoglycan (GAG) and RT-PCR analysis of the expression of chondrocyte phenotypic markers, including sox 9, type IX collagen and Indian hedgehog.

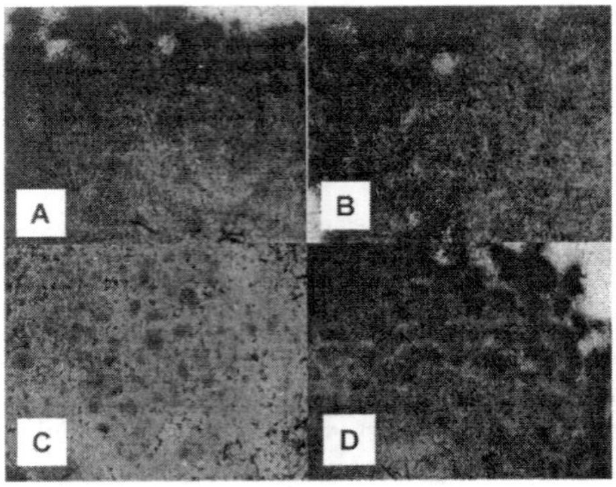

Figure 2. Effects of Frzb-1 and Wnt-8 on proteoglycan accumulation in limb bud cell cultures. Alcian blue staining of limb bud cultures infected with Frzb-1 (B), Wnt-8 (C) or both (D) viruses. Control culture (A) was infected with insert-less virus.

Forced Expression of Frzb-1 and Wnt-8 in Chondrocyte Cultures

To investigate the role of Frzb-1 and Wnts in chondrocyte differentiation, we next introduced Wnt-8 and Frzb-1 into primary cultured chondrocytes. The Frzb-1-virus-infected cells became rounder than the control cells, whereas Wnt-8 virus-infected cells became flatter. Cells infected with both Frzb-1 and Wnt-8 virus were similar to the control cells. To evaluate the phenotype, we measured the GAG content and ALPase activity of the cultures. Wnt-8-virus-infected cultures showed lower level of GAG content than that of the control cultures (Fig. 3A). Frzb-1-virus-infected cultures contained more GAG than the control cultures (Fig. 3A). Wnt-8 virus infected cultures showed a much higher level of ALPase activity than controls (Fig. 3B), while Frzb-1 virus infection inhibited ALPase activity (Fig. 3B). Wnt-8 also strongly promoted calcification, while Frzb-1 inhibited matrix mineralization (data not shown).

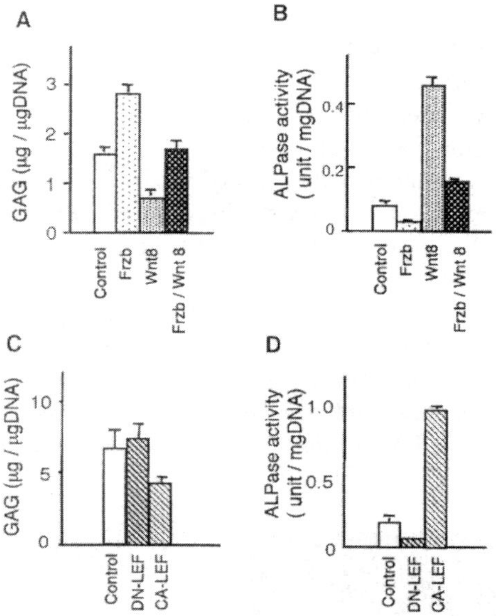

Figure 3. Effects of Frzb-1, Wnt-8 and LEF-β-catenin mutants on chondrocyte maturation in culture. A, GAG content in 2 week-old cultures infected with Frzb-1, Wnt-8 or both viruses. B, ALPase activity in 2 week-old cultures infected with Frzb-1, Wnt-8 or both viruses. C, GAG content in 2 week-old cultures infected with DN-LEF or CA-LEF viruses. D, ALPase activity in 2 week-old cultures infected with DN-LEF or CA-LEF viruses. Control cultures were infected with insert-less virus.

Expression of DN-LEF and CA-LEF Chimeric Molecule in Chondrocyte Cultures

We next asked whether the effect of Wnt-8 on chondrocytes are mediated by LEF-1/TCF-β-catenin signaling pathway. To activate or inactivate this signaling pathway, we virally introduced two kinds of LEF-1 mutants into chondrocytes. One is a dominant negative form of LEF-1 (DN-LEF) which lacks the N-terminal portion of mouse LEF-1, the binding site to β-catenin, while the other is a constitutive active form of LEF-1-β-catenin, a fusion construct consisting of murine β-catenin amino acids 695-781 to the C-terminus of a truncated form of mouse LEF-1 (CA-LEF) [23]. In Xenopus, CA-LEF generates the same response to Wnt-8 signals and DN-LEF blocks the Wnt-8-response [23]. Chondrocytes infected with virus encoding DN-LEF became round in shape; in contrast, cells infected with the virus encoding CA-LEF became flatter. Infection with the CA-LEF-virus decreased GAG accumulation and stimulated ALPase activity in chondrocytes, whereas infection with the DN-LEF-virus inhibited ALPase activity (Fig. 3C and D). Furthermore, the CA-LEF cultures induced matrix calcification, while the DN-LEF cultures showed less calcification than the control.

β-Catenin Nuclear Translocation During Chondrocyte Maturation

Since Wnt-8 and the β-catenin-dependent pathways favor chondrocyte maturation, one would expect to see changes in the distribution of β-catenin during chondrocyte maturation. To test this possibility, we determined immunohistochemically the distribution of β-catenin in chondrocytes undergoing maturation over time. A striking change in β-catenin distribution was seen in cultured chondrocytes (Fig. 4). The protein was largely present in the cytoplasm of proliferating and matrix-synthesizing chondrocytes on Days 5 and 14 of the cultures (Fig. 4A and C, respectively), but shifted to a nuclear distribution in the older hypertrophic mineralizing Day-28 cultures (Fig. 4E, arrows). We also determined the distribution of β-catenin in the growth plates of developing chick long bones. Stained chondrocytes was observed in the proliferative, prehypertrophic and hypertrophic zones of the growth plate. At higher magnification, however, it became clear that the bulk of β-catenin in articular, proliferative and prehypertrophic cells was located in the cytoplasm, whereas in hypertrophic chondrocytes much of the β-catenine was nuclear.

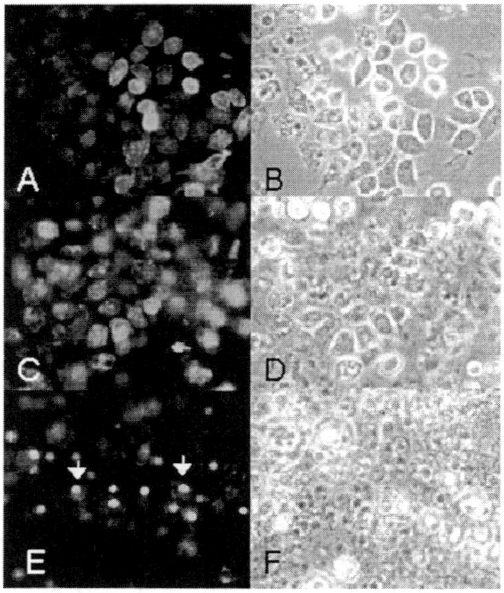

Figure 4. Localization of β-catenin in cultured chondrocytes. Day 5 (A and B), Day 14 (C and D) and Day 28 (E and F) cultures were processed for immunostaining with β-catenin antibodies. A, C and E were immunofluorescence images. B, D and F were phase contrast images corresponding to A, C and E, respectively.

Discussion

Regulation of Chondrogenesis by Wnts and Frzb-1

It has been reported that Frzb-1 is selectively expressed in precartilage and bovine cartilage extracts. Because Frzb-1 antagonizes Wnt signals, it has been hypothesized that during chondrogenic differentiation, Frzb-1 may play some role in the modulation of Wnt action. Indeed, Wnt-1 and Wnt-7a have been shown to inhibit chondrogenesis in limb bud cultures

[17, 18]. In the present study, we demonstrated that Wnt-8 also inhibited chondrogenic differentiation of limb bud cell cultures as determined by cartilage-specific proteoglycan synthesis and expression of sox 9, type IX collagen and Ihh (not shown). We also showed that co-expression of Frzb-1 counteracted the inhibitory effects of Wnt-8. It is noteworthy that in the induction of secondary axes in Xenopus, Frzb-1 can compete with Wnt-1 and Wnt-8, but not Wnts-3a, 5a and 11 [26]. Importantly, our data indicates that there is a competitive relationship in the induction of chondrogenesis between Frzb-1 and Wnt-8. We conclude that as a result of release from the Wnt signal control that inhibits chondrogenic differentiation up-regulation of Frzb-1 in condensing mesenchyme stimulates differentiation of mesenchymal cells into chondrocytes.

β-Catenin-Mediated-Wnt Signals Stimulate Chondrocyte Maturation

Our finding that Frzb-1 was sharply down-regulated in hypertrophic chondrocytes is in good agreement with previous reports [8]. The reduction in Frzb-1 expression in hypertrophic chondrocytes was a common feature at any stage or site in developing limbs that we examined. Overexpression of Wnt-8 stimulated expression of the hypertrophic phenotype of chondrocytes, characterized by an increase in ALPase activity and matrix calcification. Conversely, Frzb-1 forced-expression inhibited the elevation in phosphatase activity and mineral deposition. These findings suggest that Wnt signals promote hypertrophy and matrix calcification of chondrocytes, and Frzb-1 regulates the rate of chondrocyte maturation and prevents precocious matrix calcification. When chondrocytes Frzb-1 was overexpressed, the cells displayed a hypertrophic phenotype including elevation of ALPase activity and raised type X collagen expression. On this basis, inhibition of matrix calcification by Frzb-1 is not due to complete suppression of the hypertrophic phenotype. We have also found that Wnt-8 stimulates expression of MMP 2 and MMP13, while Frzb-1 inhibits expression of these genes. We have observed an increase in MMP activity at a late stage in endochondral ossification. Since the elevation in activity is a requirement for matrix calcification [27-29], Frzb-1 function may be closely related to the inhibition of protease activity.

Wnt-1 and Wnt-8 signals are mediated by interaction of β-catenin and members of the LEF/TCF family of transcription factors [23,30,31]. Further, it has been suggested that Wnt-4 signal activates the LEF-1/TCF-β-catenin pathway and promotes chondrocyte maturation [15]. In this study we demonstrated that the actions of Wnt-8 and Frzb-1 on chondrocytes can be mimicked by the activation or inactivation of LEF-1-β-catenin signaling, respectively. Further, in the hypertrophic region of the growth plate, β-catenin is translocated to the nucleus, while Frzb-1 expression is simultaneously downregulated. Accordingly, it is suggested that a Wnt signal, possibly not a Wnt-8 signal, mediated by LEF/TCF-β-catenin may play an important role in matrix remodeling and calcification and that this signal may be negatively regulated by Frzb-1.

Acknowledgement

We thank Dr. J. Nathans for providing murine sfrp3 cDNA, Dr. P. Chambon for mouse Wnt-8 cDNA, Dr. C. L. Cepko for RCASBP(A) and RCASBP(B) retroviral vectors and Dr. A. Hecht for LEF mutant constructs. This work was supported by grants-in-aid for scientific research from the Ministry of Education, Science and Culture of Japan (to M.E.-I. and M.I.) and by NIH grants AR40833 (to M.P.).

References

[1] Cadigan KM, Nusse R 1997 Wnt signaling: a common theme in animal development. Genes Dev 11(24):3286-305.

[2] Moon RT, Brown JD, Torres M 1997 WNTs modulate cell fate and behavior during vertebrate development. Trends Genet. 13:157-162.

[3] Nusse R, Varmus HE 1992 Wnt genes. Cell 69(7):1073-87.

[4] Eastman Q, Grosschedl R 1999 Regulation of LEF-1/TCF transcription factors by Wnt and other signals. Curr Opin Cell Biol 11(2):233-40.

[5] Miller JR, Hocking AM, Brown JD, Moon RT 1999 Mechanism and function of signal transduction by the Wnt/beta-catenin and Wnt/Ca2+ pathways. Oncogene 18(55):7860-72.

[6] Finch PW, He X, Kelley MJ, Uren A, Schaudies RP, Popescu NC, Rudikoff S, Aaronson SA, Varmus HE, Rubin JS 1997 Purification and molecular cloning of a secreted, Frizzled-related antagonist of Wnt action. Proc Natl Acad Sci U S A 94(13):6770-5.

[7] Leyns L, Bouwmeester T, Kim SH, Piccolo S, De Robertis EM 1997 Frzb-1 is a secreted antagonist of Wnt signaling expressed in the Spemann organizer. Cell 88(6):747-56.

[8] Hoang B, Moos M, Jr., Vukicevic S, Luyten FP 1996 Primary structure and tissue distribution of FRZB, a novel protein related to Drosophila frizzled, suggest a role in skeletal morphogenesis. J Biol Chem 271(42):26131-7.

[9] Rattner A, Hsieh JC, Smallwood PM, Gilbert DJ, Copeland NG, Jenkins NA, Nathans J 1997 A family of secreted proteins contains homology to the cysteine-rich ligand-binding domain of frizzled receptors. Proc Natl Acad Sci U S A 94(7):2859-63.

[10] Ladher RK, Church VL, Allen S, Robson L, Abdelfattah A, Brown NA, Hattersley G, Rosen V, Luyten FP, Dale L, Francis-West PH 2000 Cloning and expression of the Wnt antagonists Sfrp-2 and Frzb during chick development. Dev Biol 218(2):183-98.

[11] Wada N, Kawakami Y, Ladher R, Francis-West PH, Nohno T 1999 Involvement of Frzb-1 in mesenchymal condensation and cartilage differentiation in the chick limb bud. Int J Dev Biol 43(6):495-500.

[12] Wang S, Krinks M, Lin K, Luyten FP, Moos M, Jr. 1997 Frzb, a secreted protein expressed in the Spemann organizer, binds and inhibits Wnt-8. Cell 88(6):757-66.

[13] Baranski M, Berdougo E, Sandler JS, Darnell DK, Burrus LW 2000 The dynamic expression pattern of frzb-1 suggests multiple roles in chick development. Dev Biol 217(1):25-41.

[14] Duprez D, Leyns L, Bonnin MA, Lapointe F, Etchevers H, De Robertis EM, Le Douarin N 1999 Expression of Frzb-1 during chick development. Mech Dev 89(1-2):179-83.

[15] Hartmann C, Tabin CJ 2000 Dual roles of wnt signaling during chondrogenesis in the chicken limb [In Process Citation]. Development 127(14):3141-59.

[16] Kawakami Y, Wada N, Nishimatsu SI, Ishikawa T, Noji S, Nohno T 1999 Involvement of Wnt-5a in chondrogenic pattern formation in the chick limb bud. Dev Growth Differ 41(1):29-40.

[17] Parr BA, McMahon AP 1995 Dorsalizing signal Wnt-7a required for normal polarity of D-V and A-P axes of mouse limb. Nature 374(6520):350-3.

[18] Rudnicki JA, Brown AM 1997 Inhibition of chondrogenesis by Wnt gene expression in vivo and in vitro. Dev. Biol. 185:104-118.

[19] Hartmann C, Tabin CJ 2001 Wnt-14 plays a pivotal role in inducing synovial joint formation in the developing appendicular skeleton. Cell 104:341-352.

[20] Koyama E, Leatherman JL, Shimazu A, Nah HD, Pacifici M 1995 Syndecan-3, tenascin-C, and the development of cartilaginous skeletal elements and joints in chick limbs. Dev Dyn 203(2):152-62.

[21] Bouillet P, Oulad-Abdelghani M, Ward SJ, Bronner S, Chambon P, Dolle P 1996 A new mouse member of the Wnt gene family, mWnt-8, is expressed during early embryogenesis and is ectopically induced by retinoic acid. Mech Dev 58(1-2):141-52.

[22] Hughes SH, Greenhouse JJ, Petropoulos CJ, Sutrave P 1987 Adaptor plasmids simpllify the insertion of foreign DNA into helper-independent retroviral vectors. J Virol 61:3004-3012.

[23] Vleminckx K, Kemler R, Hecht A 1999 The C-terminal transactivation domain of beta-catenin is necessary and sufficient for signaling by the LEF-1/beta-catenin complex in Xenopus laevis. Mech Dev 81(1-2):65-74.

[24] Enomoto-Iwamoto M, Iwamoto M, Mukudai Y, Kawakami Y, Nohno T, Higuchi Y, Takemoto S, Ohuchi H, Noji S, Kurisu K 1998 Bone morphogenetic protein signaling is required for maintenance of differentiated phenotype, control of proliferation, and hypertrophy in chondrocytes. J Cell Biol 140(2):409-18.

[25] Tanda N, Kawakami Y, Saito T, Noji S, Nohno T 1995 Cloning and characterization of Wnt-4 and Wnt-11 cDNAs from chick embryo. DNA Seq 5(5):277-81.

[26] Wang S, Krinks M, Moos M, Jr. 1997 Frzb-1, an antagonist of Wnt-1 and Wnt-8, does not block signaling by Wnts -3A, -5A, or -11. Biochem Biophys Res Commun 236(2):502-4.

[27] Vu TH, Shipley JM, Bergers G, Berger JE, Helms JA, Hanahan D, Shapiro SD, Senior RM, Werb Z 1998 MMP-9/gelatinase B is a key regulator of growth plate angiogenesis and apoptosis of hypertrophic chondrocytes. Cell 93(3):411-22.

[28] Holmbeck K, Bianco P, Caterina J, Yamada S, Kromer M, Kuznetsov SA, Mankani M, Robey PG, Poole AR, Pidoux I, Ward JM, Birkedal-Hansen H 1999 MT1-MMP-deficient mice develop dwarfism, osteopenia, arthritis, and connective tissue disease due to inadequate collagen turnover. Cell 99(1):81-92.

[29] Inada M, Yasui T, Nomura S, Miyake S, Deguchi K, Himeno M, Sato M, Yamagiwa H, Kimura T, Yasui N, Ochi T, Endo N, Kitamura Y, Kishimoto T, Komori T 1999 Maturational disturbance of chondrocytes in Cbfa1-deficient mice. Dev Dyn 214(4):279-90.

[30] Behrens J, von Kries JP, Kuhl M, Bruhn L, Wedlich D, Grosschedl R, Birchmeier W 1996 Functional interaction of beta-catenin with the transcription factor LEF-1. Nature 382(6592):638-42.

[31] Papkoff J, Aikawa M 1998 WNT-1 and HGF regulate GSK3 beta activity and beta-catenin signaling in mammary epithelial cells. Biochem Biophys Res Commun 247(3):851-8.

The Growth Plate
I.M. Shapiro et al. (Eds.)
IOS Press, 2002

Chondrocyte Kinetics in the Growth Plate

Cornelia E. Farnum[1] and Norman J. Wilsman[2]
[1]*Department of Biomedical Sciences, Cornell University, Ithaca, N.Y., 14853;* [2]*Department of Comparative Biosciences, University of Wisconsin-Madison, Madison, WI, 53706*

Abstract. In this review a summary is made of studies in the last 15 years that have used stereological approaches to examine the multiple chondrocytic kinetic parameters during the proliferative and hypertrophic stages of the chondrocytic differentiation cascade that quantitatively contribute to the total bone growth achieved by a given growth plate. The focus is primarily on understanding differential growth of multiple growth plates at one point in time, or through time. Secondly, three significant transition points during the differentiation cascade are examined – reserve/proliferative, proliferative/hypertrophic and hypertrophic/ apoptosis. Currently there is significant understanding of the multiple 'players' at these transitions and their upstream and downstream regulators. This knowledge has been gained primarily during the past five years through analyses of transgenic animals with ever increasing subtleties to the specific constructs made. However, there is still minimal understanding of what regulates the *kinetics* of transition at these points. Finally, a consideration is given to possibilities of how growth plate chondrocytes integrate the multiple cues from their environment to effectively complete their differentiation resulting in longitudinal growth.

During postnatal bone development the differentiation cascade of growth plate chondrocytes is translated into longitudinal growth. Although there have been variations in the number of explicit stages of differentiation described, most simply one can define a stage of clonal expansion as proliferative chondrocytes followed by terminal differentiation characterized by a significant volume increase historically described as hypertrophy. In mammalian growth plates, the terminal hypertrophic chondrocyte of each column dies*, leaving both a space for the penetration of endothelial cells and osteoprogenitor cells, as well as calcified longitudinal septae that are the scaffold for new bone deposition (recently reviewed in [2,3,4]). The fundamental unit of growth is the column of differentiating chondrocytes, arranged spatially such that the alignment mirrors the temporal sequence of differentiation, parallel to the direction of growth of the bone. A territorial matrix unites a column as a unit, separated from adjacent columns by an interterritorial matrix which calcifies in the distal hypertrophic cell zone. In addition, each individual chondrocyte is surrounded by a pericellular matrix which is an interface between the cell membrane and the external matrix environment.

In this paper three questions are addressed: 1) What is known about cellular kinetics during proliferation and hypertrophy that result quantitatively in longitudinal growth? 2) Do columns of chondrocytes represent a differentiation cascade that is a continuum of development in which each cell differs both in space *and* time from its adjacent neighbors, or are individual columns best thought of as populations of cells separated from adjacent

* It is recognized that there is still considerable controversy about this point. For the purposes of this paper, it will be assumed that death is, in fact, the fate of the terminal hypertrophic chondrocyte. An excellent recent paper [1] reviews the conflicting interpretations of data on this point.

populations with gates and/or switches? 3) What information do we have – from experimental data, from the analysis of transgenic animals, or from our understanding of the pathogenesis of disease – that allows discrimination between these models of the spatial/temporal organization of the growth plate?

One can study and understand multiple aspects of the neuron without ever being able to deduce the property of memory, which is an example of an emerging property of a complex system. Likewise, longitudinal growth should be considered as a property of a complex system – of chondrocytic activity in the growth plate as an organ with a functional blood supply in long bones of a living, moving animal. Multiple aspects of the cellular and molecular biology of growth plate chondroyctes can be studied effectively in *ex vivo* experimental systems, and similarly, analyses of the chondrocytic extracellular matrix can be made *ex vivo*, with or without the presence of the chondrocytes themselves However, longitudinal growth, and especially differential longitudinal growth, is best understood if analyses can be done maintaining the complexity of the system as a whole. The tools of quantitative stereology have long been an appropriate experimental method to achieve this goal [5]. Recently, the availability of transgenically modified animals with ever increasing subtleties of the specific modifications made has generated what can only be described as 'an explosion' of data for analysis of the regulatory mechanisms operating in the whole animal that result in longitudinal bone growth.

Cellular Kinetics During Proliferation and Hypertrophy Resulting in Longitudinal Growth

An intriguing aspect of postnatal longitudinal growth is that the instantaneous elongation rate of growth plates of long bones of the body is differential – that is, both at any one time, as well as through time, each growth plate is growing at a different rate. One important consequence of this is that the two growth plates at the end of a given long bone contribute differentially to the overall growth of the bone, both because at any one time they are growing at different rates and because through time they grow for different lengths of time. The differential contributions can be dramatically different – the distal ulnar growth plate in children contributes ~95% of the total length of the ulna [6]. Stereological analysis of differential growth in multiple growth plates at one point in time [7], or in the same growth plate through time [8], has provided a powerful experimental system for understanding how the kinetics of differentiation during proliferation and hypertrophy result quantitatively in longitudinal growth in the whole animal with no perturbation of the systemic or local environment in which growth is occurring. In these analyses, rate of growth over a defined period can be assessed by giving the animal a fluorescent calcium chelator, and characteristics of the cell cycle can be analyzed by the use of thymidine autoradiography or bromodeoxyuridine (BrdU) labeling [9,10].

Fig. 1 shows two murine growth plates growing at a 2.9X difference in rate. Although differences in cellular populations are immediately apparent, the two dimensional appearance of growth plates can be misleading. Stereological analysis allows the two dimensional image to be understood quantitatively as a three-dimensional representation, either of the entire growth plate volume or of the volume of an idealized cylinder of the growth plate [5,7]. This is the only way that the significance of parameters such as cellular volumes, cellular densities, and matrix volumes can be understood. A summary of multiple stereological studies of growth plates growing at different rates [7-18] yields the following generalizations as to the proportions of growth contributed during the proliferative *vs.* the hypertrophic phase, and the relative significance of cellular *vs.* matrix contributions to growth:

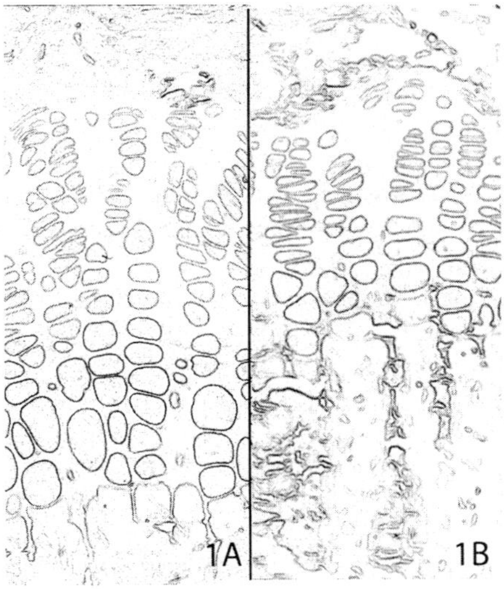

Figure 1. This is a comparison of the proximal tibial growth plate from mice of two different ages. The growth plate in 1A is from a five-week-old animal and is growing at ~70μm/day; 1B is from a nine-week-old animal growing at ~24μm/day. Although a difference in zonal heights and cellularity can clearly be noted, a three dimensional analysis is required to fully understand the kinetic parameters involved with converting the differentiation cascade into longitudinal growth. X450

Proliferative Cell Population

1. There is a high positive correlation between the number of cells in the proliferative pool and rate of growth;
2. Growth plates growing at different rates have significant differences in their cell cycle time, with the greatest difference being in the G1 phase; although cycle time in general has a positive correlation with rate of growth, growth plates can have nearly identical cell cycle times and yet have very different rates of growth;
3. Proliferative cell volume and density are different in growth plates growing at different rates – the slower the rate of growth the smaller the cellular volume and the higher the cellular density.

Hypertrophic Cell Population

1. There is a positive correlation between the number of cells in the hypertrophic phase and rate of growth; however, this is not as dramatic as the correlation for proliferative cells;
2. There is a strongly positively correlation between final hypertrophic cell volume and rate of growth;
3. The faster the rate of growth, the faster the rate of hypertrophy and the larger the final volume;
4. Hypertrophy is explosive once initiated and continues to increase until at least the penultimate cell;
5. A shape change accompanies the volume change such that cells increase disproportionately in height compared to width; this shape change is a critical

component of the hypertrophy process; in rapidly growing growth plates, ~80% of growth can be accounted for by the sum of the incremental height changes of cells during hypertrophy.

Matrix/Cellular Contributions

1. Cells modulate their activity primarily by altering the level of activity of intracellular organelles (RER, Golgi, mitochondria) rather than by an increase in the surface area or volume of organelle membranes;
2. Accompanying the volume change, there is a significant increase in microtubules, suggesting this might be significant either in signaling mechanisms or in effecting the shape change accompanying hypertrophy;
3. At all rates of growth, the contributions from cellular size changes are greater than the contributions from matrix production;
4. The slower the rate of growth, the more significant is the contribution from matrix;
5. Hypertrophic cells have a significantly larger matrix domain than proliferative cells, meaning that net matrix synthesis increases as the cells enter the hypertrophic phase;
6. Increased pericellular/territorial matrix volume makes a greater contribution to growth than does increased interterritorial matrix volume.

Proliferative/Hypertrophic Phase Contributions During Differential Growth

1. At all rates of growth, there are contributions during both the proliferative and the hypertrophic phases;
2. At all rates of growth, contributions are greater during the hypertrophic phase;
3. The faster the rate of growth, the larger per cent growth during the hypertrophic phase;
4. Over a wide range of growth rate, during a twenty-four hour period the number of new cells born in the proliferative cell zone equals the number of hypertrophic cells turned over at the chondro-osseous junction;
5. The faster the rate of growth, the faster the rate of differentiation – that is, significantly more cells are 'run through the system' each day.

In summary, there is significant understanding of chondrocytic kinetics in the proliferative and hypertrophic cell populations of mammalian growth plates and how they contribute quantitatively to differential growth.

As a conclusion, three points should be emphasized. First, to date almost all studies of this kind have been done on rodent growth plates, primarily the proximal tibial growth plate. This growth plate is ideal for the systematic sampling required for stereological studies because it is small, it is relatively flat, and different areas of the growth plate are growing essentially at the same rate [19]. However, in the young rat the proximal tibia is growing at an exceedingly rapid rate. Therefore, its growth probably is through a different mix of the relative significance of cellular hypertrophy *vs.* matrix production than occurs in growth plates of large mammals such as ourselves who grow over a long period of time. Secondly, this approach is a powerful way to analyze growth plate abnormalities either in specific diseases – including chondrodysplasias, metabolic or nutritional disease – or altered biomechanical environment. Several studies using this have recently appeared in the literature [19-21]. However, because of the time intensive nature of stereological analyses, they should not be undertaken in instances where derangement of the growth plate is extreme and a descriptive analysis suffices to understand the pathology involved. Third, to date the majority of studies of this kind have been done on mammalian growth plates. Avian growth plates, which can grow almost an order of magnitude faster than rodent growth plates, use a different 'strategy' to achieve this growth, in which the rate of differentiation is the primary variable, not the rate and extent of hypertrophy [22]. Therefore, generalizations derived from the study of rodent growth plates should be used cautiously in the interpretation of experiments involving avian bone growth, and even of *in vitro* experiments using avian cells in culture.

Spatial Organization of the Growth Plate and Transitions Between Populations

Longitudinal bone growth is achieved during clonal expansion following a stem cell division, represented spatially in the growth plate as a column of cells. The spatial alignment of cells within a column mirrors the temporal 'life history' of an individual chondrocyte in its differentiation cascade. Significant transition points can be identified during this differentiation cascade, and one can conceptualize that there are 'kinetics of transition' at these points, analogous to the cellular kinetics that have just been described within the proliferative and hypertrophic cell populations. In other words, there is evidence that there are points during the differentiation cascade that are controlled as gates of transition; the rate of transition and hence the amount of bone growth over a given period can be modulated by controlling the rates at which these transitions occur. In Fig. 2, significant transition points to consider have been identified as arrows between populations of cells.

The first transition, which is the point of initiation of stem cell expansion, can be defined morphologically by the change in cellular size, shape, and degree of columnation. However, essentially nothing is known about regulation at this transition and such fundamental questions exist as 1) what combination of hormonal/ paracrine/ autocrine/ environmental signals are significant to initiate clonal expansion; 2) does a given stem cell have a set or a variable number of divisions; 3) to what extent is the so-called 'reserve cell zone' the source of stem cells, given that, especially in large species, this zone contributes to growth of the secondary center and may have a structural role as well [23]; 4) to what extent is the *rate* of initiation of stem cell division significant, *i.e.*, are there controls that regulate the kinetics of this transition point [24]? Until the specific cells which are the true stem cells that create the daughter cells that represent the proliferative cells of a given column can be uniquely identified on tissue sections and can be uniquely isolated for study in cell culture, it is unlikely that this critical transition point will be able to be studied either experimentally or in disease states.

The second transition represents the point at which cellular proliferation ceases and cells initiate the volume increase and shape change associated with hypertrophy. Conceptually one could consider this as two steps – the end of cellular division followed by the initiation of volume increase – that may or may not be linked. Alternatively, one could consider this as a broad transition where the changes in gene expression associated with ending proliferation and beginning hypertrophy are inexorably linked [24]. The depiction in Fig. 2 begs the question between these alternatives with large arrows representing the former (two steps, on the left) and small arrows representing the latter (a broad transition, on the right). In the latter case, a 'pre-hypertrophic stage' can be defined, and this terminology is often used in the literature. In either case, in the last five years there has been an explosion in the understanding of regulators of this transition point that is comparable to the explosion of volume increase of the cells during this transition, beginning with the original paper by Vortkamp *et al.* in Science in 1996 and continuing unabated to the present [25]. Control at this point has been described as a 'gate' in which the *rate* of transition to hypertrophy is regulated by a complex negative feedback loop involving PTH/PTHrP, their receptors, and Ihh as major players and a host of upstream and down stream regulators and modifiers. The significance of this gate in prenatal development and perhaps in postnatal development as well has been clarified primarily through the analysis of transgenic animals. Recent reviews integrate the current knowledge, and the significance of this feedback loop in understanding several chondrodysplasias has been demonstrated [26-28]. The point to made here is that, analogous to the kinetic behavior of cells within the proliferative population, conceptually one must consider the kinetics of the transition to hypertrophy, the significance of the more than 50 genes that might be upregulated at this transition [29,30], and the potential of modulation of the kinetics at this transition point in affecting the overall amount of growth achieved.

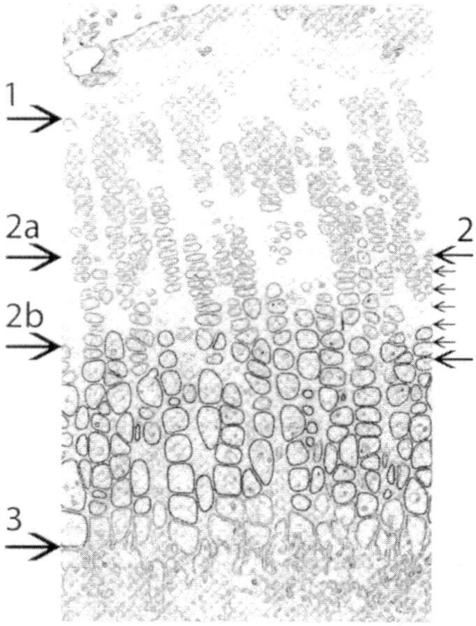

Figure 2. Arrows on the left side of this micrograph of the proximal tibial growth plate of a five-week-old rat indicate three major transitions between cellular populations in the differentiation cascade. Arrows on the right indicate that the transition between the end of proliferation and the beginning of hypertrophy can be conceptualized as a broad transition of a 'pre-hypertrophic' stage. X225

A final point of transition occurs at the chondro-osseous junction where the terminal hypertrophic cell dies, and endothelial cells accompanied by osteoprogenitor cells invade through the territorial matrix, eventually leading to remodeling of the calcified longitudinal septae and new bone formation. Clearly there is a 'kinetics of turnover' at this transition studied most thoroughly by a serial section analysis of swine growth plates in which it was demonstrated that, in a growth plate growing at ~140μm /day, ~5.4 hypertrophic cells turned over in 24 hours [31]. Just as rate of proliferation is a critical variable in understanding the amount of growth achieved by a given growth plate, a rate of turnover at the chondro-osseous junction also is critical; in the steady state, production and turnover cancel each other [7]. However, it should also be noted that production and turnover rates can be identical and yet very different total amounts of growth can be achieved, if the volume and height changes of hypertrophic cells in the two growth plates are different. In two growth plates each producing and turning over eight cells a day, if the proliferative cell heights (5μm) are identical but the final hypertrophic heights are 30μm and 25μm respectively, then $\sum\Delta$ height (hypertrophic – proliferative) X 8 = 200μm for one and only 160μm for the other, leading to significantly different amounts of daily growth.

As recently as three years ago, and continuing at an accelerating rate into the present, regulators of the rate of turnover at the chondro-osseous junction have been described, with a primary role played by VEGF and MMPs (see [32-35] for recent studies). Again, the power of the analyses comes primarily from the study of transgenic animals, although other approaches have also provided important information [36]. It had been known for several decades that vascularization can be delayed in some disease states, sometimes coupled with an absence of matrix calcification such as in rickets [37], and sometimes even in the presence of heavy mineralization of the matrix such as in the osteochondroses [38]. In the experimental models which have been described to date, the evidence is clear that chondrocyte turnover can be severely delayed by aberrations in the signaling pathways involving VEGF and the MMPs; whether it continues at a slow rate and whether there are significant prenatal/ postnatal differences is less clear. The point to be made is that at this transition there is a complex system of major regulators coupled with upstream and down stream modifiers that potentially control the *rate* at which cells die and vascularization proceeds (reviewed in [27]). The kinetics of this control point has significant effects on the total amount of growth that can be achieved by a given growth plate.

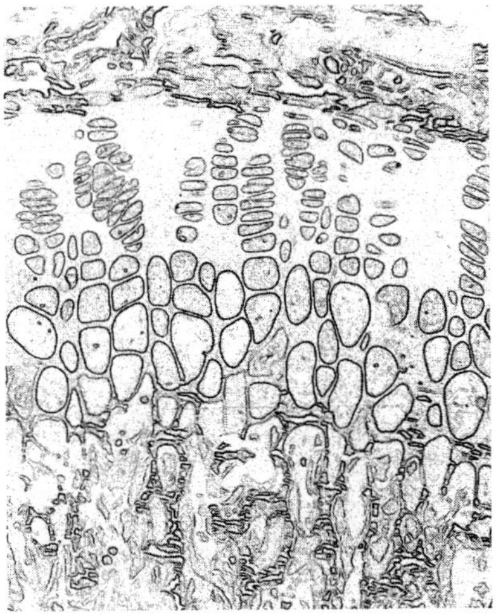

Figure 3. In this micrograph of a caudal vertebral growth plate from a six-week-old rat, the transition from proliferation to hypertrophy appears as an abrupt increase in cellular volume over the space of only one or two cells. X450

A Continuum of Differentiation or Discrete Stages with Gates?

In the preceding section evidence has been presented that would suggest that chondrocytic differentiation as seen in columns representing the clonal expansion of a stem cell can be conceptualized as two populations of cells, proliferative and hypertrophic, separated from each other, and from the stem cell population proximally and from apoptosis at the metaphyseal junction distally, by control points with complex gates and/or switches. It also has been emphasized that, when considering the potential of a growth plate to produce a given amount of daily longitudinal growth, the kinetics of activity *during* the stages of

Figure 4. In this micrograph of the distal radial growth plate of a seven-week-old puppy, the transition between cellular zones appears morphologically indistinct. Without additional specific markers there is an indistinct transition between proliferative and hypertrophic cell populations. X80

proliferation and hypertrophy are significant, *as well as* the kinetics of regulators of the transition points between populations. As visualized morphologically, the distinction of two populations of cells with abrupt transitions between them can be dramatic as seen in the vertebral rat growth plate in Fig. 3. Chondrocytes in this growth plate seem to explode into a volume increase over a transition of one or two cells. In contrast, when one looks at the puppy growth plate in Fig. 4, chondrocytes of a given column seem to lazily enjoy their transition to hypertrophy. If one were to draw transition points for this growth plate without the use of specific markers such as PTHrP or collagen X expression, it would be arbitrary where the population divisions would be marked. The growth plate in Fig. 2 is between these two extremes.

Since the columns of growth plate chondrocytes are linearily arranged, it is logical to think that each position of a cell in the column is unique and that the progression of the cells within the column is a spatial representation of the temporal differentiation of all chondrocytes. If this were true, it would be logical to think of essentially linearly acting regulatory signaling pathways. But if one were to ask the question, "Does each cell in a column differ both in space *and* time from its adjacent neighbors?", the answer would be different for proliferative and hypertrophic cells. During hypertrophy it seems apparent that, once hypertrophy is initiated, each cell continues on a course of volume increase that reaches a maximum just before apoptosis. In the absence of appropriate timing of vascular penetration, however, cellular volume does not continue to increase, but rather remains stable at the level characteristic of that growth plate at that rate of growth. This provides evidence that during hypertrophy volume increase is regulated by mechanisms that are independent of the signaling for apoptosis and/or vascular penetration. The point is that it can be assumed that the spatial representation of chondrocytes in columns in the hypertrophic cell zone is a mirror of the temporal terminal differentiation of an individual cell.

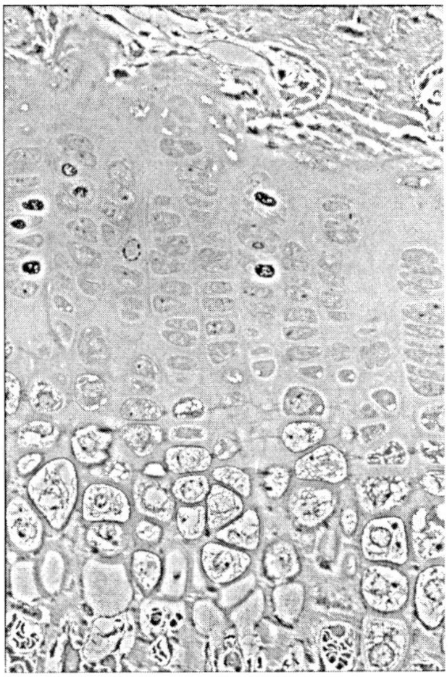

Figure 5. This is a phase micrograph of the proximal tibial growth plate of a five-week-old rat labeled 30 minutes prior to euthanasia with BrdU, then localized with a monoclonal antibody. Note that positive cells (darkly labeled) are more numerous proximally in the growth plate and there is a lack of reaction product in many cells which, by their axial ratio of width to height, would still be considered to be in the proliferative zone. X450

By contrast, the spatial presentation of cells in the proliferative population is *not* a mirror of the temporal differentiation of an individual cell. Rather, it is an interspersion of cells of different ages since new cells are born every time a proliferative cell undergoes a division. There is good evidence from studies using either single or repeated pulse labeling of BrdU and following the pulse(s) over time [9,10] that 1) all cells in a column are capable of division; 2) cells positioned proximally in the proliferative cell zone divide more rapidly than cells positioned distally (see Fig. 5; and [3]) distal neighbors of the most distal labeled cell have not yet started to hypertrophy. The conclusion is that there is variation in the number of times individual daughter cells divide, resulting in an 'age distribution' of proliferative cells in the column. Some cells may have divided three or four times before initiating hypertrophy; at the other extreme, after the last division of a given cell, the daughter cell produced may never divide before initiating hypertrophy.

Given these differences in spatial/temporal positioning of cells in these two populations, one can ask the question, how do individual chondrocytes integrate cues that ultimately result in the synchronized differentiation of the entire column resulting in longitudinal growth? Growth plate chondrocytes are influenced by a complex panoply of players, some stimulatory and some inhibitory. Multiple systemic hormones affect longitudinal growth of which the most significant are growth hormone and IGF-1, thyroxine, glucocorticoids, insulin, and the sex steroids. The list of paracrine/ autocrine regulators is almost endless but includes as major players the BMPs, FGFs, TGF-β, IGFs, VEGF, and PTH/PTHrP. Multiple transcription factors with profound effects on longitudinal growth include Bcl2, cbfa1, A-CREB, sox9, p21, and the Smads. Growth plate chondrocytes are in a complex extracellular matrix that differs in composition not only proximally to distally within the growth plate, but also with lateral distance from the chondrocyte. Not only is the structural integrity of the multiple collagens, proteoglycans, and minor proteins of this matrix important, but also the timely activation of latent enzymes in this matrix including the MMPs and growth factors such as members of the TGF-β superfamily [39,40]. Long bone growth is altered significantly with changes in levels and/or quality of nutrition, and with manipulations of the biomechanical environment of the growth plate. Additionally, experimental evidence from studies of 'catch-up growth' as well as therapeutic interventions involving growth hormone administration following multiple causes of short stature suggest that to some extent chondrocytes in the growth plate are playing out a pre-programmed differentiation cascade that can be delayed or accelerated, but not significantly changed from the final amount of growth which is programmed as the 'sense of size' [24,41,42].

In a recent review [43] it was suggested that in multiple tissues the characteristic patterns specific for the final shape, form, and function of that tissue require that, during differentiation, individual cells be switched between different phenotypes or 'fates', and that neighboring cells may appear to act independently of each other when assuming these so-called fates. In the broadest sense, switching is determined by an interplay of systemic hormones, soluble growth factors, influences of the extracellular matrix, and biomechanical forces which are part of the local organ environment. Using theoretical computer simulations, the authors present the idea that there are a limited number of 'fates' for a given cell during the entire differentiation process which, in the case of growth plate chondrocytes, would include stem cell division, division as a proliferative cell, volume increase during hypertrophy, and apoptosis. In the general information processing of the cell, both the very general stimuli (mechanical forces, nutrient availability, levels of systemic hormones), coupled with specific molecular clues (autocrine levels of IGF-1, Ihh, VEGF, etc) elicit signals that follow multiple trajectories but ultimately will converge into one of the 'fates' which in the case of growth plate chondrocyte, as described above, potentially is only four. Fig. 6 is from this paper and gives three typical states of multiple cellular lineages as an example. This conceptualization of progress through a differentiation program that leads to a specific outcome, in our case longitudinal growth, does not rely on a series of linear signaling pathways.

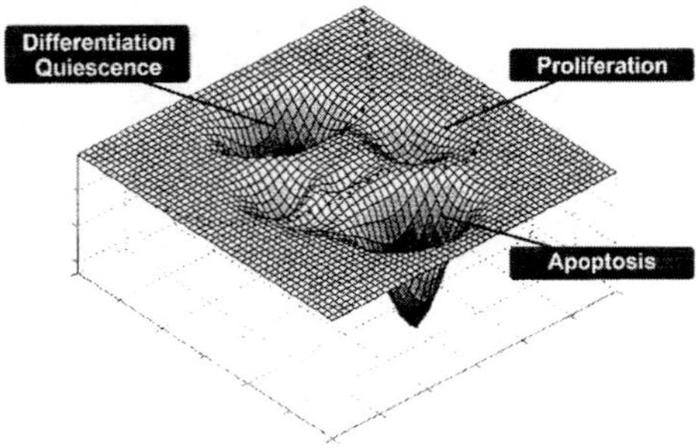

Figure 6. Attractor landscape representation of cellular fate, described in Huang and Ingber, 2000 [43] and used by permission of Academic Press. This model presents a hypothetical "potential landscape" representing an N-dimensional space framework compressed into two dimensions, in which multiple potential fates encountered in a differential cascade are presented as attractors.

This model presents a way for thinking about the complexity of the differentiation cascade of growth plate chondrocytes. The growth plate, as an organ, is as sophisticated as any in the body in the complexity of the environment – molecular, cellular, extracellular, biomechanical, blood supply – that *in toto* presents competing signals to the chondrocytes. These signals are presented to individual chondrocytes of the column within the context of an organ that may have intrinsic spatial polarity [44], and one in which the nutrient blood supply enters from the metaphyseal side, thus providing a gradient of availability of blood-derived factors across the growth plate [45]. The growth plate is a dynamic organ: at a growth rate of 300 μm/day, the initial daughter cell of the clonal expansion will have a life span of only about six days [9]. The growth plate is an organ that temporally is self renewing, such as skin epithelium or gut epithelium; however, unlike the latter two, the rate of self renewal alters over time, and will only continue only until the animal reaches adulthood. Finally, and perhaps most intriguing, at any given time, each growth plate is more or less 'marching to its own drummer' in the sense that it is growing at its own characteristic rate, despite the fact that the systemic hormonal environment in which it functions is presumably the same as for all other growth plates of the body.

Currently we have a good understanding of the kinetics of cellular activity within the two major differentiation states of the chondrocyte – proliferative and hypertrophic. We also have models of the essential variables that demonstrate predictably how chondrocytic kinetic activity within these two populations of cells quantitatively account for the amount of differential longitudinal growth achieved in multiple growth plates. We also can identify key transition points in the differentiation cascade and are developing a good understanding of the important regulators at two of these points – transition to hypertrophy and turnover at the chondro-osseous junction. However, we lack understanding of what regulates the *kinetics* of these transitions. Finally, our understanding of the complexity of signals that the chondrocyte needs to integrate is fast outpacing our knowledge of how chondrocytes within the column actually integrate these signals to change their gene expression.

Acknowledgments

The authors would like to thank Ellen Leiferman, Andea Lee, and Michelle Lenox for technical help and with preparation of the illustrations. The work was supported by NIH grant AR-35155.

List of Abbreviations

A-CREB: cAMP response element binding protein
BCl2: pro-survival/pro-apoptotic protein
BMP: bone morphogenic protein
BrdU: bromodeoxyuridine
Cbfa1: core binding factor alpha (osteoblast transcription factor)
FGF: fibroblast growth factor
IGF-1: insulin-like growth factor 1
Ihh: Indian hedgehog
MMP: matrix metalloproteinase
P21: CIP1/WAF1: a cyclin-dependent kinase inhibitor
PTH/PTHrP: parathyroid/parathyroid-related protein
RER: rough endoplasmic reticulum
Sox9: SRY-like HMG-box
Smads: plasma membrane serine/threonine kinase receptors and cytoplasmic effectors

TGFß: transforming growth factor beta
VEGF: vascular endothelial growth factor

References

[1] Bianco P, Cancedda FD, Riminucci M, Cancedda R 1998 Bone formation via cartilage models: The "borderline" chondrocyte. Matrix Biol 17:185-192.

[2] Farnum CE 1994 Differential growth rates of long bones. In: Hall BK (ed) Mechanisms of Bone Development and Growth, Vol 8, Bone. CRC Press, Boc Raton, pp. 193-222.

[3] Farnum CE, Wilsman NJ 1998 Growth plate cellular function. In: Buckwalter J, Andersen HC (eds.) Advances in the Basic and Clinical Understanding of the Growth Plate. American Academy of Orthopaedic Surgeons, Chicago, pp. 203-224.

[4] Hunziker EB 1994 Mechanism of longitudinal bone growth and its regulation by growth plate chondrocytes. Microsc Res Tech 28:505-519.

[5] Pritchett JW 1991 Growth plate activity in the upper extremity. Clin Orthop 268: 235-242.

[6] Cruz-Orive LM, Hunziker EB 1986 Stereology for anisotropic cells: application to growth cartilage. J Microsc 143:47-80.

[7] Hunziker EB, Schenk RK 1989 Physiological mechanisms adopted by chondrocytes in regulating longitudinal bone growth in rats. J Physiol 414:55-71.

[8] Wilsman NJ, Farnum CE, Leiferman EM, Fry M, Barreto C 1996 Differential growth by growth plates as a function of multiple parameters of chondrocytic kinetics. J Orthop Res 14:927-936.

[9] Farnum CE, Wilsman NJ 1993 Determination of proliferative characteristics of growth plate chondrocytes by labeling with bromodeoxyuridine. Calcif Tissue Int 52:110-119.

[10] Wilsman NJ, Farnum CE, Green EM, Lieferman EM, Clayton MK 1996 Cell cycle analysis of proliferative zone chondrocytes in growth plates elongating at different rates. J Orthop Res 14:562-572.

[11] Breur GJ, VanEnkevort BA, Farnum CE, Wilsman NJ 1991 Linear relationship between the volume of hypertrophic chondrocytes and the rate of longitudinal bone growth in growth plates. J Orthop Res 9:348-359.

[12] Buckwalter JA, Mower D, Ungar R, Schaeffer J, Ginsberg B 1986 Morphometric analysis of chondrocyte hypertrophy. J Bone Joint Surg Am 68-A:243-255.

[13] Hunziker EB, Schenk RK,Cruz-Orive L-M 1987 Quantitation of chondrocyte performance in growth-plate cartilage during longitudinal bone growth. J Bone Joint Surg Am 69-A:162-173.

[14] Hunziker EB, Kapfinger E, Saager C 1999 Hypertrophy of growth plate chondrocytes in vivo is accompanied by modulations in the activity state and surface area of their cytoplasmic organelles. Histochem Cell Biol 112:115-123.

[15] Noonan KJ, Hunziker EB, Nessler J, Buckwalter JA 1998 Changes in cell, matrix compartment and fibrillar collagen volumes between growth-plate zones. J Orthop Res 16:500-508.

[16] Breur GJ, Lapierre MD, Kazmierczak K, Stechuchak KM, McCabe GP 1997 The domain of hypertrophic chondrocytes in growth plates growing at different rates. Calcif Tissue Int 61:418-425.

[17] Farquharson C, Lester D, Seawright E, Jefferies D, Houston B 1999 Microtubles are potential regulators of growth-plate chondrocyte differentiation and hypertrophy. Bone 25:405-412.

[18] Breur GJ, Turgai J, Vanenkevort BA, Farnum, Wilsman NJ 1994 Stereological and serial section analysis of chondrocytic enlargement in the proximal tibial growth plate of the rat. Anat Rec 239:255-268.

[19] Farnum CE, Nixon A, Lee AO, Kwan DT, Belanger L, Wilsman NJ 2000 Quantitiative three-dimensional analysis of chondrocytic kinetic response to short-term stapling of the rat proximal tibial growth plate. Cells Tissues Organs 167:247-258.

[20] Álvarez J, Balbín M, Fernández M, López JM 2001 Collagen metabolism is markedly altered in the hypertrophic cartilage of growth plates from rats with growth impairment secondary to chronic renal failure. J Bone Miner Res 16:511- 524.

[21] Cobo A, López JM, Carbajo E, Santos F, Álvarez J, Fernández M, Weruaga A 1999 Growth plate cartilage formation and resorption are differentially depressed in growth retarded uremic rats. J Am Soc Nephrol 10: 971-979.

[22] Barreto C, Wilsman NJ 1994 Hypertrophic chondrocyte volume and growth rates in avian growth plates. Res Vet Sci 56:53-61.

[23] Farnum CE, Wilsman NJ 1988 Lectin-binding histochemistry of intracellular and extracellular glycoconjugates of the reserve cell zone of growth plate cartilage. J Orthop Res 6: 166-179.

[24] Conlon I, Raff M 1999 Size control in animal development. Cell 96:235-244.

[25] Vortkamp A, Lee K, Lanske B, Segre GW, Kronenberg HM, Tabin CJ 1996 Regulation of rate of cartilage differentiation by Indian hedgehog and PTH-related protein. Science 273:613-622.
[26] Amizuka N, Henderson JE, White JH, Karaplis AC, Goltzman D, Sasaki T, Ozawa H 2000 Recent studies on the biological action of parathyroid hormone (PTH)-related peptide (PTHrP) and PTH/PTHrP receptor in cartilage and bone. Histol Histopathol 15:957-970.
[27] Karsenty, G 1999 The genetic transformation of bone biology. Genes Dev 13:3037-3051.
[28] Olsen BR, Reginato AM, Wang W 2000 Bone development. Annu Rev Cell Dev Biol 16:191-220.
[29] Linsenmayer TF, Nurminskaya M 1998 Cartilage hypertrophy: Isolation and characterization of upregulated genes. In: Ninomiya Y, et al. (eds.) Extracellular Matrix-Cellular Interaction: Molecules to Diseases. Japan Sci Soc Press, Tokoyo, pp. 339-351.
[30] Volk SW, Leboy PS 1999 Regulating the regulators of chondrocyte hypertrophy. J Bone Min Res 14:483-486.
[31] Farnum CE, Wilsman NJ 1989 Cellular turnover at the chondro-osseous junction of growth plate cartilage: analysis by serial sections at the light mircoscopical level. J Orthop Res 7:654-666.
[32] Holmeck K, Bianco P, Caterina J, Yamada S, Kromer M, Kuznetsov SA, Mankani M, Robey PG, Poole AR, Pidoux I, Ward JM, Birkendal-Hansen H 1999 MT1-MMP-deficient mice develop dwarfism, osteopenia, arthritis, and connective tissue disease due to inadequate collagen turnover. Cell 99:81-92.
[33] Engsig MT, Chen TH, Vu A, Pedersen A, Therkidsen B, Lund LR, Kenricksen K, Lenhard T, Foged NT, Werb Z, Delaisse J 2000 Matrix metalloproteinase 9 and vascular endothelial growth factor are essential for osteoclast recruitment into developing long bones. J Cell Biol 151:879-889.
[34] Zhou Z, Apte SS, Soininen R, Cao R, Baaklini GY, Rauser RW, Wang J, CaoY, Tryggvason K 2000 Impaired endochondral ossification and angiogenesis in mice deficient in membrane-type matrix metalloproteinase I. Proc Natl Acad Sci USA 97:4052-4057.
[35] Vu TH, Shipley JM, Bergers G, Helms JA, Hanahan D, Shapiro SD, Senior RM, Werb Z 1998 MMP-9/gelatinase B is a key regulator of growth plate angiogenesis and apoptosis of hypertrophic chondrocytes. Cell 93:411-422.
[36] Gerber HP, Vu TH, Ryan AM, Kowalski J, Werb Z, Ferrara N 1999 EGF couples hypertrophic cartilage remodeling, ossification, and angiogenesis during endochondral bone formation. Nature Medicine 5:623-628.
[37] Dean DD, Muniz OE, Berman I, Pita JC, Carreno MR, Woessner, Jr. JF, Howell DS 1985 Localization of collagenase in the growth plate of rachitic rats. J Clin Invest 76:716-722.
[38] Farnum CE, Wilsman NJ, Hilley HD 1984 An ultrastructural analysis of osteochondritic growth plate cartilage in growing swine. Vet Pathol 21:141-151.
[39] Ray RP, Wharton KA 2001 Twisted perspective: new insights into extracelluar modulation of BMP signalling during development. Cell 104:801-804.
[40] Sage EH 2001 Regulation of interactions between cells and extracellular matrix: a command performance on several stages. J Clin Invest 107:781-783.
[41] Boersma B, Wit JM 1997 Catch-up growth. Endocr Rev 18:646-661.
[42] Stevens DA, Hasserjian RP, Robson H, Siebler T, Shalet SM, Williams GR 2000 Thyroid hormones regulate hypertrophic chondrocyte differentiation and expression of parathyroid hormone related peptide and its receptor during endochondral bone formation. J Bone Miner Res 15:2431-42.
[43] Huang S, Ingber DE 2000 Shape-dependent control of cell growth, differentiation and apoptosis: switching between attractors in cell regulatory networks. Exp Cell Res 261:91-103.
[44] Abad V, Uyeda J, Temple HT, DeLuca F, Baron J 1999 Determinants of spatial polarity in the growth plate. Endocrinology 140:958-962.
[45] Trueta T 1968 Studies of the Development and Decay of the Human Frame, William Heinemann Medical Books Ltd, London.

Localization of Bone Morphogenetic Proteins and their Intercellular Signaling Components (Smads) in the Growth Plate

Yuichirou Yazaki, Shunji Matsunaga, Takashi Sakou, Yasuhiro Ishidou, and Setsurou Komiya
Department of Orthopaedic Surgery, Faculty of Medicine, Kagoshima University, 35-1 Sakuragaoka, Kagoshima, Japan

Abstract. Bone morphogenetic proteins (BMPs) that belong to transforming growth factor-β (TGF-β) super family, transduce signals from the cell membrane to the nucleus via specific type I and type II receptors and Smads. Smad1 and Smad5 are specific mediators for intra cellular signaling of BMPs, whereas Smad4 is a common mediator. In this study, we studied immunohistochemically the spatial and temporal localization of BMP-2/4, osteogenic protein-1 (OP-1, also termed BMP-7), and BMP receptors (BMPRs), i.e. BMPR-IA, BMPR-IB and BMPR-II in the epiphyseal plate of growing rats. At 12 weeks after birth, in the proximal tibia, BMP-2/4 and OP-1 were expressed markedly in proliferating and maturing chondrocytes. BMPR-IA, -IB and -II were clearly co-expressed in proliferating and maturing chondrocytes. A lower level of expression was observed in hypertrophic chondrocytes. At 24 weeks, the expression of BMP-2/4 and OP-1 was decreased, but BMPRs were still well-expressed in proliferating chondrocytes. We also examined the expression of Smad1, 4 and 5 in the epiphyseal plate using immunohistochemical techniques. The expression of Smad was correlated with the expression of BMPs and BMPRs. Smad proteins were localized to the cytoplasm, but partially accumulated in the nucleus of proliferating and maturing chondrocytes. The temporal and spatial expression of BMPs, BMPRs and Smads suggests that BMP signaling play a role in the multistep cascade of events that lend to endochondral ossification in the epiphyseal growth plate.

Introduction

Bone morphogenetic proteins (BMPs) were originally identified as the growth factors that can induce endochondral ossification at ectopic sites [1, 2]. BMPs exert pleiotropic biological effects in developmental processes and regulate growth, differentiation, and apoptosis of various cell types: ostoblasts, chondrocytes, neural cells, and epithelial cells [3, 4]. BMPs belonging to the transforming growth factor-β (TGF-β) super family, transduce signals through two different types of serine/threonine kinase receptors termed type I and type II. Up to the present, two type I receptors and a type II receptor, specific for BMPs have been identified in mammals [5, 6, 7]. BMPs binds to BMP receptor (BMPR) type IA (BMPR-IA, also termed activin receptor-like kinase [ALK]-3), BMPR-IB (also termed ALK6) and BMPR-II. BMPs also can bind to activin type II, activin type IIB receptors, and activin type I receptor (also termed ALK2) [8, 9]. After ligand binding, BMPRs transmit intracellular signaling to the nucleus by Smad proteins. Smads are signaling molecules of the TGF-β superfamily including TGF-βs, Activins and BMPs [10].

Eight Smads are known in mammals and are classified into three groups based on their structure and function: receptor- regulated Smads (R-Smads), common mediator Smad (Co-Smad) and an Anti-Smad that interferes with signaling by R-Smad and Co-Smad. Smad 1, 5 and 8 are activated by BMPR-IA or BMPR-IB, whereas Smad 2 and 3 are activated by TGF-β type I receptor (TβR-I) or activin type IB receptor (ActR-IB). Furthermore, activin type IA receptor (ActR-IA) that was originally identified as a receptor for activin, can bind OP-1 and activate Smad1 and 5. Smad 1, 5 or 8 are phosphorylated by BMPR-IA or BMPR-IB and form heteromeric complexes with Smad 4. These heteromeric complexes translocate from the cytoplasm into the nucleus where they regulate transcription of target genes in cooperation with other transcriptional factors [10, 11].

The development of the long bones is regulated by diverse systemic and local factors. Longitudinal growth of long bones is dependent on endochondral bone formation in the epiphyseal growth plate. Various cytokines including BMPs plays important roles in the physiology of the growth plate. In this study, in order to elucidate the participation of BMPs in the development of long bone, we studied immunohistochemically the spatial and temporal localization of BMP-2/4, osteogenic protein-1 (OP-1, also termed BMP-7), and BMPR-IA, BMPR-IB and BMPR-II. We also examined the expression and localization of Smad1, 4 and 5 in the epiphyseal plate to confirm that there was activation of BMP signaling in vivo.

Materials and Methods

Tissue Preparation

Fifteen male Wistar rats aged 6, 12 and 24 weeks were sacrificed for this study. The tissues were fixed by cardiac perfusion with 10% neutral buffered formalin under intraperitoneal anesthesia using pentobarbiturate. Proximal parts of the tibiae were removed and fixed in 10% neutral buffered formalin for 24h at 4°C. After decalcification with 0.36 M ethylenediamine tetraacetic acid (pH 7.0-7.2) for 3-4 weeks, the samples were embedded in paraffin and 3-5 μm thick sections were prepared. They were subjected to hematoxylin and eosin staining, alcian blue staining and immunohistochemistry using the specific antibodies for BMP-2/4, OP-1, BMPR-IA, BMPR-IB, BMPR-II, Smad1, Smad4 and Smad 5.

Antibodies

Anti-Ligands: Polyclonal rabbit IgG against BMP-2/4 that recognized both BMP-2 and BMP-4 and monoclonal mouse IgG against OP-1 that reacted with OP-1 but not with BMP-2 and BMP-4 were generated and used as previously reported [12].

Anti-BMP receptors: Polyclonal rabbit antisera against BMPR-IA, BMPR-IB and BMPR-II were prepared and used as previously described [12, 13]. The synthetic peptides corresponding to the intracellular juxtamembrane parts of BMPR-IA, -IB and -II were used as immunogens.

Anti-Smads: Anti-Smad polyclonal rabbit antisera were raised against synthetic peptides corresponding to amino acid sequences of variable proline-rich linker regions of Smad1, Smad4 and Smad5 [14,15].

Immunohistochemistry

Immunohistochemistry was performed by the avidin-biotin peroxidase complex method using an Elite ABC RABBIT IgG KIT and an *Elite* ABC MOUSE IgG KIT (Vector Laboratories, Burlingame, CA, U.S.A.), [12,13]. Color was developed using 3,3'-diaminobenzidine tetrachloride (Dojindo Chemical Laboratories, Kumamoto, Japan). For negative controls, phosphate-buffered saline (PBS), normal rabbit IgG or normal mouse IgG were used instead of the primary antibodies.

Results

The cartilage of the epiphyseal growth plate can be divided into four different zones with distinct cellular morphologies: resting, proliferating, maturing, and calcifying cartilage. At 6 weeks after birth, the epiphyseal growth plate is well developed, and the four distinct zones are evident. At 12 weeks after birth, the epiphyseal growth plate width had decreased. At 24 weeks after birth, zones of proliferating and maturing chondrocytes had markedly decreased. We assessed the expression and localization of BMPs, BMPRs and Smads in the four zones of chondrocytes, using hematoxylin and eosin and alcian blue staining.

Expression of BMP-2/4 and OP-1(BMP-7)

BMP-2/4 and OP-1 were well expressed in proliferating and maturing chondrocytes at 6 and 12 weeks after birth. Expression of BMP-2/4 was more intense than that of OP-1 in proliferating chondrocytes. Expression of OP-1 was dominant in maturing and hypertrophic chondrocytes. At 24 weeks after birth, the expression level had decreased.

Expression of BMPR-IA, BMPR-IB and BMPR-II

BMPR-IA, -IB and -II were clearly expressed in proliferating and maturing chondrocytes at 6 and 12 weeks after birth. At 24 weeks, expression of BMP-2/4 and OP-1 was very weak as the number of chondrocytes in the epiphyseal growth plate had decreased, but those of BMPR-IA, -IB and -II were still well expressed. The expression of BMPRs decreased in hypertrophic chondrocytes.

Expression and Subcellular Localization of Smad 1, 4 and 5

Smad1 and 5 were well expressed in proliferating and maturing chondrocytes at 6 and 12 weeks after birth. Smad proteins were mainly localized to the cytoplasm; they had partially accumulated in the nucleus of proliferating and maturing chondrocytes. Smad4 was expressed in chondrocytes of all zones.

Discussion

The cartilage of the epiphyseal growth plate was characterized by four different zones with distinct cellular morphologies: resting, proliferating, maturing, and calcifying cartilage. In the present study, BMP-2/4 and OP-1 were well expressed in proliferating and maturing chondrocytes. However, expression of BMP-2/4 was more intense in proliferating chondrocytes, while OP-1 was dominant in maturing and hypertrophic chondrocytes. It was previously reported that BMP-2 and 4 were expressed at an early stage whereas OP-1 was expressed by hypertrophic chondrocyte in embryonic cartilage [16, 17, 18]. These distinct BMP expression patterns suggest that members of BMPs family have overlapping, but critically different functions during proliferation and differentiation of chondrocytes [12]. BMPRs, i.e. BMPR-IA, -IB and -II, were clearly coexpressed in proliferating and maturing chondrocytes at all stages. Although the zones of proliferating and maturing chondrocytes displayed a remarkably decreased expression of BMP ligands at 24 weeks after birth, expression of BMPR-IA, -IB and -II remained high.

BMPs can induce chondrocyte differentiation and permit the development of long bones [19, 20]. A decrease in expression of BMPs at 24 weeks after birth may reduce BMP signaling in epiphyseal growth cartilage development. Interestingly, expression of BMP-4 was transient during fracture healing [21], but in late stage healing there was expression of BMPR-IA, -IB and -II [12]. With respect to quantitative regulation of BMPs signaling, the change in expression of BMPs may be more critical than that of BMPRs.

In our study, localization patterns of BMPR-IA, BMPR-IB and BMPR-II in chondrocytes of the epiphyseal growth plate of growing rats were almost the same. Coexpression of BMPR-Is and BMPR-II in chondrocytes may be an *in vivo* example of use of both type I and type II receptors for BMP signaling. Lack of BMPR- IA, BMPR-IB or BMPR-II caused a loss of phenotypic expression in chondrocytes [22, 23, 24]. Exogenous expression of dominant negative (DN) type of BMPR-IB and BMPR-II were potent in suppressing chondrocyte maturation [22, 24], whereas expression of DN-BMPR-IA inhibited early-phase differentiation into chondrocytes [23]. We could not clearly detect differences in expression patterns of BMPR-IA and -IB in epiphyseal growth plate cells. These results suggest that the functional roles of BMPR-IA and -IB may be redundant and that BMPR-IA and –IB may have similar, but not identical, ligand binding properties [6]. It is possible that these BMPR-Is transmit BMP signals from different ligands to each other.

After ligand binding, BMPRs transmit intracellular signaling to the nucleus by Smad proteins. Smad 1, 5 or 8, phosphorylated by BMPR-IA or BMPR-IB. They form heteromeric complexes with a common-mediator Smad 4, and translocate from the cytoplasm into the nucleus where they regulate transcription of target genes. Smad1 and 5 were well expressed in proliferating and maturing chondrocytes at 6 and 12 weeks after birth. Futhermore, immunohistochemically Smad 1 and 5 accumulated in the nucleus of proliferating and maturing chondrocytes. These nuclear accumulations of Smad proteins indicate that Smads may be phosphorylated by BMPR-Is and translocated from the cytoplasm into the nucleus. Indeed, it was confirmed that BMP signaling was activated in chondrocytes in the epiphyseal growth plate in vivo.

Chondrocytes express specific genes such as type II and X collagen, alkaline phosphatase and proteoglycan during proliferation, maturation and hypertrophy. Smads may regulate these genes in cooperation with other transcriptional factors. In our study, localization of Smads protein were broad. The additional cell specific transcriptional factors, including Runx2 (Cbfa-1) and SOX9 may be critical for the phased expression of phenotypic genes [25, 26, 27]. However, the mechanism of transcriptional control associated with Smad 1 and 5 is still unknown. Recent studies have revealed the mechanism by which Smad-mediated BMP signals associate with Runx 2 [28, 29].

In conclusion, BMPs, BMPRs and Smads participate in the regulation of proliferation and differentiation of chondrocytes in the epiphyseal growth plate. Further studies are needed to elucidate the detailed mechanism of BMPs signaling in endochondral bone formation in the growth plate.

References

[1] Urist MR 1965 Bon: Formation by autoinduction. Science 150:893-899.

[2] Wozney JM, Rosen V, Celeste AJ, Mitsock LM, Whitters MJ, Kritz RW, Hewick RM, Wand EA 1988 Novel regulators of bone formation: molecular clones and activities. Science 242:1528-1534.

[3] Reddi AH 1997 Bone morphogenetic proteins: an unconventional approach to isolation of first mammalian morphogens. Cytokine Growth Factor Rev 8:11-20.

[4] Sakou T 1998 Bone morphogenetic proteins: From basic studies to clinical approaches. Bone 22:591-603.

[5] Koenig BB, Cook JS, Wolsing DH, TingJ, Tiesman JP, Corres PE, Olson CA, Pecquet AL, Ventura F, Grant RA, Chen G-X, Wrana JL, Massague J, Rosenbaum JS 1994 Characterization and cloning of a receptor for BMP-2 and BMP-4 from NIH3T3 cells. Mol Cell Biol 14:5961-5974.

[6] ten Dijke P, Yamashita H, Sampath TK, ReddiAH, Estevez M, Riddle DL, Ichijo H, Heldin C-H, Miyazono K 1994 Identification of type I receptors for osteogenic protein-1 and bone morphogenetic protein-4. J Biol Chem 269:16985-16988.

[7] Rosenzweig BL, Imamura T, Okadome T, Cox GN, Yamashita H, ten Dijke P, Heldin C-H, Miyazono K 1995 Cloning and characterization of a human type II receptor for bone morphogenetic proteins. Proc Natl Acad Sci USA 92:7632-7636.

[8] Yamashita H, ten Dijke P, Heldin CH, Miyazono K 1996 Bone morphogenetic protein receptors. Bone 19:569-574.

[9] Kawabata M, Imamura T, Miyazono K 1998 Signal transduction by bone morphogenetic proteins. Cytokine Growth Factor Rev 9:49-61.

[10] Massagu• J 1998 TGF-β signal transduction. Annu Rev Biochem 67:753-791.

[11] Miyazono K, ten Dijke P, Heldin CH 2000 TGF-β signalling by Smad proteins. Adv Immunol 75:115-157.

[12] Onishi T, Ishidou Y, Nagamine T, Yone K, Imamura T, Kato M, Sampath TK, ten Dijke P, Sakou T 1998 Distinct and overlapping patterns of localization of bone morphogenetic protein (BMP) family members and a BMP type II receptor during fracture healing in rats. Bone 22:605-612.

[13] Ishidou Y, Kitajima I, Obama H, Maruyama I, Murata F, Imamura T, Yamada N, ten Dijke P, Miyazono K, Sakou T 1995 Enhanced expression of type I receptors for bone morphogenetic proteins during bone formation. J Bone Miner Res 10:1651-1659.

[14] Nakao A, Imamura T, Souchelnytskyi S, Kawabata M, Ishisaki A, Oeda E, Tamaki K, Hanaki J, Heldin C-H, Miyazono K, ten Dijke P 1997 TGF-§ receptor mediated signalling through Smad2, Smad3 and Smad4. EMBO J 16:5353-5362.

[15] Imamura T, Takase M, Nishihara A, Oeda E, Hanai J, Kawabata M, Miyazono K 1997 Smad6 inhibits signalling by TGF-β superfamily. Nature 389:622-626.

[16] Lyons KM, Pelton RW, Hogan BLM 1989 Patterns of expression of murine Vgr-1 and BMP-2a RNA suggest that transforming growth factor-§-like genes coordinatly regulate aspects of embryonic development. Gene Dev 3:1657-1668.

[17] Jones CM, Lyons KM, Hogan BLM 1991 Involvement of bone morphogenetic protein-4 (BMP-4) and Vgr-1 in morphogenesis and neurogenesis in the mouse. Development 111:531-542.

[18] Vukicevic S, Latin V, Chen P, Batorsky R, Reddi AH, Sampath TK 1994 Localization of osteogenic protein-1 (bone morphogenetic protein-7) during human embryonic development: high affinity binding to basement membranes. Biochem Biophys Res Commun 198:693-700.

[19] Asahina I, Sampath TK, Hauschka PV 1996 Human osteogenic protein-1 induces chondroblastic, osteoblastic and/or adipocytic differentiation of clonal murine target cells. Exp Cell Res 222:38-47.

[20] Duprez D, Bell EJdH, Richardson MK, Archer CW, Wolpert L, Brickell PM, Fracis-West PH 1996 Overexpression of BMP-2 and BMP-4 alters the size and shape of developing skeletal elements in the chick limb. Mech Dev 57:145-157.

[21] Nakase T, Nomura S, Yoshikawa H, Hashimoto J, Hirota S, Kitamura Y, Oikawa S, Ono K, Takaoka K 1994 Transient and localized expression of bone morphogenetic protein 4 messenger RNA during fracture healing. J Bone Miner Res 9:651-659.

[22] Enomoto-Iwamoto M, Iwamoto M, Mukudai Y, Kawakami Y, Nohno T, Higuchi Y, Takemoto S, Ohuchi H, Noji S, Kurisu K 1998 Bone morphogenetic protein signaling is required for maintenance of differentiated phenotype, control of proliferation, and hypertrophy in chondrocytes. J Cell Biol 140:409-418.

[23] Shukunami C, Akiyama H, Nakamura T, Hiraki Y 2000 Requirement of autocrine signaling by bone morphogenetic protein-4 for chondrogenic differentiation of ATDC5 cells. FEBS Letters 469:83-87.

[24] Volk SW, D'Angelo M, Diefenderfer D, Leboy PS 2000 Utilization of bone morphogenetic protein receptors during chondrocyte maturation. J Bone Miner Res 15:1630-1639.

[25] Takeda S, Bonnamy JP, Owen MJ, Ducy P, Karsenty G 2001 Continuous expression of Cbfa1 in nonhypertrophic chondrocytes uncovers its ability to induce hypertrophic chondrocyte differentiation and partially rescues Cbfa1-deficient mice. Gens Dev 15:467-481.

[26] Kim IS, Otto F, Zabel B, Mundlos S 1999 Regulation of chondrocyte differentiation by Cbfa1. Mech Dev 80:159-170.

[27] Zehentner BK, Dony C, Burtscher H 1999 The transcription factor Sox9 is involved in BMP-2 signaling. J Bone Miner Res 14:1734-1741.

[28] Lee KS, Kim HJ, Li QL, Chi XZ, Ueta C, Komori T, Wozney JM, Kim EG, Choi JY, Ryoo HM, Bae SC 2000 Runx2 is a common target of transforming growth factor beta1 and bone morphogenetic protein 2, and cooperation between Runx2 and Smad5 induces osteoblast-specific gene expression in the pluriopotent mesenchymal precursor cell line C2C12. Mol Cell Biol 20:8783-8792.

[29] Leboy P, Grasso-Knight G, D'Angelo M, Volk SW, Lian JV, Drissi H, Stein GS, Adams SL 2001 Smad-Runx interactions during chondrocyte maturation. J Bone Joint Surg Am 83A:S15-22.

Author Index